Bio-Synthetic Hybrid Materials and Bionanoparticles

A Biological Chemical Approach Towards Material Science

RSC Smart Materials

Series Editors:
Professor Hans-Jörg Schneider, *Saarland University, Germany*
Professor Mohsen Shahinpoor, *University of Maine, USA*

Titles in this Series:

How to obtain future titles on publication:
A standing order plan is available for this series. A standing order will bring delivery of each new volume immediately on publication.

For further information please contact:
Book Sales Department, Royal Society of Chemistry, Thomas Graham House, Science Park, Milton Road, Cambridge CB4 0WF, UK
Telephone: +44 (0)1223 420066, Fax: +44 (0)1223 420247
Email: booksales@rsc.org
Visit our website at www.rsc.org/books

Bio-Synthetic Hybrid Materials and Bionanoparticles

A Biological Chemical Approach Towards Material Science

Edited by

Alexander Böker
University of Potsdam, Fraunhofer-Institut für Angewandte Polymerforschung IAP, Potsdam, Germany
Email: alexander.boeker@iap.fraunhofer.de

and

Patrick van Rijn
University of Groningen, University Medical Center Groningen, Groningen, The Netherlands
Email: p.van.rijn@umcg.nl

THE QUEEN'S AWARDS
FOR ENTERPRISE:
INTERNATIONAL TRADE
2013

RSC Smart Materials No. 16

Print ISBN: 978-1-84973-822-4
PDF eISBN: 978-1-78262-210-9
ISSN: 2046-0066

A catalogue record for this book is available from the British Library

Published by the Royal Society of Chemistry,
Thomas Graham House, Science Park, Milton Road,
Cambridge CB4 0WF, UK

Registered Charity Number 207890

For further information see our web site at www.rsc.org

Preface

New developments with respect to bionanoparticles and bionanoparticle hybrid systems and their use in composite materials have become increasingly important. The recognition of the particle character and behavior of proteins and protein-based particles has a major impact on the development of novel nanoparticle systems and materials, adding new functions and possibilities. The types of particles and systems discussed here span the full spectrum required to achieve a final functional system or develop novel and interesting applications:

Part I: Design and synthesis of bionanoparticles and biohybrid structures (Chapters 1–4)
Part II: Bionanoparticle behavior in solution and at interfaces (Chapters 5–7)
Part III: Bionanoparticle functional materials and systems (Chapters 8–11)

The approaches considered relate to current developments but are not exhaustive. They represent the general approaches taken to redesign proteins and protein-based particles and to utilize their assembly behavior and physical properties both in solution and at various interfaces. These approaches permit the newly developed protein-based biohybrid materials to be used in novel functional systems and applications.

Nature is a continuous source of inspiration and this includes using proteins and protein-based systems as presented in this book. The newly formed bionanoparticles behave differently from the native species and the newly added properties and assembly/adsorption behavior give access to new functional materials, not only in (bio)medical and biotechnological applications but also in electronic devices and possibly membrane technology. Although a large variety of systems have already been

RSC Smart Materials No. 16
Bio-Synthetic Hybrid Materials and Bionanoparticles: A Biological Chemical Approach Towards Material Science
Edited by Alexander Böker and Patrick van Rijn
© The Royal Society of Chemistry 2015
Published by the Royal Society of Chemistry, www.rsc.org

developed, many more remain to be discovered: significant advances are expected when more (sub-)disciplines in chemical and biological science are combined. In *Bio-Synthetic Hybrid Materials and Bionanoparticles*, this interdisciplinarity is embraced and exploited, stressing that combined efforts are needed for the advancement of bio-based and biohybrid (nano) materials.

Contents

RSC Smart Materials No. 16
Bio-Synthetic Hybrid Materials and Bionanoparticles: A Biological Chemical Approach Towards Material Science
Edited by Alexander Böker and Patrick van Rijn
© The Royal Society of Chemistry 2015
Published by the Royal Society of Chemistry, www.rsc.org

Part II: Bionanoparticle Behavior in Solution and at Interfaces

Part III: Bionanoparticle Functional Materials and Systems

Synthetic Modifications of Proteins

ULRICH GLEBE[a], BARBARA SANTOS DE MIRANDA[b],
PATRICK VAN RIJN*[b,c], AND ALEXANDER BÖKER*[a,d]

[a]Fraunhofer Institute for Applied Polymer Research IAP, Geiselbergstr. 69,
14476 Potsdam-Golm, Germany; [b]Department of Biomedical Engineering,
FB-40, University Medical Center Groningen, University of Groningen,
A. Deusinglaan 1, 9713 AV Groningen, The Netherlands; [c]W. J. Kolff Institute
for Biomedical Engineering and Materials Science, FB-41, University
Medical Center Groningen, University of Groningen, A. Deusinglaan 1,
9713 AV Groningen, The Netherlands; [d]Lehrstuhl für Polymermaterialien
und Polymertechnologie, Universität Potsdam, 14476 Potsdam-Golm,
Germany
*E-mail: p.van.rijn@umcg.nl, alexander.boeker@iap.fraunhofer.de

1.1 Introduction: General Approaches to Protein Modification

Proteins are fascinating building blocks with outstanding functions in Nature. Bioinspired material synthesis utilizes these unique properties of proteins when integrated into synthetic structures. The properties of both the protein and the synthetic compound, inorganic or organic, can be combined to create novel hybrid materials for a broad range of applications.[1] Proteins provide

RSC Smart Materials No. 16
Bio-Synthetic Hybrid Materials and Bionanoparticles: A Biological Chemical Approach
Towards Material Science
Edited by Alexander Böker and Patrick van Rijn
© The Royal Society of Chemistry 2015
Published by the Royal Society of Chemistry, www.rsc.org

a multitude of applications, *e.g.* as catalysts, recognition sites, transporters, nanochannels and much more. Proteins are defined by their three-dimensional structures with distinct shape and size dimensions. Under suitable conditions, these structures self-regulate and can be re-formed into their natural shape. Unfortunately, this nearly perfect system in Nature is a disadvantage for the synthesis of biohybrid materials. Protein self-assembly is a sensitive feature and protein stability can decrease strongly under unsuitable conditions. To preserve the protein shape, including the resulting functions mentioned above, satisfactory conditions need to be found that allow the integration of synthetic components into protein-based systems in order to form biohybrid structures. In many cases, conditions that mimic the natural environment provide a suitable foundation. Generally, proteins are stable in aqueous media. However, not all chemical reactions needed for protein modification can be realized in pure aqueous systems. In some cases, organic cosolvents can be used to dissolve a wider range of chemical reactants. Nevertheless, the specific ratio for each organic solvent that conserves functionality and the overall folded structures of the protein requires attention. Additionally, heat, drastic changes in pH and certain chemicals cause denaturation of proteins.

The size, folding and function of proteins are predetermined by the precise sequence of amino acids. Several approaches are feasible to modify proteins selectively without influencing their functionality. The use of biotechnological methods allows for the exchange of specific amino acids or the deletion of some parts of the peptide sequence while retaining the protein shape and stability. Such changes can be used to add functional groups and also binding sites, to stabilize a protein or to alter the polarity of certain parts.[1] Proteins can be integrated into organic or inorganic structures. By forming covalent or non-covalent bonds, small molecules, specific functional groups or even much larger nanoparticles can be attached to proteins. Particularly important are polymers, which are frequently used to modify proteins. The residues of amino acids which are oriented to the outside of the protein cage can be used for direct and covalent chemical modification. If polarity, charge and steric demand are not altered significantly, the protein remains comparatively stable with respect to its natural occurrence.

Hybrid materials offer the possibility to combine the beneficial properties of inorganic, bio- and polymeric structures to generate materials with novel functions.[2] Applications and objectives for protein-based hybrid structures are drug/gene delivery, biosensing and bioimaging, in addition to creating materials for electronic devices and functional membranes.[1,3,4] Furthermore, the conjugation to polymers may improve protein stability, solubility and biocompatibility and also increase the interfacial activity, decrease the amount of denaturation of the protein on an interface or render the proteins more compatible with other polymer structures.[1,5] Over the past few years, bioconjugation strategies have been investigated intensively.[6]

Protein–polymer systems are attractive for medical applications in the fields of diagnostics, therapeutics and theragnostics.[7] Biocompatible or even biodegradable polymers could result in applicable hybrid systems. Although not yet ready for practical use, polymeric templates with enzymes and other proteins are envisaged to act as reaction spaces in nanoscale dimensions.[7] Here, the insertion and encapsulation of biological compounds into polymer systems are possible, thereby protecting enzymatic reactions through the surrounding polymer structure.[7] Such systems combine the stability and robustness of polymer structures with the specificity and selectivity that the biological compounds provide.[7] Polymer conjugation can stabilize therapeutic proteins and protect them, especially from enzymatic degradation.[8] It has recently been shown that a proline-specific endopeptidase (PEP) was stabilized to such an extent that it retained function even when presented with conditions as harsh as those found in the stomach. This was accomplished through the covalent attachment of a positively charged dendronized polymer.[8,9]

For the conjugation of proteins to synthetic structures, complementary functional groups are needed. Two types of reactions are commonly used: polymerization and 'click' reactions. Polymers can be attached in one step (grafting-to approach) or the polymerization can be performed directly from initiator sites on the protein (grafting-from approach). Section 1.3 addresses the conjugation of polymers to proteins.

'Click' chemistry is a still fairly recent field that comprises bond-forming reactions with the aim of linking molecular building blocks *via* heteroatoms. In general, 'click' reactions are highly selective, easy to perform with a high tolerance towards the presence of other functional groups and solvents and lead to products that can easily be isolated in high yield with little or no by-product formation.[10,11] Many 'click' reactions are biocompatible and can be performed in water, which makes them ideal tools for protein functionalization.[5,12-14] The applicability to bionanoparticle modification was demonstrated shortly after the initial development of 'click' chemistry.[13] The reaction between azides and alkynes to form a 1,2,3-triazole ring is one reaction of the 'click' chemistry pool, known as copper-catalysed azide–alkyne cycloaddition (CuAAC). Biological structures normally do not have alkyne and azide groups; however, they can easily be incorporated into proteins.[5,15] CuAAC is a very suitable reaction for a specific ligation strategy for proteins and the most commonly applied 'click' reaction for bioconjugation [Scheme 1.1(a)].[16,17] However, the correct selection of the catalyst–ligand system, because of potential damage to biomolecular structures, and a comparably slow reaction rate at the low concentrations typically required for bioconjugation purposes must be emphasized.[18] The copper-free variant of the reaction, strain-promoted azide–alkyne cycloaddition (SPAAC), provides a metal-free alternative for protein conjugation [Scheme 1.1(b)].[15] The activation barrier of the reaction is lowered by taking advantage of the inherent ring strain of cyclooctynes.[18]

Copper(I)-catalysed 1,3-dipolar cycloaddition is the best known reaction, but other 'click' reactions are also commonly used and several of the reactions presented in this chapter for protein modifications indeed fulfil the requirements of 'click' reactions. Among them are the Diels–Alder reaction [Scheme 1.1(c)], Staudinger ligation [Scheme 1.1(d)], pyridyl disulfide,

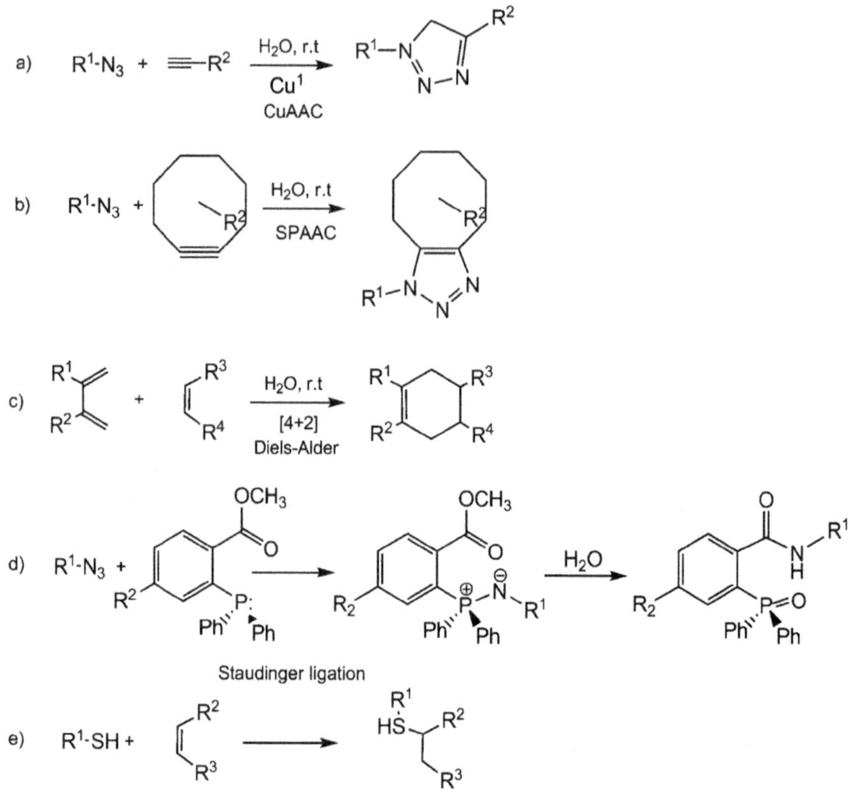

Scheme 1.1   General scheme of 'click' reactions. Reproduced from ref. 16.

Michael addition and thiol–ene reactions [Scheme 1.1(e)], and also hydrazine and oxime formation.[5,16,17,19–22] These copper-free 'click' reactions have the advantage that the toxicity and the possible denaturing character of Cu^I are avoided.[5,15,19] Furthermore, Cu ions readily promote the generation of reactive oxygen species (ROS) that often harm biological structures.[18] Sometimes proteins have to be functionalized first in order to apply a 'click' reaction. In other cases, native proteins can be used directly without modification, an example being the SH group of cysteine. The reaction of thiols with non-activated alkenes can proceed *via* a radical (thiol–ene chemistry) or anionic (thiol Michael addition) pathway.[16,23] Additionally, 'click' reactions are well suited for the surface immobilization of small molecular components, including fluorescent labels, and also polymers on protein structures.[16,22]

1.2   Amino Acid Targeting for Synthetic Protein Modification

Since the functional groups of chemical compounds, including polymers, are highly diverse, the chemical strategy for the synthesis of protein hybrid conjugates is predetermined by the amino acid residues of the protein. First,

10 out of the 20 natural amino acids bear functional groups that are suitable for ligation chemistry.[24] Second, it is important to consider whether or not these amino acids are accessible on the protein surface for chemical modifications. Lysine and cysteine are the two amino acids most commonly used for conjugation. In addition, tyrosine, glutamine, tryptophan, histidine, aspartic acid, glutamic acid, arginine and phenylalanine plus the N- and C-termini of the peptide chain are also possible reaction centres.[24] For the other amino acids, almost no specific reaction that proceeds without damaging the native state of the protein or competing with functional groups of other amino acids is known.[24] Furthermore, the modification of a specific amino acid residue should not influence the conformation and function of the protein.

Amino acids that are embedded in the globular structure of a protein are not accessible for chemical reactions. The average surface accessibility (ASA) indicates whether amino acids are addressable by chemical reactions. On average, aspartic acid, lysine, glycine, glutamic acid, glutamine and serine are mostly located on the surface of proteins.[24,25] If one amino acid is accessible multiple times on the protein surface, this can lead to undefined and polydisperse products with varying degrees of modification. As a result, the monodisperse character that is a defining feature of proteins is lost.

The lysine residues and the N-terminus of proteins or peptides are the most popular sites for polymer conjugation.[21,26] Lysines not only have a common location on the surface of proteins, but additionally can be used in a range of chemical reactions. The nucleophilicity of the amine group is higher than those of nucleophilic groups of other amino acids.[24] Reactions with activated carboxylic acids are commonly used. For this purpose, *N*-hydroxysuccinimidyl (NHS) esters are widely utilized in addition to other carboxylic acid derivatives such as NHS carbonates, NHS carbamates, anhydrides and acid halides. In addition to such acylations, alkylation is also used for lysine modification.[24] Here, a charge on the amine group can be retained, for example through amidation with imido esters.[21,24] This may be advantageous if the charge on the residue is important for the stability and structure of the protein. Aldehydes react reversibly with lysine groups to form imines that can be reduced to amines under mild conditions.[21,24] Scheme 1.2 gives an overview of common products of lysine modifications. In general, most proteins possess many accessible lysine residues. This, however, is accompanied with the loss of monodispersity due to differing degrees of amine modification, as mentioned earlier.[26,27] For different proteins, the reactivity of lysine residues varies significantly owing to accessibility and pK_a values.[28] By adjusting the pH, one can influence whether the reaction takes place at the amine group of lysine ($pK_a \approx 10.5$) or at the N-terminus of the protein ($pK_a \approx 7.8$).[21,24] However, in practice this is very difficult to achieve.[5]

Cysteine is also often used for conjugation and usually does not have the disadvantage of forming polydisperse bioconjugates.[24] Cysteine is present in only small amounts on the protein surfaces and there are especially few cysteine residues that do not participate in disulfide bonds.[5] In combination with the possibility of using reactions that are not feasible with other amino

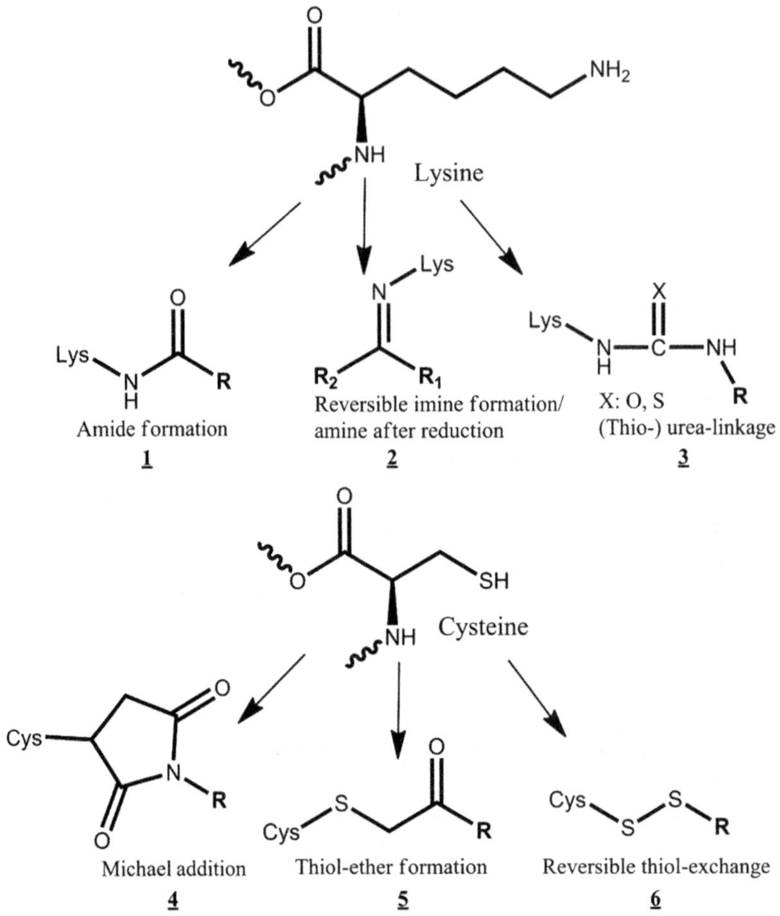

Scheme 1.2 Frequently used reactions for chemical modification of proteins with synthetic compounds. Most convenient is the modification of either lysines (reactions 1–3) or cysteines (reactions 4–6). Reproduced from ref. 1.

acids, the small amount of surface amino acids can be exploited for selective modification. Two reaction pathways are common for cysteine modifications: the formation of disulfide bonds and alkylation during a Michael addition or thiol–ene coupling (Scheme 1.2).[24] Activated disulfide groups such as orthopyridyl disulfides or methanethiosulfonates react with cysteine residues with the formation of disulfide bonds.[21,24] Reducing agents can cleave the disulfide linkages formed, which can be an advantage or disadvantage depending on the desired application.[5,21] Activated alkenes such as maleimides and vinyl sulfones form stable thioether bonds with thiol groups in Michael additions.[21,24,26] Recently, it was shown that also mono- and dibromomaleimides and electron-deficient alkynes can be used for conjugation to cysteine residues.[24,26] Sometimes disulfide bonds have to be cleaved to make them accessible for ligation. It is possible to reduce selectively the surface-accessible disulfide bridges only.[5] In many cases, the removal of a disulfide bond in a protein influences the overall three-dimensional structure. However, possibilities exist for preserving linkages in these positions by linking the two sulfur atoms *via* a three-carbon bridge. After cleavage of the disulfide bridge, a reaction with a bis(thiol)-specific reagent stabilizes the correct tertiary structure of the protein.[5,21,24]

In addition to the common protein modification approaches using lysine and cysteine residues, further chemical reactions addressing the residues of other amino acids should be mentioned that may be needed especially when the surface of the protein structures is devoid of any conventional addressable functionality. Tyrosine can be modified by palladium-catalysed alkylation of the hydroxyl group, by an electrophilic aromatic substitution such as a Mannich-type reaction or by a diazonium coupling.[21,24] However, these approaches have rarely been used up to now.[24] For glutamine, only enzymatic reactions mediated by transglutaminase are available for modification.[21,24] The indole ring of tryptophan can be used for rhodium-catalysed reactions in which a rhodium carbene is formed *in situ*.[21,24] Histidine residues form stable complexes with divalent transition metal ions. So-called His tags are composed of six histidine residues in a row and are used as a common binding motif for metal ions in peptide chemistry.[24] In order to bind chemical compounds *via* this motif, these compounds need also to be equipped with a metal binding site. Carboxylic acid groups of aspartic acid and glutamic acid and the C-terminus of proteins are reacted with amines in reverse to lysine modification. This is achieved by enzymatic and biosynthetic approaches.[24] The carboxylic acid group can be converted to very different chemical groups that may be suitable for further modification.

For a more detailed overview regarding the synthetic modification of amino acids in proteins, the reader is referred to reviews by Jung and Theato,[24] Canalle *et al.*[5] and Gauthier and Klok.[21]

1.3 Synthetic Approaches for Polymer–Protein Hybrid Structures

The modification of proteins with polymers is one of the most widespread approaches to create hybrid structures and offers tremendous possibilities for functional systems.[29,30] Both polymerization and coupling reactions are used to achieve polymer structures on protein surfaces.[31] First, three major strategies have to be distinguished (Scheme 1.3). The grafting-to approach involves the synthesis of polymer chains with suitable end-groups that are then linked to proteins. Protein-reactive groups can be introduced to polymers in a post-polymerization step or protein-reactive initiators are used directly for the polymerization reaction. The latter guarantees that each polymer chain indeed has a protein-reactive end-group.[17] The biomolecules are only involved in the last synthetic step as the polymers are synthesized

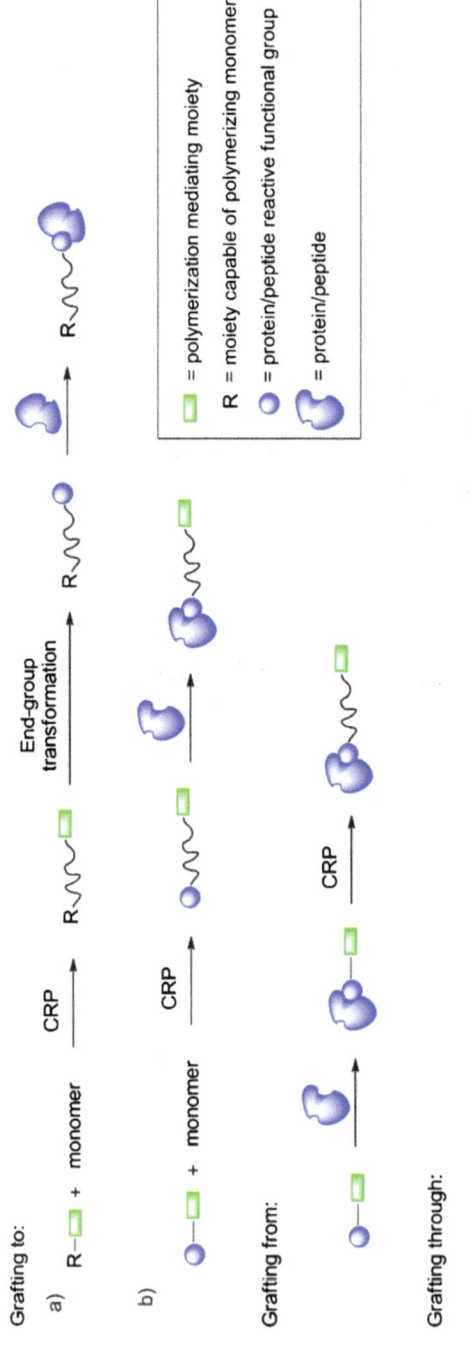

Scheme 1.3 Methods for the preparation of protein–polymer conjugates. Reproduced from ref. 26.

separately. An excess of polymer is needed for the conjugation reaction, which hampers purification of the protein–polymer hybrid. Unreacted polymer and maybe even protein have to be removed from the conjugate.[26] This disadvantage can be avoided with the grafting-from approach. Here, the polymer chains grow directly from a protein. First, small initiator molecules are linked to the protein surface able to initiate the polymerization from these so-called macroinitiators. Hence the purification of the product is easier because only small molecules, unreacted monomer and the catalyst, have to be removed.[26,27] Furthermore, the location of the polymers is defined by the initiator positions and steric hindrance is negligible.[26] Particularly in cases where it is intended to link a high density of polymer chains to a protein, grafting-from is superior to grafting-to because the steric hindrance of the growing, initially short, polymer chains is much smaller.[27] Additionally, the synthesis of hydrophobic polymer–protein conjugates and the preparation of conjugates with high molecular weight homopolymers or block copolymers are easier *via* grafting-from.[26,32] The drawback associated with the grafting-from approach is that polymerization conditions have to be chosen that do not affect the activity and folded structure of the protein. The third possibility is the grafting-through strategy.[5,21,27] Here, monomers that contain proteins/peptides are polymerized so that every repeat unit of the comb-shaped polymer structure has a protein/peptide moiety. Grafting-through potentially allows the incorporation of a high amount of proteins/peptides into the conjugates. Alternatively, monomers with a protein/peptide binding site can be polymerized and the biocomponents attached to the polymer backbone in a post-polymerization step. More details and examples of the use of the different grafting strategies are given below.

Controlled radical polymerization (CRP) techniques allow the synthesis of a wide range of polymers with a narrow molecular weight distribution.[27] The two most widely employed methods for synthesizing polymer–protein hybrid structures are atom transfer radical polymerization (ATRP) and reversible addition fragmentation chain transfer (RAFT) polymerization.[26,27] During an ATRP reaction, the polymer chains grow by the addition of intermediate radicals to monomers.[33,34] ATRP utilizes a Cu^I catalyst to establish an equilibrium state between an active and a dormant species [Scheme 1.4(b)]. The active species, the radical of the growing chain to which monomer is added, is formed by reaction of the dormant species, the halogenated polymer chain, with Cu^I ions. The oxidized transition metal atom rapidly deactivates the propagating polymer chain and reforms the dormant species.[34] By keeping the concentration of the active species low, termination reactions are suppressed and highly monodisperse polymer chains can be achieved. Therefore, radical polymerizations behave in a nearly living or controlled manner.[34,35] Mild reaction conditions, tolerance towards various chemical functionalities, experimental simplicity and the commercial availability of most initiators and catalysts make ATRP the most studied and applied CRP method for the synthesis of bioconjugates.[6] RAFT polymerization does not require a metal catalyst. The RAFT group, typically a thiocarbonylthio group, reacts reversibly with propagating polymer radicals to form the dormant species [Scheme 1.4(c)]. Here, a degenerate chain-transfer process establishes reaction control.[36] In view of the grafting-from approach, the RAFT chain-transfer agent (CTA) can be linked to amino acid residues *via* its R- or Z-group. With the R-group approach, polymer chains are grown between the protein surface and the CTA, which allows post-synthetic modification of the polymer end-group and also chain extension. The Z-group approach is not strictly grafting-from as the polymer chains do not grow from the protein, but to the CTA directly attached to the biomolecule. However, the advantages are that termination products remain in solution and the possible separate characterization of the polymers because of the labile bonding to the protein *via* the CTA.[32] In addition to the ATRP and RAFT reactions, the third major method of controlled radical polymerizations, nitroxide-mediated radical polymerization (NMP),[37] has also been used occasionally for the synthesis of protein hybrid structures. Here, nitroxides are persistent radicals that act as mediators for the polymerization. The nitroxides $(N-O^{\cdot})$ react reversibly with the growing polymer with a radical chain-end $(P^{\cdot})$ to give the dormant species $P-ON$ [Scheme 1.4(a)].

Especially ATRP and RAFT have huge potential in the field of polymer–protein bioconjugates and have brought major improvements in this area.[27,32] CRP methods made it possible to synthesize polymers with a high degree of control over architecture and insertion of functional groups. CRP allows for diverse chemical compositions since many monomers and initiators can be used in addition to organic and aqueous solvents. However, *in vivo* applications usually require the complete removal of the RAFT group and of the ATRP catalyst. Many proteins exhibit binding pockets for copper or iron ions and their biological integrities may be affected after performing a metal-catalysed polymerization reaction.[38] Copper ions are known to be cytotoxic[26,27,31] and the use of some ATRP catalysts is limited owing to their influence on protein structure and function. The method used for purification was shown to be critical in this context.[26] Nevertheless, protein–polymer conjugates could have numerous medical applications. Many examples show that the biological activity of proteins in bioconjugates is preserved to a high degree.[27]

ATRP has been extensively studied in the grafting-from approach and especially biological-compatible reaction conditions can be used such as aqueous or buffered solutions, near-ambient temperature and catalyst systems that do not significantly harm protein structures.[32] However, ATRP under aqueous conditions is challenging, mainly owing to high radical concentrations, and hence increased rate of termination and catalyst stability.[39–42] Halide dissociation from Cu^{II}, competitive coordination of solvent, destabilization of the Cu–ligand complex, disproportionation or oxidation of Cu^{I} can occur in addition to hydrolysis of the alkyl halide initiators and chain ends.[32] Proper reaction conditions need to be found that are successful in combination with protein structures. Some studies have systematically investigated ATRP-based polymerization reactions under biological-compatible conditions in the last few years.[32,41,43] Several variants of CRP techniques have been investigated for polymerization in aqueous media in recent years and are promising

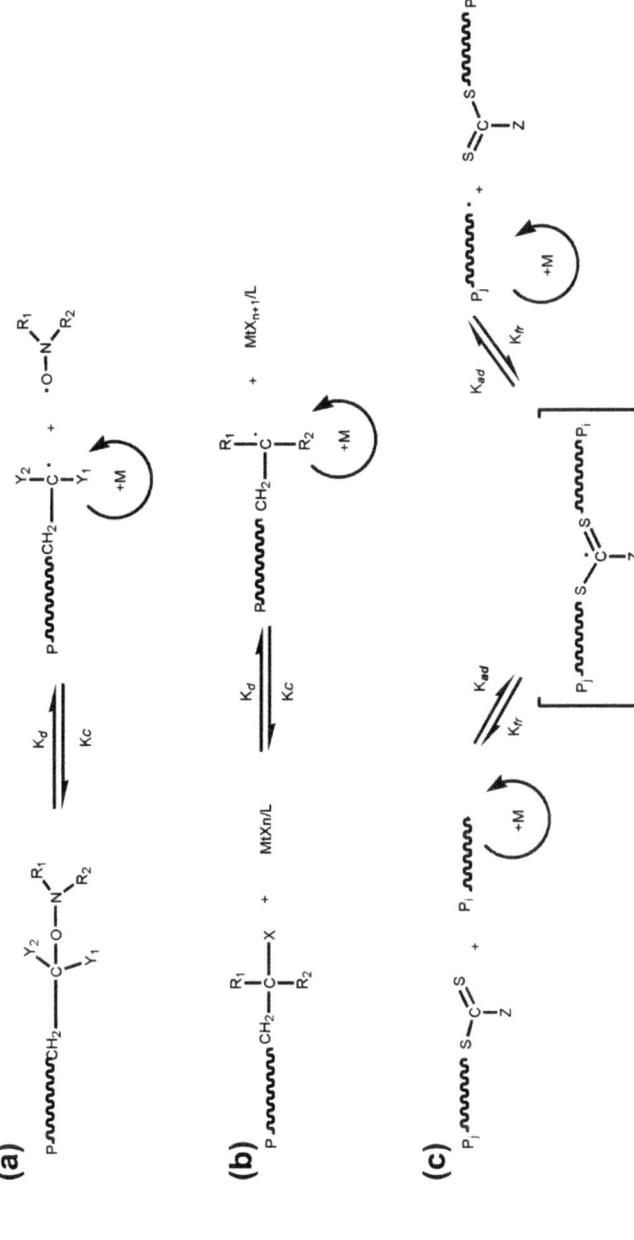

Scheme 1.4 Reversible termination in nitroxide-mediated polymerization (NMP) (a), reversible termination in atom-transfer radical polymerization (ATRP) (b) and exchange mechanism in reversible addition–fragmentation transfer (RAFT) polymerization (c). Reprinted from ref. 6, R. P. Johnson, *et al.*, *Euro. Polym. J.*, **49**, 2925–2948, Copyright 2013, with permission from Elsevier.

for the modification of proteins. Some of these, namely ARGET ATRP and AGET ATRP, were successfully used for grafting polymers from proteins.[41,43] For these ATRP variants, a reduced amount of copper salt is needed and they provide a more stable catalyst precursor, aimed at improving the polymerization from proteins. The substantially lower content of copper favours these types of polymerization for biomedical applications.

In the following, several examples of the synthesis of protein–polymer conjugates are presented. As only a few examples of particular proteins and recent developments are presented, the reader is referred to reviews for a more comprehensive overview.[5,6,17,21,26,27,37,38,44,45]

The grafting-to technique is the first and most widespread approach for the synthesis of protein–polymer bioconjugates.[27] Well-defined polymers were synthesized containing functional groups—at either the α- or ω-terminus, or both[21,27]—that were chosen for conjugation to proteins. Appropriate functional initiators are used in their native form or in a protected state for α-functionalization through radical, anionic and cationic polymerizations.[21] For ω-functionalization, labile end-groups of polymers are modified into useful functionalities.[21] Polymers with functionalities at the α- and ω-positions can be used to connect proteins and form dimers or multimers of higher orders.[27] Some reviews provide an overview of protein-reactive initiators, terminators and mediating agents used for polymerization reactions prior to grafting-to.[5,17,21,27] The preformed polymers are then linked to selected amino acid residues. PEGylation, the modification of proteins with poly(ethylene glycol) (PEG) polymers, is a typical example of grafting-to and was complemented with other PEG-containing polymers.[17,27] The formation of PEGylated protein drugs was reviewed by Maynard and co-workers.[46] Proteins such as lysozyme and bovine serum albumin (BSA) are typical examples that have been used in many grafting-to conjugation approaches. The amine groups provided by the lysine residues were often coupled with NHS esters or aldehyde groups of polymers whereas thiol groups of cysteine residues were reacted with maleimide or pyridyl disulfide α-functionalized polymers.[27]

CuAAC 'click' chemistry has also been used in combination with the grafting-to approach. Both combinations of azide-functionalized polymers and alkyne-bearing proteins or *vice versa* have been utilized.[27] For polymer synthesis, ATRP initiators and RAFT agents with azide or alkyne functions can be used, in addition to post-synthetic substituents of halide atom chain ends.[27] Especially the combination of CuAAC and ATRP has been used for the construction of hybrid bionanoparticles. The halogen end-groups of polymers prepared by ATRP can be easily functionalized and the catalyst–ligand system is very similar for both reactions.[6]

The grafting-from strategy is a straightforward, but not trivial, approach that has not been as widely investigated as grafting-to.[6] Polymerization reactions on proteins cannot be initiated directly from amino acid residues. Hence amino acids are converted to polymerization initiators *via* functionalization with appropriate small molecular components. Proteins modified in this way are called macroinitiators. The strategies for the synthetic modification of amino acids presented in Section 1.2 are utilized for this purpose. For example, ATRP initiators such as 2-bromoisobutyric acid are coupled to amine groups *via* activated ester chemistry.[24] Polymerization from protein macroinitiators was successful with ATRP, NMP and RAFT methods.[21] CRP techniques are well suited for use in the presence of biological structures in view of their tolerance towards many functional groups. Among the first examples of grafting-from were the ATRP reaction of *N*-isopropylacrylamide (NIPAAm) on streptavidin and polymerization from BSA by Maynard and co-workers.[47,48]

An example of a protein to show various modification strategies is ferritin. Ferritin is an iron storage protein with a well-defined globular morphology that can be found in many animals, plants and prokaryotes. For ferritin, it was shown that several approaches for protein modification are successful, including grafting-to, grafting-from and 'click' reactions. Ferritin is stable when exposed to small amounts of organic solvents during synthesis. Wang and co-workers modified the lysine residues of apo-horse spleen ferritin (apo-HSF) with NHS ester chemistry.[49] On one side, an alkyne moiety for subsequent CuAAC 'click' reaction was introduced [Scheme 1.5(a)]. A macroinitiator for ATRP was formed on the other side and PEG methacrylate was polymerized in this grafting-from approach, used for the first time with ferritin [Scheme 1.5(b)]. The lysine residues on the outside of the HSF nanocage were similarly used for modification by the groups of Russell and Emrick.[50] A 2-bromoisobutyryl initiator unit was conjugated by reaction of the incorporated NHS ester bond with the amine groups of the lysine residues. Grafting-from polymerization of methacryloyloxyethyl phosphorylcholine (MPC) or PEG methacrylate (PEGMA) was performed by ATRP in order to control the polymer molecular weight through the monomer-to-initiator ratio. Alternatively, NHS-terminated poly(PEGMA) was synthesized in an ATRP reaction and conjugated to ferritin in a grafting-to approach. The polymer–ferritin particles displayed a different affinity towards polystyrene-*b*-polyethylene oxide (PS-*b*-PEO) block copolymer thin films and also a different interaction with antibodies compared with native ferritin.[50] Another example of the synthesis of hybrid materials with ferritin is the construction of new membrane materials (see below) and complex networks, as developed by Böker and co-workers using the increased affinity of the hybrid particles towards polar/apolar interfaces.[29,51–53]

A few examples demonstrate the synthesis of protein–polymer conjugates containing block copolymers using a grafting-from approach. Sumerlin and co-workers showed that two consecutive RAFT polymerizations can be performed from the surfaces of lysozyme and BSA.[54,55] A RAFT CTA was conjugated to both proteins *via* its R-group. This allowed the formation of block copolymers as the polymer chains grew between the protein surface and the RAFT group. The CTAs were linked to the lysine residues of lysozyme by reaction of the NHS-activated ester bond of the CTA or to a cysteine residue of BSA *via* Michael addition. After the first well-controlled polymerization of NIPAAm, the RAFT end-group was retained. Thus, the polymer chains could be extended by polymerization

(a)

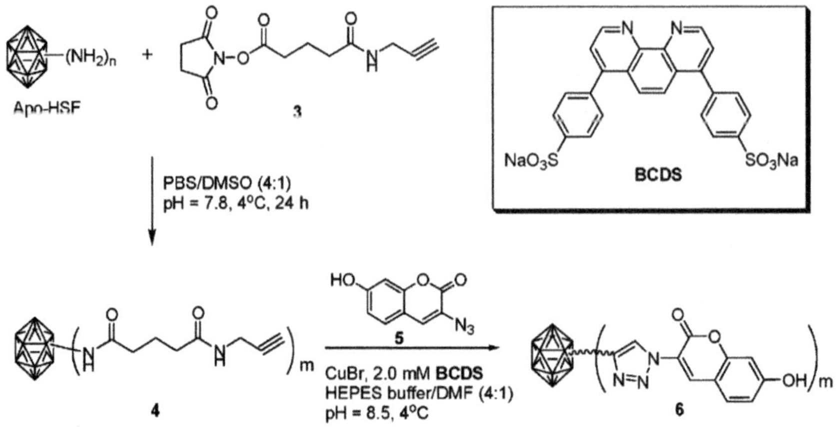

(b)

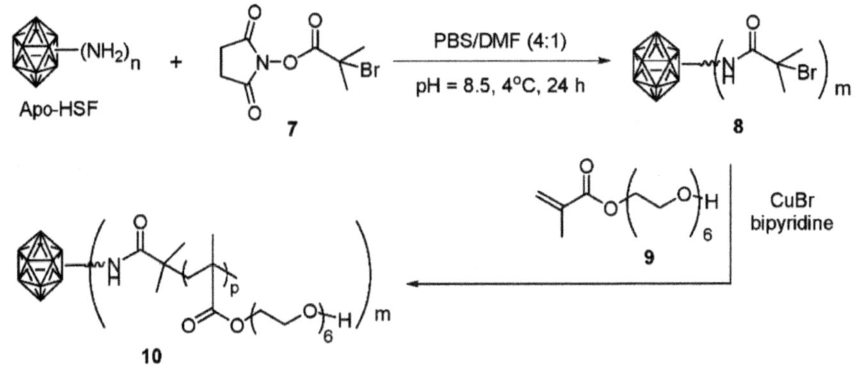

Scheme 1.5 Modification of apo-HSF with NHS ester chemistry and subsequent CuAAC 'click' reaction (a) or ATRP reaction (b). Reproduced from ref. 49.

of *N,N*-dimethylacrylamide to yield the thermoresponsive conjugates containing block copolymer chains (Scheme 1.6). In addition to RAFT polymerization, block copolymers were also successfully grafted from a protein by ATRP. An ATRP initiator was linked to the lysine residues of chymotrypsin and the monomers sulfobetaine methacrylamide (SBAm) and NIPAAm were polymerized successively.[56]

The conjugates mentioned in the last paragraph contain responsive polymers that are occasionally used to create 'smart' hybrid conjugates. The polymer chains show responsiveness towards external stimuli such as pH, temperature, enzymes or ion/cofactor binding and undergo a change in size or structure.[3] Poly(NIPAAm) is commonly used as a thermoresponsive polymer. As an example, Scheme 1.7 shows thermoresponsive conjugates that can be exploited for the separation of proteins from a complex mixture.[57]

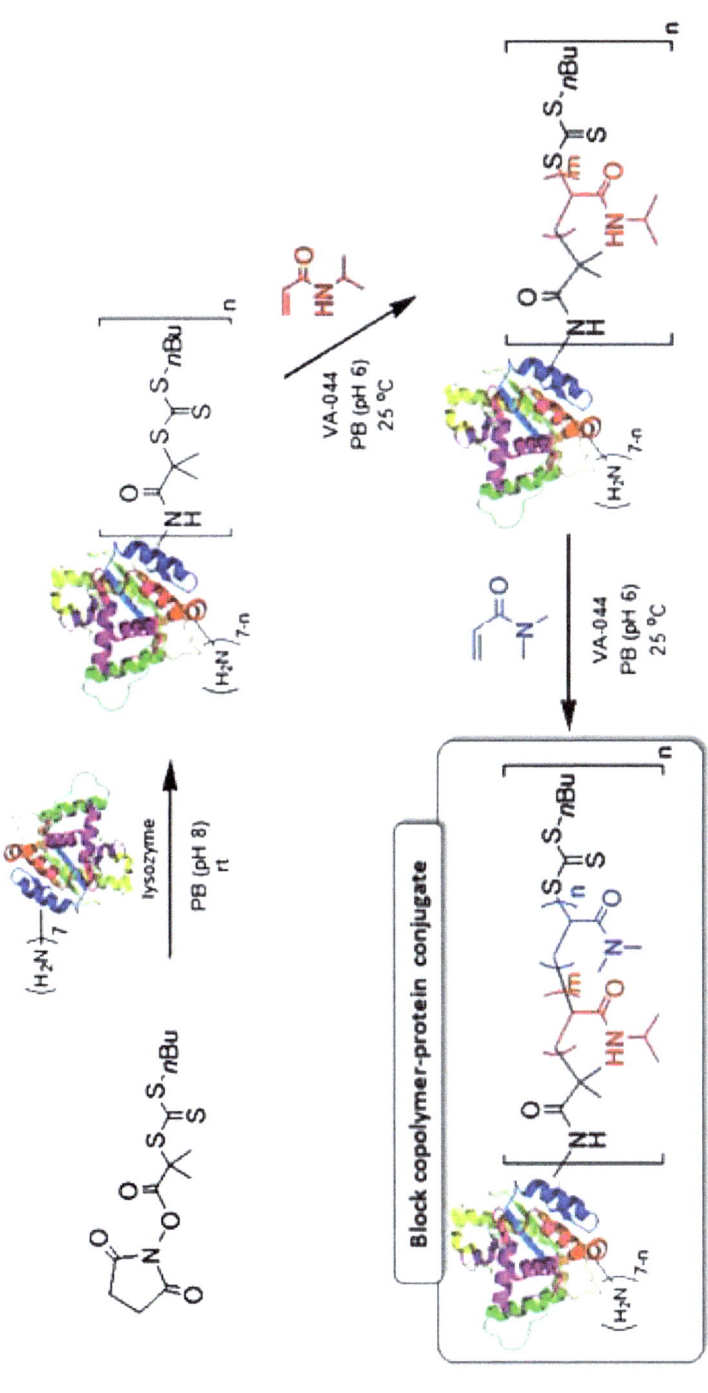

Scheme 1.6 Synthesis of lysozyme–PNIPAAm-*b*-PDMA conjugates by sequential RAFT polymerizations from CTA-modified lysozyme. Reproduced from ref. 54.

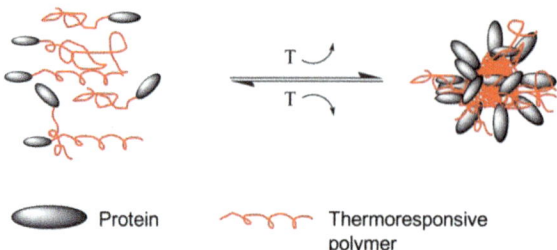

Scheme 1.7   Formation of thermoresponsive protein–polymer nanoparticle conjugates. Reproduced from ref. 57.

Russell's group recently discovered that the responsiveness of attached polymers could be used to tailor enzyme activity and stability.[56,58,59] Chymotrypsin, a serine protease, was used as a model enzyme; 12 out of 14 surface lysine residues were modified by reaction with a water-soluble ATRP initiator [Scheme 1.8(a)].[58] The subsequent grafting-from polymerization reaction conjugated poly[2-(dimethylamino)ethyl methacrylate] [poly(D-MAEMA)] to chymotrypsin. With this temperature- and pH-responsive polymer, the polymer shell surrounding the protein could be changed in order to achieve higher stability and activity of the enzyme, even in a pH range where it is usually inactive or unstable [Scheme 1.8(b)].[58] In addition, chymotrypsin conjugates with the thermoresponsive polymers poly(NIPAAm) and poly[N,N'-dimethyl(methacryloylethyl)ammonium propane sulfonate] [poly(DMAPS)] were synthesized and investigated in the context of this so-called 'polymer-based protein engineering'.[59] The conjugate of chymotrypsin with the polymer chains poly(SBAm-*b*-NIPAAm) introduced in the last paragraph responds to both low and high temperatures because of the different UCST or LCST values of the polymer blocks. The enzymes surrounded by a dual-zone shell showed higher stability even to harsh conditions of temperature, pH and protease degradation.[56]

Maynard and co-workers developed a strategy not to modify amino acids of formed proteins, but to design artificial amino acids with initiator units that will subsequently be incorporated into peptide strands.[26,60] In this way, it is possible to run polymerization reactions precisely from predetermined amino acids even when the protein contains large amounts of the chosen amino acid (Scheme 1.9). The peptide macroinitiator was prepared by solid-phase peptide synthesis (SPPS). Matyjaszewski and co-workers recently reported a variation of this strategy for grafting from green fluorescent protein (GFP) that allows the direct analysis of the grafted polymer.[61] A non-canonical amino acid with an ATRP initiator unit that has a base-labile ester bond was genetically incorporated into GFP. After polymerization, the polymer chain formed could be cleaved from the protein through hydrolysis of the ester bond and analysed by gel permeation chromatography (GPC).

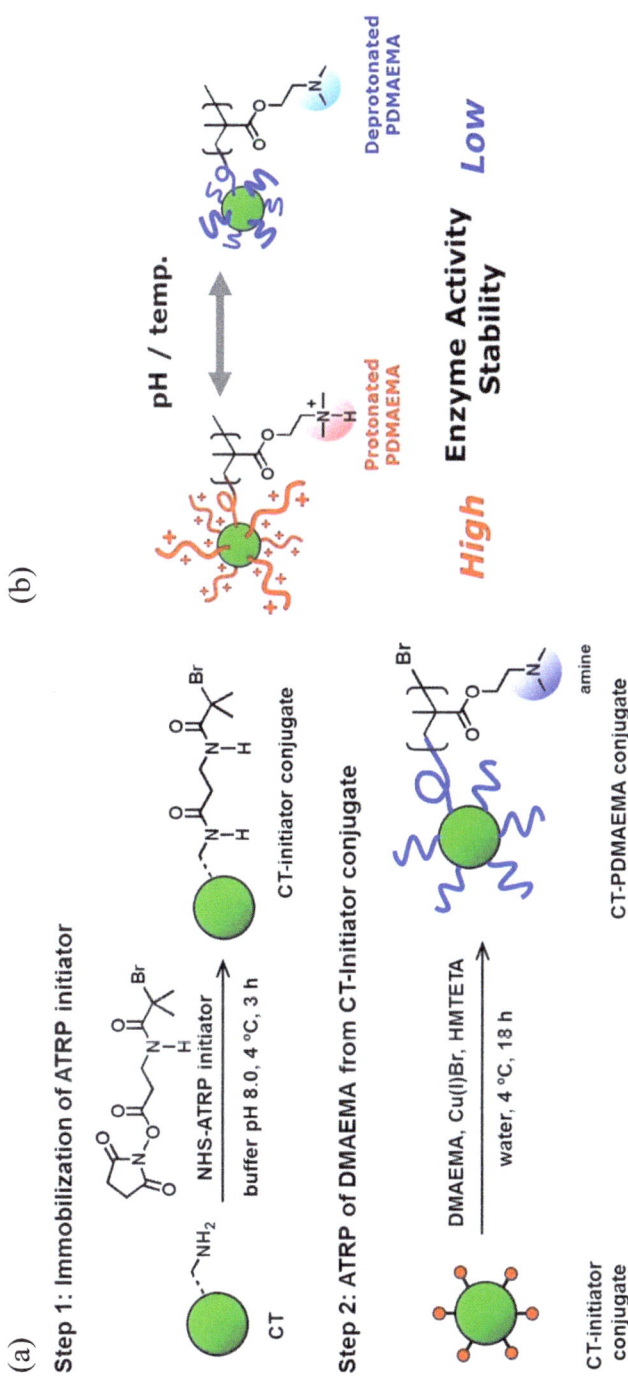

Scheme 1.8 Synthesis of chymotrypsin macroinitiator and responsive hybrid conjugate (a), tailoring of enzyme activity and stability of the conjugate through pH or temperature (b). Reprinted with permission from ref. 58, H. Murata, *et al.*, *Biomacromolecules*, 2013, **14**, 1919, Copyright 2013 American Chemical Society.

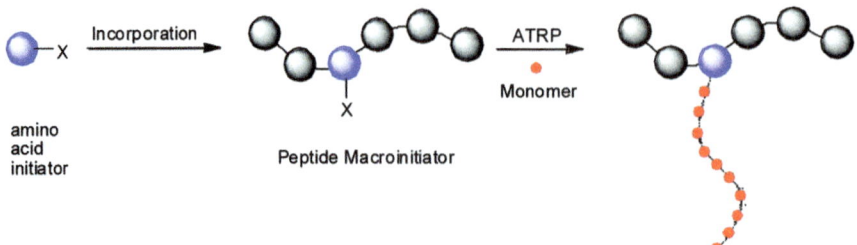

Scheme 1.9   Incorporation of amino acid initiator into peptide chain for precise
             biohybrid synthesis. Reprinted with permission from ref. 60, R. M.
             Broyer, *et al., J. Am. Chem. Soc.,* 2008, **130**, 1041, Copyright 2008 Amer-
             ican Chemical Society.

The grafting-through technique leads to bioconjugates with increased
local concentrations of peptides or proteins. Their observed biological activity can be significantly enhanced through possible multivalent interactions.[5]
Several systems have been developed with successful application of the grafting-through strategy.[27] The polymerization of peptide/protein-based macro-
monomers has the advantage that all repeat units of the conjugate bear a
peptide or protein unit. Peptides have been prepared that contain a broad
range of functional groups used for polymerization reactions.[21] However,
synthesizing peptide-functional monomers is not trivial.[5] For the alternative,
the post-polymerization modification of a side-chain functional unit, it is difficult to achieve quantitative functionalization with peptide units.

Sometimes it is necessary initially to synthesize a protein. Possible reasons for this can be a low abundance of a natural protein, difficulties related
to its isolation or purification, in addition to the aim of having a particular
amino acid residue at a specific position in the peptide chain.[21] There are
three main methods for synthesizing peptide or protein structures, namely
polymerization of amino acid *N*-carboxyanhydrides (NCAs), solid-phase
peptide synthesis and protein biosynthesis. Non-canonical amino acids can
also be incorporated into the structures.[21] Particularly for protein–polymer
conjugates, typically two different approaches are used to synthesize proteins by means of the above-mentioned methods. First, polymer and protein are synthesized separately and linked subsequently, or second, one of
the compounds is synthesized on the other one which is prepared first.[45] The
combination of a peptide synthesis method with a CRP technique allows the
preparation of well-defined hybrid conjugates. The polymerization of NCAs
has been combined with either a CRP method or 'click' chemistry to construct polymer–peptide bioconjugates.[27,45,62–64] The drawback of NCA ring-
opening polymerization is the lack of precise control over the chain length
and monomer sequence.[45] Recent efforts have been aimed at the preparation
of polypeptides with well-defined structures *via* NCAs.[6] Solid-phase peptide
synthesis is a routine technique with precise control over chain length and
monomer sequence, but the maximum chain length is restricted to about 50

amino acids. The tailor-made peptide is built *via* stepwise addition of amino acids on a resin support.[38] Here, the desired side-groups can be introduced at a specific step in the peptide synthesis. To synthesize conjugates, traditional peptide coupling conditions are frequently used as described before.[21,45] Moreover, it is possible to synthesize the peptide domain from a soluble or solid-supported polymer and even the whole bioconjugate on a resin support.[21,27,38] Longer peptides or proteins are conveniently prepared using biosynthetic methods that overcome the limitations of the NCA and SPPS methods. Although not a trivial method, through protein biosynthesis high molecular weight proteins with precisely controlled chain length and monomer sequence can be achieved. Again, polymers can be added by grafting-to in addition to ATRP or RAFT polymerization on protein macroinitiators.[21,45]

Several proteins are used as components in polymer-based membrane materials. Protein building blocks are of potential interest for membrane formation, because they form well-defined and uniform pores owing to their defined structure and also their variability in size and shape.[2] In membranes, proteins can either act as templates for holes after denaturation or act as channels in the polymer structure on their own. The possibility of not only generating size-selective pores, but also selective ion or molecular transport is especially interesting.[2] Two different approaches are used for the creation of polymer–protein membrane materials. In the first, the proteins are not covalently linked to the polymer material, but inserted into polymer membranes.[65] In the second, bioconjugates are synthesized by covalently linking proteins to polymers to construct membranes with improved properties such as chemical and mechanical stability.

The first approach focuses on the construction of vesicle structures that can function as nanoreactors. These are usually composed of amphiphilic triblock copolymers that form spherical hollow vesicles, the so-called polymersomes, in aqueous solution that are advantageous compared with the traditionally used liposomes with respect to their higher stability.[7,66,67] In principle, various supramolecular assemblies can be formed with amphiphilic block copolymers, ranging from dendrimers, spherical micelles, cylindrical micelles and capsules to polymersomes.[7,68] The polymersome membrane is composed of the hydrophobic polymer part inside the two hydrophilic layers. The flexibility of the polymer systems permits the insertion of proteins into vesicle membranes if the hydrophobic part of the polymer membrane has dimensions that fit the hydrophobic part of the channel protein used. Such proteins are especially suitable because the polymer membrane mimics their natural environment and stabilizes the protein. If the hydrophobic parts of the protein and of the polymer membrane do not have exactly the same dimensions, this so-called hydrophobic mismatch renders it inefficient to incorporate a large amount of proteins into the membranes.[2,67,69] Not only have suitable polymers been constructed, but also a protein has been engineered to match the polymer material.[69] Polymersomes with incorporated proteins are nanocompartments that can be used for enzyme catalysis or as traps for compounds and hence are called synthosomes.[70] The nanocontainer walls should be as impermeable as possible, so that the flux through the membrane is controlled only by the channel proteins. Examples include the insertion of the outer membrane protein F (OmpF),[71–76] Aquaporin Z (AqpZ),[77,78] Tsx,[76] bacterial channel-forming protein (LamB),[79] ferric hydroxamate uptake protein component A (FhuA),[69,70,73,80] claudin-2,[81] reduced nicotinamide adenine dinucleotide (NADH):ubiquinone reductase (complex I)[82] and adenosine triphosphate (ATP) synthase together with bacteriorhodopsin[83] into polymer vesicles. In nearly all examples, poly(2-methyloxazoline)-*b*-polydimethylsiloxane-*b*-poly(2-methyloxazoline) (PMOXA-*b*-PDMS-*b*-PMOXA) was used exclusively, which in part relates to the extreme flexibility of the hydrophobic block that adapts best to the requirements of the membrane proteins.[84] Figure 1.1 illustrates how a reaction can be catalysed in a polymersome with incorporated protein channels enabling permeability. The examples given show that proteins can function as selective gates, chemical reactions take place exclusively inside the vesicles and compounds can be entrapped for specific release. Polymersomes are especially envisaged as confined reaction spaces and delivery vehicles for medical applications.[7,85,86] Additionally, a biomimetic membrane with incorporated AqpZ channels was formed on a porous planar support by vesicle spreading.[77]

Other studies have focused on the direct ligation of polymers to proteins and subsequent formation of the membrane by cross-linking the covalently bound polymer chains. By this approach, the amount of incorporated proteins will be significantly higher than in the approach for synthosomes. As an example, ferritin was modified in a copolymerization with NIPAAm

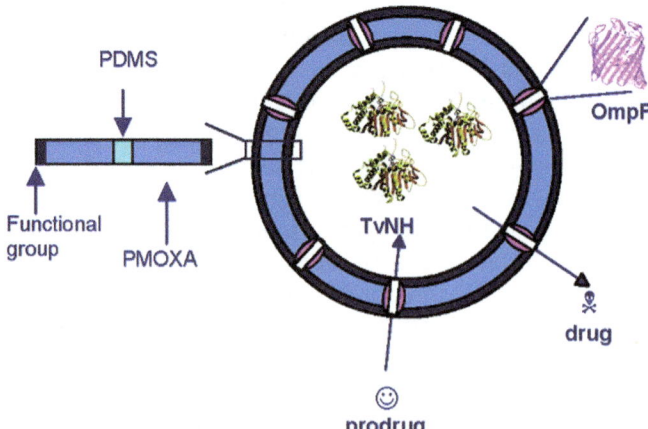

Figure 1.1   Schematic representation of a completely functionalized nanoreactor built up from poly(2-methyloxazoline)-*block*-polydimethylsiloxane-*block*-(2-methyloxazoline) (PMOXA–PDMS–PMOXA), permeabilized by the bacterial outer membrane protein OmpF and encapsulated with *Trypanosoma vivax* nucleoside hydrolase (TvNH). Reprinted with permission from ref. 76, A. Ranquin, *et al.*, *Nano Lett.*, 2005, **5**, 2220, Copyright 2005 American Chemical Society.

and cross-linkable 2-(dimethylmaleimido)-*N*-ethylacrylamide (DMIAAm) *via* 2-bromoisobutyryl-functionalized lysine residues on the surface of the protein and hybrid membranes were constructed from Pickering emulsions at liquid/liquid interfaces.[29,51–53] After self-assembly and cross-linking of the conjugates at an air/water interface, a polymer membrane was formed in which ferritin acted as a template for uniform holes that were created through denaturation of the protein.[87]

A similar approach was utilized by Mann and co-workers, who ligated poly(NIPAAm) to BSA and linked the proteins after self-assembly of the conjugates at the oil/water interface.[88] The amount of amine residues of BSA was increased by reaction of the carboxylic acid groups of aspartic and glutamic acid with 1,6-hexanediamine followed by grafting-to of functionalized poly(NIPAAm) [Figure 1.2(a)]. Proteins were cross-linked by ligation of unreacted amine groups with PEG-bis(*N*-succinimidyl succinate) after self-assembly at an oil/water interface [Figure 1.2(b)]. The assembly did not occur with the unmodified protein, indicating that the amphiphilicity of the protein–polymer construct was critical for stabilizing the water micro-droplets. These 'proteinosomes' exhibit protocellular properties.[88]

The examples described so far mostly consist of conjugates with one protein and one or multiple polymer chains. With either homo- or heterotelechelic polymers and also star-shaped structures, conjugates were formed with multiple biomolecules and even multiple proteins of different types.[26]

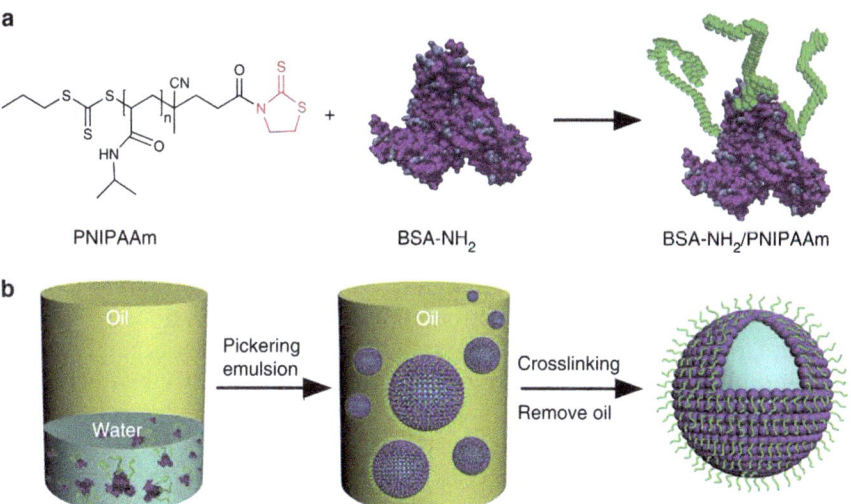

Figure 1.2 (a) Coupling of mercaptothiazoline-activated poly(NIPAAm) polymer chains with primary amine groups of cationized BSA–NH$_2$ to produce protein–polymer nanoconjugates [BSA–NH$_2$(poly(NIPAAm))]. (b) Use of protein–polymer building blocks for the spontaneous assembly of proteinosome micro-compartments in oil and their transfer into a bulk water phase. Reproduced from ref. 88, X. Huang, *et al.*, *Nat. Commun.*, 2013, **4**, 2239, with permission from Macmillan Publishing Ltd.

Higher binding affinities, efficient anchoring to surfaces and improved bio-sensors could be achieved by multiple anchoring.[26]

1.4 Non-Covalent Approaches for Polymer–Protein Conjugates

In addition to the described strategies to form covalently bound protein–polymer bioconjugates, apart from the polymersome approach, strategies with a non-covalent binding motif were developed. This is usually done by the use of cofactors. These are small organic molecules that fit perfectly into the active site of a certain protein usually having great importance for the activity of the protein. By covalently attaching the cofactor to a polymer, bio-conjugates with non-covalent binding sites can be constructed.[5] The best known example is the system of (strept)avidin and the cofactor biotin. Both avidin and streptavidin are constituted of four subunits, each of them having a biotin binding site.[38] The interactions between avidin and biotin ($K_a = 10^{15}$ M^{-1}) and streptavidin and biotin ($K_a = 10^{13} M^{-1}$) are actually seen as the strongest non-covalent biological interaction.[27] The complexes are very stable even under harsh conditions and the kinetics of dissociation are extremely slow compared with the time scale of most experimental methods.[38] The carboxylic acid group of biotin can easily be used for ligation to polymers and also for the incorporation of biotin into polymerization initiators.[5,17] Further examples of the construction of polymer–protein bioconjugates based on non-covalent interactions use the haem cofactor protoporphyrin IX together with (apo-)myoglobin or (apo-)horseradish peroxidase (HRP) plus the barstar–barnase system ($K_a = 10^{14} M^{-1}$).[1,5,27]

Glycopolymers are another example of the exploitation of non-covalent interactions to bind multiple proteins. Sugars are information-rich mole-cules owing to their structural diversity. Although the mechanisms are not completely clear, recognition processes are thought to proceed by spe-cific, non-covalent, carbohydrate–protein interactions.[38] Although protein–saccharide interactions are typically weak, a high affinity and a high specificity in the binding motif are reached through multivalent interactions.[27,89] Glyco-polymers, polymers bearing a saccharide moiety in every monomer, can act as a binding site for certain proteins. Lectins are such carbohydrate-bind-ing proteins. Synthetic glycopolymers can be prepared by polymerization of either glycomonomers or monomers bearing reactive sites for subsequent functionalization with sugar moieties.[27] The multivalency of glycopolymers provides the opportunity to bind lectins. Glycopolymers can be obtained with many incorporated functional groups, for example, biotin, maleimide, pyridyl disulfide or azide moieties.[27] These functionalities were used for con-jugation onto several proteins.[90] Glycopolymers also proved applicable for protein stabilization.[91,92]

Electrostatic binding motifs are less specific non-covalent interactions used for the conjugation of polymers to proteins. Many biological structures exhibit charged surfaces that can be utilized for electrostatic interactions.[6] Furthermore, histidine-tagged proteins can form a complex with immobilized nickel ions, for example.[1]

1.5 Protein–Nanoparticle Hybrids *via* Surface Conjugation

Nanoparticles are frequently used to construct hybrid conjugates with proteins. Biologically modified nanoparticles are an important goal in nanomedicine as the crossing of biological barriers, the delivery of therapeutic agents and the achievement of specific targeting could be improved.[22,27,38,57,93] Hence nanoparticles biofunctionalized with (poly)peptides have been synthesized and often applied for drug delivery, biosensing, bioimaging and even bioelectronic devices.[4,93] A wide variety of core materials, including metals, metal oxides and semiconductors, can be used. Protein–nanoparticle conjugates have been prepared by either covalent or non-covalent linkages. The interactions of nanoparticles with proteins may have several consequences for the biomaterial where the interface plays a crucial role, *e.g.* the probability of partial protein unfolding can be enhanced.[94] This may lead to perturbance of the biological function due to the high local protein concentration on the nanoparticle surface. Especially upon non-specific covalent conjugation onto polymer nanoparticles, proteins tend to lose their biological activity.[57] In other cases, the adsorption onto nanoparticles can stabilize the structure and activity of a protein.[95] Covalent protein–nanoparticle conjugates formed by protein side-chain modifications or engineered ligands were used to master the challenge of preserving protein structure and activity.[57,96] A biocompatible spacer, usually oligo(ethylene glycol) (OEG), is commonly used to minimize protein denaturation.[96] The adsorption to nanoparticles through non-covalent interactions can similarly influence the conformation of the protein and thereby the forces that stabilize the structure. Non-covalent conjugates can be prepared using complementary electrostatic charges between the protein and the nanoparticle, in addition to metal-mediated and hydrophobic and bioaffinity interactions [biotin–strept(avidin), carbohydrate–lectin; see Section 1.3 for details].[57,93,95,96] The reversible nature of non-covalently bound conjugates permits applications in sensing and protein delivery. In biological environments, proteins form a bionanointerface on the nanoparticle surface, the so-called 'protein-corona'.[94] This corona is composed of many proteins attached to the surface of the nanoparticle.

1.6 Biocatalytic Approaches for Biohybrid Structures

Enzymes are widely used as biocatalysts in order to produce covalent bonds, and the advantage of enzymatic reactions compared with chemical and physical methods is that they are biocompatible by nature.[97,98] Enzymes promote site-specific modifications under mild reaction conditions and generally do not produce toxic by-products, which therefore allows applications in bio-medicine such as tissue engineering, and also in drug and food processes. Certain enzymes are able to target specific amino acids and are able to modify them. This section explores the most frequently used enzymes together with some applications in protein modification for creating biohybrid materials.

Transferase is a class of enzymes able to transfer a functional group from one molecule to another and functions under conditions that ensure the preservation of the overall structure, folding and activity of most protein structures, which is a very interesting advantage. Wagner *et al.* presented a wide variety of substrates used by aminoacyl tRNA transferase to modify the N-terminus of a protein.[99] By selectively targeting the N-terminus with respect to other available amine groups found in lysine residue containing peptide strands, one can synthesize very specific Janus-like structures with a high degree of control over position and amount of modification.

Oxidoreductase is a class of widely applied enzymes that have the same modification mechanism. These enzymes act as a catalyst in electron transfer (oxyreduction) reactions. Tyrosinase acts on tyrosine residues by transforming the phenolic group in an *o*-quinone. Furthermore, the modified protein can couple non-enzymatically with other molecules containing nucleophilic groups such as primary amines.[97,100] Converting only one specific type of amino acid into a more reactive one without altering the overall three-dimensional structure is very convenient, otherwise it would mean redesigning the protein genetically in order to introduce more reactive amino acids. Another interesting enzyme is laccase, which is an oxidoreductase of the multinuclear copper-containing oxidoreductases.[101] The enzyme functions under relatively mild conditions, consumes oxygen and produces water as its only by-product.[102] It was shown that laccase can also modify small peptides in the tyrosine residues by transforming the tyrosine into a free radical, permitting a free radical polymerization.[103] Peroxidases modify the tyrosine group, generating a radical with the abstraction of a proton, they are able to perform site-specific cross-linking and they use hydrogen peroxide as cofactor. It was shown that even genetically introduced tyrosine residues were modified specifically.[104]

Enzymes are used as biocatalysts of cross-linking reactions in food processing, leather and textile fabrication, tissue engineering and biochemical and biomedical research. A lot of effort has been put into developing a better understanding of these areas because of its potential on an industrial scale. Enzymes are already widely used on an industrial scale for the synthesis of various chemical intermediates ranging from amino acids to alcohols, acids and penicillin, and are also widely applied for protein cross-linking in food processes.[97,105] This can add functionality, lower the immune-response and add nutritional value to the protein.

It is of great interest to improve an enzyme's performance, stability, activity and also identify new substrates to be targeted.[106] The immobilization of enzymes on a surfaces is one of the strategies to improve stability, allow re-use of the biomolecule and enhance the activity.[107,108] During immobilization, the enzyme can lose part of its activity owing to unfavourable positioning during binding or induce denaturing effects. Using multipoint binding can reduce such effects. Improvements in enzyme immobilization and activity are very much related to synthetic modifications, as mentioned in previous sections. Also for protein immobilization, 'click' chemistry reactions have made a significant impact, and an overview of specific developments in this area was recently published by Palomo.[109] Combining enzymatic chemical conversions with other synthetic methods broadens the overall approaches that can be adopted in order to construct new biohybrid materials. Although they are not yet as frequently used for this purpose as chemical modifications, in the near future more biocatalytic approaches will most likely emerge.

1.7 Discussion and Conclusion

While it has been recognized that biohybrid materials are of great importance, a high degree of control over particle synthesis is required to achieve control over the material properties. Various approaches ranging from non-covalent or covalent synthetic methods to biocatalytic modifications of protein/peptide structures are used for the synthesis of conjugates. Every synthetic modification has its specific conditions, which are not always compatible with delicate biological structures such as proteins. Therefore, it is pivotal that new and refined synthetic methods are developed to meet the requirements for the vast amount of protein structures available, each with their own stability, function and amino acid/chemical composition, and useful in different (bio)medical, sensing, food and cosmetic applications. So far the combination of proteins with polymers is regarded as the most versatile and hopeful approach for hybrid bioconjugates,[30] since numerous polymers/monomers are available with all kinds of different chemical functionalities that can be used in different proportions with many possibilities also to change the polymer architecture, such as linear, block copolymer, branched, dendrimer, bottle-brush forms, *etc.* Even though many combinations have already been developed, the most recent examples of which have been highlighted here, new combinations and with them new functions and materials are still under intense investigation, stressing the huge potential of the presented approaches.

References

1. P. van Rijn and A. Böker, *J. Mater. Chem.*, 2011, **21**, 16735.
2. P. van Rijn, M. Tutus, C. Kathrein, L. Zhu, M. Wessling, U. Schwaneberg and A. Böker, *Chem. Soc. Rev.*, 2013, **42**, 6578.
3. J. Y. Shu, B. Panganiban and T. Xu, *Annu. Rev. Phys. Chem.*, 2013, **64**, 631.
4. E. Katz and I. Willner, *Angew. Chem., Int. Ed.*, 2004, **43**, 6042.
5. L. A. Canalle, D. W. P. M. Löwik and J. C. M. van Hest, *Chem. Soc. Rev.*, 2010, **39**, 329.

6. R. P. Johnson, J. V. John and I. Kim, *Eur. Polym. J.*, 2013, **49**, 2925–2948.
7. C. G. Palivan, O. Fischer-Onaca, M. Delcea, F. Itel and W. Meier, *Chem. Soc. Rev.*, 2012, **41**, 2800.
8. H. D. Maynard, *Nat. Chem.*, 2013, **5**, 557.
9. G. Fuhrmann, A. Grotzky, R. Lukić, S. Matoori, P. Luciani, H. Yu, B. Zhang, P. Walde, A. D. Schlüter, M. A. Gauthier and J.-C. Leroux, *Nat. Chem.*, 2013, **5**, 582.
10. H. C. Kolb, M. G. Finn and K. B. Sharpless, *Angew. Chem., Int. Ed.*, 2001, **40**, 2004.
11. P. Wu, A. K. Feldman, A. K. Nugent, C. J. Hawker, A. Scheel, B. Voit, J. Pyun, J. M. J. Fréchet, K. B. Sharpless and V. V. Fokin, *Angew. Chem., Int. Ed.*, 2004, **43**, 3928.
12. J. E. Hein and V. V. Fokin, *Chem. Soc. Rev.*, 2010, **39**, 1302.
13. B. Le Droumaguet and K. Velonia, *Macromol. Rapid Commun.*, 2008, **29**, 1073.
14. M. Meldal and C. W. Tornøe, *Chem. Rev.*, 2008, **108**, 2952.
15. J. C. M. van Hest and F. L. van Delft, *ChemBioChem*, 2011, **12**, 1309.
16. J. M. Palomo, *Org. Biomol. Chem.*, 2012, **10**, 9309.
17. K. L. Heredia and H. D. Maynard, *Org. Biomol. Chem.*, 2007, **5**, 45.
18. E. Lallana, R. Riguera and E. Fernandez-Megia, *Angew. Chem., Int. Ed.*, 2011, **50**, 8794.
19. M. F. Debets, J. C. M. van Hest and F. P. J. T. Rutjes, *Org. Biomol. Chem.*, 2013, **11**, 6439.
20. S. S. van Berkel, M. B. van Eldijk and J. C. M. van Hest, *Angew. Chem., Int. Ed.*, 2011, **50**, 8806.
21. M. A. Gauthier and H.-A. Klok, *Chem. Commun.*, 2008, 2591.
22. R. K. Iha, K. L. Wooley, A. M. Nyström, D. J. Burke, M. J. Kade and C. J. Hawker, *Chem. Rev.*, 2009, **109**, 5620.
23. A. Dondoni and A. Marra, *Chem. Soc. Rev.*, 2012, **41**, 573.
24. B. Jung and P. Theato, *Adv. Polym. Sci.*, 2013, **253**, 37.
25. S. Moelbert, E. Emberly and C. Tang, *Protein Sci.*, 2004, **13**, 752.
26. R. M. Broyer, G. N. Grover and H. D. Maynard, *Chem. Commun.*, 2011, **47**, 2212.
27. B. Le Droumaguet and J. Nicolas, *Polym. Chem.*, 2010, **1**, 563.
28. B. S. Lele, H. Murata, K. Matyjaszewski and A. J. Russell, *Biomacromolecules*, 2005, **6**, 3380.
29. P. van Rijn, N. C. Mougin and A. Böker, *Polymer*, 2012, **53**, 6045.
30. P. van Rijn, *Polymers*, 2013, **5**, 576.
31. H. G. Börner, *Macromol. Chem. Phys.*, 2007, **208**, 124.
32. B. S. Sumerlin, *ACS Macro Lett.*, 2012, **1**, 141.
33. K. Matyjaszewski, *Macromolecules*, 2012, **45**, 4015.
34. K. Matyjaszewski and J. Xia, *Chem. Rev.*, 2001, **101**, 2921.
35. H. Fischer, *Chem. Rev.*, 2001, **101**, 3581.
36. *Handbook of RAFT Polymerization*, ed. C. Barner-Kowollik, Wiley-VCH, Weinheim, 2008.
37. J. Nicolas, Y. Guillaneuf, C. Lefay, D. Bertin, D. Gigmes and B. Charleux, *Prog. Polym. Sci.*, 2013, **38**, 63.

38. J. Nicolas, G. Mantovani and D. M. Haddleton, *Macromol. Rapid Commun.*, 2007, **28**, 1083.
39. N. V. Tsarevsky, T. Pintauer and K. Matyjaszewski, *Macromolecules*, 2004, **37**, 9768.
40. W. A. Braunecker, N. V. Tsarevsky, A. Gennaro and K. Matyjaszewski, *Macromolecules*, 2009, **42**, 6348.
41. A. Simakova, S. E. Averick, D. Konkolewicz and K. Matyjaszewski, *Macromolecules*, 2012, **45**, 6371.
42. Y. Qi and A. Chilkoti, *Polym. Chem.*, 2014, **5**, 266.
43. S. Averick, A. Simakova, S. Park, D. Konkolewicz, A. J. D. Magenau, R. A. Mehl and K. Matyjaszewski, *ACS Macro Lett.*, 2012, **1**, 6.
44. G. N. Grover and H. D. Maynard, *Curr. Opin. Chem. Biol.*, 2010, **14**, 818.
45. H.-A. Klok, *Macromolecules*, 2009, **42**, 7990.
46. S. N. S. Alconcel, A. S. Baas and H. D. Maynard, *Polym. Chem.*, 2011, **2**, 1442.
47. D. Bontempo and H. D. Maynard, *J. Am. Chem. Soc.*, 2005, **127**, 6508.
48. K. L. Heredia, D. Bontempo, T. Ly, J. T. Byers, S. Halstenberg and H. D. Maynard, *J. Am. Chem. Soc.*, 2005, **127**, 16955.
49. Q. Zeng, T. Li, B. Cash, S. Li, F. Xie and Q. Wang, *Chem. Commun.*, 2007, 1453.
50. Y. Hu, D. Samanta, S. S. Parelkar, S. W. Hong, Q. Wang, T. P. Russell and T. Emrick, *Adv. Funct. Mater.*, 2010, **20**, 3603.
51. P. van Rijn, N. C. Mougin, D. Franke, H. Park and A. Böker, *Chem. Commun.*, 2011, **47**, 8376.
52. P. van Rijn, H. Park, K. Özlem Nazli, N. C. Mougin and A. Böker, *Langmuir*, 2013, **29**, 276.
53. N. C. Mougin, P. van Rijn, H. Park, A. H. E. Müller and A. Böker, *Adv. Funct. Mater.*, 2011, **21**, 2470.
54. H. Li, M. Li, X. Yu, A. P. Bapat and B. S. Sumerlin, *Polym. Chem.*, 2011, **2**, 1531.
55. M. Li, H. Li, P. De and B. S. Sumerlin, *Macromol. Rapid Commun.*, 2011, **32**, 354.
56. C. Cummings, H. Murata, R. Koepsel and A. J. Russell, *Biomacromolecules*, 2014, **15**, 763.
57. C. Boyer, X. Huang, M. R. Whittaker, V. Bulmus and T. P. Davis, *Soft Matter*, 2011, **7**, 1599.
58. H. Murata, C. S. Cummings, R. R. Koepsel and A. J. Russell, *Biomacromolecules*, 2013, **14**, 1919.
59. C. Cummings, H. Murata, R. Koepsel and A. J. Russell, *Biomaterials*, 2013, **34**, 7437.
60. R. M. Broyer, G. M. Quaker and H. D. Maynard, *J. Am. Chem. Soc.*, 2008, **130**, 1041.
61. S. E. Averick, C. G. Bazewicz, B. F. Woodman, A. Simakova, R. A. Mehl and K. Matyjaszewski, *Eur. Polym. J.*, 2013, **49**, 2919.
62. M. A. Quadir, M. Martin and P. T. Hammond, *Chem. Mater.*, 2014, **26**, 461.
63. J. Huang and A. Heise, *Chem. Soc. Rev.*, 2013, **42**, 7373.

64. T. J. Deming, *Adv. Polym. Sci.*, 2006, **202**, 1.
65. J. Kowal, X. Zhang, I. A. Dinu, C. G. Palivan and W. Meier, *ACS Macro Lett.*, 2014, **3**, 59.
66. C. LoPresti, H. Lomas, M. Massignani, T. Smart and G. Battaglia, *J. Mater. Chem.*, 2009, **19**, 3576.
67. K. Kita-Tokarczyk, J. Grumelard, T. Haefele and W. Meier, *Polymer*, 2005, **46**, 3540.
68. A. Blanazs, S. P. Armes and A. J. Ryan, *Macromol. Rapid Commun.*, 2009, **30**, 267.
69. N. Muhammad, T. Dworeck, M. Fioroni and U. Schwaneberg, *J. Nanobiotechnol.*, 2011, **9**, 8.
70. M. Nallani, S. Benito, O. Onaca, A. Graff, M. Lindemann, M. Winterhalter, W. Meier and U. Schwaneberg, *J. Biotechnol.*, 2006, **123**, 50.
71. S. Ihle, O. Onaca, P. Rigler, B. Hauer, F. Rodríguez-Ropero, M. Fioroni and U. Schwaneberg, *Soft Matter*, 2011, **7**, 532.
72. P. Tanner, O. Onaca, V. Balasubramanian, W. Meier and C. G. Palivan, *Chem.–Eur. J.*, 2011, **17**, 4552.
73. M. Nallani, O. Onaca, N. Gera, K. Hildenbrand, W. Hoheisel and U. Schwaneberg, *Biotechnol. J.*, 2006, **1**, 828.
74. C. Nardin, J. Widmer, M. Winterhalter and W. Meier, *Eur. Phys. J. E: Soft Matter Biol. Phys.*, 2001, **4**, 403.
75. W. Meier, C. Nardin and M. Winterhalter, *Angew. Chem., Int. Ed.*, 2000, **39**, 4599.
76. A. Ranquin, W. Versées, W. Meier, J. Steyaert and P. Van Gelder, *Nano Lett.*, 2005, **5**, 2220.
77. H. Wang, T.-S. Chung, Y. W. Tong, K. Jeyaseelan, A. Armugam, Z. Chen, M. Hong and W. Meier, *Small*, 2012, **8**, 1185.
78. M. Kumar, M. Grzelakowski, J. Zilles, M. Clark and W. Meier, *Proc. Natl. Acad. Sci. U. S. A.*, 2007, **104**, 20719.
79. A. Graff, M. Sauer, P. Van Gelder and W. Meier, *Proc. Natl. Acad. Sci. U. S. A.*, 2002, **99**, 5064.
80. O. Onaca, P. Sarkar, D. Roccatano, T. Friedrich, B. Hauer, M. Grzelakowski, A. Güven, M. Fioroni and U. Schwaneberg, *Angew. Chem., Int. Ed.*, 2008, **47**, 7029.
81. M. Nallani, M. Andreasson-Ochsner, C.-W. D. Tan, E.-K. Sinner, Y. Wisantoso, S. Geifman-Shochat and W. Hunziker, *Biointerphases*, 2011, **6**, 153.
82. A. Graff, C. Fraysse-Ailhas, C. G. Palivan, M. Grzelakowski, T. Friedrich, C. Vebert, G. Gescheidt and W. Meier, *Macromol. Chem. Phys.*, 2010, **211**, 229.
83. H.-J. Choi and C. D. Montemagno, *Nano Lett.*, 2005, **5**, 2538.
84. J.-F. Le Meins, O. Sandre and S. Lecommandoux, *Eur. Phys. J. E: Soft Matter Biol. Phys.*, 2011, **34**, 14.
85. G.-Y. Liu, C.-J. Chen and J. Ji, *Soft Matter*, 2012, **8**, 8811.
86. O. Onaca, R. Enea, D. W. Hughes and W. Meier, *Macromol. Biosci.*, 2009, **9**, 129.
87. P. van Rijn, M. Tutus, C. Kathrein, N. C. Mougin, H. Park, C. Hein, M. P. Schürings and A. Böker, *Adv. Funct. Mater.*, 2014, **24**, 6762.

88. X. Huang, M. Li, D. C. Green, D. S. Williams, A. J. Patil and S. Mann, *Nat. Commun.*, 2013, **4**, 2239.
89. J. J. Lundquist and E. J. Toone, *Chem. Rev.*, 2002, **102**, 555.
90. V. Vázquez-Dorbatt, J. Lee, E.-W. Lin and H. D. Maynard, *ChemBioChem*, 2012, **13**, 2478.
91. J. Lee, E.-W. Lin, U. Y. Lau, J. L. Hedrick, E. Bat and H. D. Maynard, *Biomacromolecules*, 2013, **14**, 2561.
92. R. J. Mancini, J. Lee and H. D. Maynard, *J. Am. Chem. Soc.*, 2012, **134**, 8474.
93. M. De, P. S. Ghosh and V. M. Rotello, *Adv. Mater.*, 2008, **20**, 4225.
94. M. Mahmoudi, I. Lynch, M. R. Ejtehadi, M. P. Monopoli, F. B. Bombelli and S. Laurent, *Chem. Rev.*, 2011, **111**, 5610.
95. A. A. Shemetov, I. Nabiev and A. Sukhanova, *ACS Nano*, 2012, **6**, 4585.
96. S. Rana, Y.-C. Yeh and V. M. Rotello, *Curr. Opin. Chem. Biol.*, 2010, **14**, 828.
97. T. Heck, G. Faccio, M. Richter and L. Thöny-Meyer, *Appl. Microbiol. Biotechnol.*, 2013, **97**, 461.
98. K. Morihara, *Trends Biotechnol.*, 1987, **2**.
99. A. M. Wagner, M. W. Fegley, J. B. Warner, C. L. J. Grindley, N. P. Marotta and E. J. Petersson, *J. Am. Chem. Soc.*, 2011, **133**, 15139.
100. C. Thalmann and T. Lötzbeyer, *Eur. Food Res. Technol.*, 2002, **214**, 276.
101. S. Riva, *Trends Biotechnol.*, 2006, **24**, 219.
102. F. Hollmann, Y. Gumulya, C. Tölle, A. Liese and O. Thum, *Macromolecules*, 2008, **41**, 8520.
103. M.-L. Mattinen, K. Kruus, J. Buchert, J. H. Nielsen, H. J. Andersen and C. L. Steffensen, *FEBS J.*, 2005, **272**, 3640.
104. K. Minamihata, M. Goto and N. Kamiya, *Bioconjugate Chem.*, 2011, **22**, 74.
105. M. Færgemand, J. Otte and K. B. Qvist, *J. Agric. Food Chem.*, 1998, **8561**, 1326.
106. J. M. Woodley, *Curr. Opin. Chem. Biol.*, 2013, **17**, 310.
107. D. Brady and J. Jordaan, *Biotechnol. Lett.*, 2009, **31**, 1639.
108. T. Davids, M. Schmidt, D. Böttcher and U. T. Bornscheuer, *Curr. Opin. Chem. Biol.*, 2013, **17**, 215.
109. J. Palomo, *Curr. Org. Chem.*, 2013, **17**, 691.

Inorganic–Protein Hybrid Bionanostructures

PATRICK VAN RIJN*[a,b]

[a]Department of Biomedical Engineering, FB-40, University Medical Center Groningen, University of Groningen, Antonius Deusinglaan 1, 9713 AV Groningen, The Netherlands; [b]W. J. Kolff Institute for Biomedical Engineering and Materials Science, FB-41, University Medical Center Groningen, University of Groningen, Antonius Deusinglaan 1, 9713 AV Groningen, The Netherlands
*E-mail: p.van.rijn@umcg.nl

2.1 Introduction

In Nature, there are many examples where inorganic species and materials are combined with proteins, and these vary in function, *e.g.* catalysis, often in oxidation reactions or storage to regulate uptake and release of ionic species, but also for the formation of composite biological materials such as bone and teeth during biomineralization processes.[1-3] Therefore, from a material science point of view, there are many interesting possibilities to be explored using proteins combined with inorganic materials to develop new applications, synthetic methods, material properties and morphologies.[4-89]

In this chapter, bionanocomposites are addressed where a protein is used either to act as a nanoreactor for particle synthesis, as a template for new nanomaterial development or for biofunctionalization rendering

RSC Smart Materials No. 16
Bio-Synthetic Hybrid Materials and Bionanoparticles: A Biological Chemical Approach Towards Material Science
Edited by Alexander Böker and Patrick van Rijn
© The Royal Society of Chemistry 2015
Published by the Royal Society of Chemistry, www.rsc.org

nanomaterials more biocompatible. For many of these approaches, multi-component protein structures such as cage proteins and virus particles are used, which will be the focus here. These multicomponent structures have various capabilities for hosting inorganic materials both on the surface [Figure 2.1(A)] and inside the assembly [Figure 2.1(B) and (C)]. These are general approaches that can be used in combination with assembled protein structures including cage proteins and virus structures.

Cage proteins such as the iron storage protein ferritin, a 24-mer protein assembly of 12 nm in diameter with an internal cavity of about 8 nm, and ferritin-like proteins such as DNA-binding proteins from starved cells (DPS) or heat shock proteins (HSP) are often used for hybrid bionanoparticle formation.[10] These proteins have the capability of reversibly disassembling and reassembling, but also the inner cavity is connected to the outside medium *via* small channels allowing specific species to enter. Many virus particles have similar structural properties and assembly behaviour, meaning that also here a multiprotein structure provides an isolated cavity that can be reached by opening the structure slightly for species to enter but also *via* disassembly/reassembly. An additional interesting feature of virus particles is that there are different sizes and shapes, which makes their use as templates also very interesting for the production of anisotropic nanostructures that

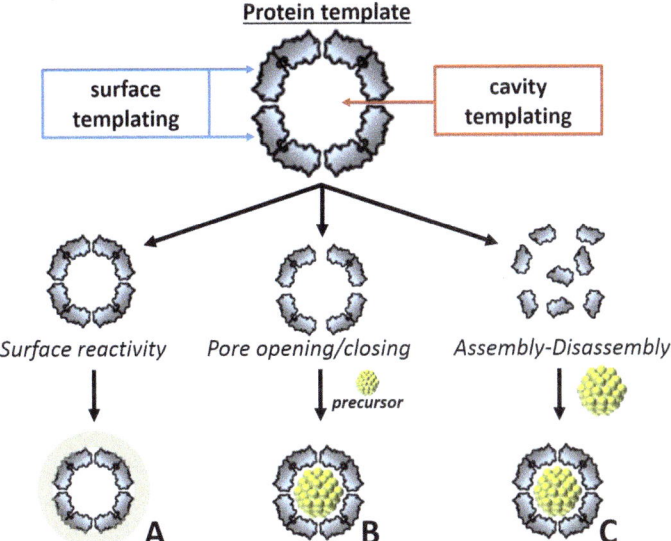

Figure 2.1 The main approaches for combining inorganic materials with multi-component protein structures. Either the surface reactivity is used to deposit inorganic materials, creating a shell around the protein structure (A); small precursor components such as ions penetrate the structure *via* channels that are large enough or that can be opened and subsequent formation of a solid core is induced (B); the protein complex is disassembled in the presence of an inorganic structure that is entrapped upon reassembly (C).

cannot be reached *via* conventional synthetic approaches.[11-14] However, it should be mentioned that nanorods and other shapes are becoming increasingly accessible without the use of templates *via* site-controlled nucleation and growth, as is seen for gold nanorod formation.

Although the inside of cage proteins and also the outside can be used as a template, an additional feature is that the outside and inside are chemically different, which is the reason why ferritin, for example, can specifically form an iron oxide core but also the reason why virus particles interact with RNA or DNA that is encapsulated within the virus capsid protein assembly. This difference in chemistry and thereby affinity can be use for specific interaction with surfaces, often *via* charge interactions. Disassembly of the protein structures and consecutively reassembly in the presence of complementary surfaces will direct the protein assembly towards that surface, thereby coating it with a single layer of protein subunits. This convenient approach of biofunctionalization renders a surface biocompatible in addition to adding new surface properties. Many metallic and other inorganic surfaces have been modified in this way, ranging from planar surfaces to micro- and nanoparticles formed from various materials such as gold, silicon dioxide and even semiconductor nanoparticles.[15-18]

For many approaches, the natural [wild-type (wt)] cage proteins or virus particles can be used. However, one of the major advantages of protein structures is that they can be genetically modified and therefore specific interactions, reactivity and affinity can be tuned to broaden the scope of protein–inorganic material combinations and the features and uses of hybrid bionanoparticles.

2.2   Protein–Inorganic Hybrid Nanostructures *via* Nanoparticle Templating

The combination of proteins with inorganic materials is an often used approach since one can combine the structural features or catalytic activity of the protein with the many different properties associated with metallic, metal oxide or semiconducting materials.[15,16,19,20] Many different approaches can be used to functionalize the surface with biomolecules such as proteins. The approaches can be either *via* covalent attachment or *via* non-covalent interactions such as charge interactions, hydrogen bonding or van der Waals interactions.[20] The type of interaction and the protein structure dictate the orientation that the proteins take at the surface. The use of multicomponent protein assemblies is advantageous with respect to protein orientation since these often have a specific interior that is significantly different from the outside of the protein assembly and there is already a strong affinity between the proteins due to complementary interactions from which the natural assembly behaviour originates (Figure 2.2).[15,20,21] Although many other approaches also with non-protein assemblies are able to add interesting functionalities and biocompatibility, in this chapter protein assemblies are

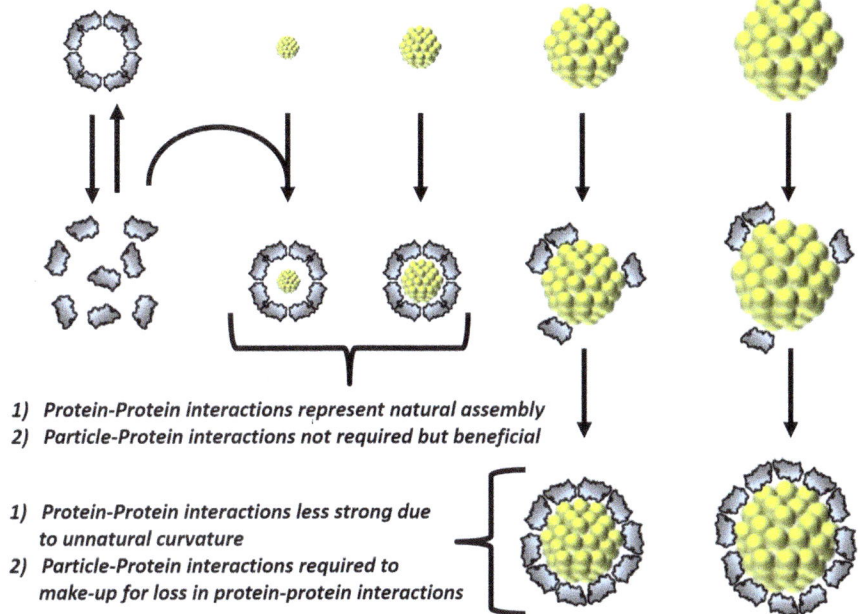

1) **Protein-Protein interactions represent natural assembly**
2) **Particle-Protein interactions not required but beneficial**

1) **Protein-Protein interactions less strong due to unnatural curvature**
2) **Particle-Protein interactions required to make-up for loss in protein-protein interactions**

Figure 2.2 Protein complexes that can disassemble/reassemble can be used as coatings for inorganic structures. The ability to encapsulate the structure depends on the size of the structure and also the protein subunit interactions with the surface of the inorganic structure.

mainly discussed. For more approaches and interface interactions concerning inorganic materials and proteins, the reader is referred to recent interesting reviews on this topic.[15,20]

For rendering surfaces biocompatible, protein structures can be used to mask the surface properties of the material. When a protein complex is used, whether it is a globular assembly or rod-like assembly, two interactions play an important role in the encapsulation/adhesion process. First there is the natural complementary interaction between protein subunits of which the complex is formed. These complementary interactions ensure complex stability and any alterations in orientation between the subunits will diminish the favourable interactions. Therefore, when inorganic structures/particles of dimensions smaller than the inner compartment are encapsulated, the inter-protein subunit interactions are maintained (Figure 2.2). However, when encapsulation of structures larger than the original cavity is attempted, the inter-protein subunit interactions will diminish and it will be less likely that a closed shell will be formed. In order to ensure proper encapsulation, favourable interactions between particle and protein subunit can overcome the loss of interactions between subunits and therefore still collect a protein shell around the particle (Figure 2.2). The remaining complementarity between the protein subunits will provide additional stability of the protein coating, but this will be lower with increasing mismatch in curvature, *i.e.* with increasing particle size. The influence of the particle size around which the protein subunits are collected has been shown to alter protein particle symmetry significantly in, *e.g.*, virus particles.[22]

To achieve optimal interaction between the protein subunit and the particle surface, one needs to determine the chemical nature of the inner surface of the protein complex and tailor the particle surface properties accordingly. In most cases this will be the surface charge. However, even though the particle surface charge and the protein subunit are complementary, additional factors such as the ionic strength of the surrounding medium also play a significant role, as is known from analysing many protein assemblies under different conditions.[23–27]

The influence of charge, pH and charge interactions was clearly demonstrated by Dragnea and co-workers.[22,25,26,28] They used the brome mosaic virus (BMV), which is a globular virus structure composed of 180 identical copies of the virus capsid protein. Conventionally, BMV assembles around the genomic template, but it is also able to assemble around other negatively charged templates. For this purpose, gold nanoparticles were modified using a carboxylate-terminated thioalkylated tetraethylene glycol ligand to mimic the interactions found between BMV capsid proteins and RNA template (Figure 2.3).[28] Although the interactions between the negatively charged surface of the particle and the BMV capsid proteins under physiological pH (7.5) assembled around the template, adjusting the pH of the assembly conditions to 4.5 resulted in a smaller particle size. The difference between the particle sizes at

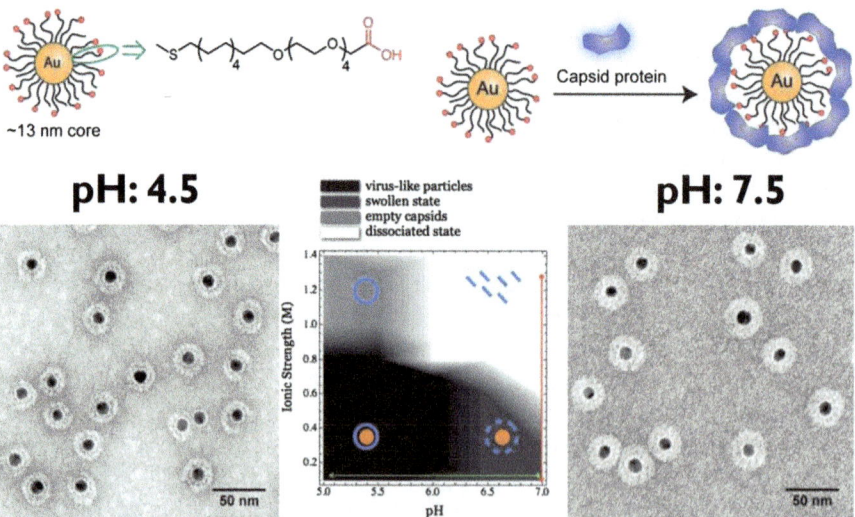

Figure 2.3 Modified gold nanoparticle having a carboxylate-containing surface that strongly associates with the BMV capsid proteins, creating a shell around the particle and creating a virus-like particle. The structure and the assembly process depend on both pH and ionic strength. Adapted from ref. 28 with permission from the Royal Society of Chemistry.

pH 7.5 (31 ± 3 nm) and pH 4.5 (28 ± 2 nm) was small but indicated a difference not only in morphology but also in the assembly process. This was further investigated using the intrinsic fluorescent properties of the proteins which originate from aromatic amino acid residues such as tryptophan, tyrosine and phenylalanine, which are quenched by the interaction with gold nanoparticles. It was found that the assembly process is much more cooperative than under physiological conditions. The explanation for this effect originates from the difference between protein–protein interactions and protein–surface interactions. At pH 7.5, 99% of the carboxylates is deprotonated and therefore the negatively charged surface has a high affinity for the lysine-containing inner face of the BMV capsid proteins. At pH 4.5, a maximum of 40% of the carboxylates is deprotonated, hence the charge interactions decrease. Upon assembly at pH 7.5, individual protein subunits are collected at the surface of the particles until they are completely covered due to the strong coulombic interactions. At pH 4.5, this interaction is not as strong as at pH 7.5, therefore initially the protein–protein interactions prevail, forming small aggregated structures. These are then collected by the particles and any increase in protein concentration is associated with further aggregation around the existing protein–particle aggregates that act as nucleation points, which contributes to the cooperative effect.

In addition to pH, it was also demonstrated that ionic strength plays an important role in the process. From the phase diagram (Figure 2.3), at low ionic strength both low and high pH resulted in nanoparticle-encapsulated structures where at high pH the shell was more diffuse. At higher ionic strength, charges will be screened, meaning that the specific coulombic interactions between particle and protein will be strongly diminished by the presence of high concentrations of other ions. This results in an empty virus particle at low pH and a completely non-aggregating system at high pH in combination with high ionic strength. This kind of assembly behaviour is not specific for a particle virus capsid protein assembly, and is also observed with many native virus species such as tobacco mosaic virus (TMV) and cowpea chlorotic mosaic virus (CCMV), and also the native structure of BMV.[29,30]

The approach discussed above is a general one and will result in a uniform coating around the inorganic particle, and any particle material bearing similar surface-confined functional groups can be used. It should be noted that the generality of the approach depends on the curvature of the surface (size of the particle), since any strong deviation of the natural virus particle curvature will result in a larger mismatch between protein subunit–protein subunit interactions, which will influence the packing and stability.[31] The interactions are based on coulombic interactions and the assembly will follow the direction of the surface of the particle. This means that when a spherical particle is used, a spherical assembly will be formed despite the original shape of the virus particle, be it globular or rod-like, although this is not a general rule since too great a deviation of the natural shape may strongly influence the packing density.

The native assembly process with retention of virus morphology but still occurring at the particle surface is an approach that was recently designed and developed by Wege and co-workers.[32] They made use of the natural assembly process around the native genomic template, thereby maintaining the virus particle morphology. However, instead of performing the assembly with the RNA in solution, the assembly process was carried out at the surface by attaching the RNA with one side *via* covalent coupling. The assembly then commences around the RNA strand, creating a surface covered with fully assembled virus particles; in this case TMV was used. The approach was performed initially on planar surface, but there should not be any restrictions on this provided that the RNA strand can be attached to the surfaces, which can be on nano/microparticle, rod and planar surfaces.[33–35] This approach was demonstrated in combination with gold nanoparticles onto which the RNA strands were attached and subsequently TMV was assembled, creating very interesting assembled coatings around nanoparticles but also as a complete particle assembly presenting new virus-like structures.[32]

So far adhesion of subunits onto the surface has been discussed, but it is also possible to rely on the assembly in the presence of nanoparticles that are small enough to be encased by simple reassembly of the protein cage. As mentioned earlier, complementary interactions between protein subunits and particles aid in the inclusion but are not absolutely necessary. When there are no specific interactions, there will be a distribution between filled and empty cages and any increase in complementary interactions will ensure a higher encapsulation efficiency.[36] A protein often used for this purpose is ferritin, which conveniently disassembles and reassembles under the influence of pH (Figure 2.4) and is able to include various structures such as small molecular components, but also nanoparticles of different properties other than the native ferrous oxide.

When ferritin is dialysed at pH 2, the cage structure is compromised and thereby the native core will be released. However, the assembly process is completely reversible and therefore upon increasing the pH to 8.5, the cage is re-formed.[10,37] When the reassembly process is performed in combination with nanoparticles, these can be encapsulated. Without any additional interactions to facilitate the encapsulation process, the efficiency is directed by the concentration and the relative ratio between cage protein and nanoparticle.[20,21,38,39] In order to achieve much higher encapsulation efficiencies, other strategies are more reliable, such as synthesizing the particle directly inside the cage structure. This will not only increase the encapsulation efficiency but also allows for the creation of very well-defined anisotropic structures

Figure 2.4   Schematic representation of the removal of the inner core of ferritin *via* disassembly and the possible routes for the incorporation of a new synthetic core *via* reassembly. Adapted from ref. 4 with permission from Elsevier.

that are generally not possible *via* conventional synthetic approaches, *e.g.* using the 4 nm cavity within tubular virus particles such as TMV. Using the internal space of cage proteins and virus particles as a nanoreactor and in the process creating biofunctionalization of the inorganic structure is therefore an interesting alternative to the use of a pre-existing inorganic template around which the protein subunits are assembled.

2.3 Protein-Templated Synthesis of Inorganic Structures

In Section 2.2, inclusion of particles and materials that were preformed and then encapsulated *via* reconstitution of the protein cage structure was discussed and, although this is a very convenient approach, the encapsulation efficiency can still be low. To circumvent the low occupation, particles can be directly synthesized inside the protein. When soluble precursors are confined to the interior of the assembled protein structure *via* diffusion or actively encapsulating them *via* disassembly/reassembly, the nanomaterials will be formed inside when applying the appropriate reductive or oxidizing conditions (Figure 2.5).

The approach of forming the nanomaterial will have the same result as using a predefined inorganic template onto which the protein subunits will adhere, with one major advantage, namely control over the dimensions of the inorganic structure. The confinement of the nanomaterial precursors will also dictate the final size of the nanomaterial and the shape. Since bionanoparticles come in different sizes and shapes, globular and tubular, these can be utilized as such to create well-defined structures and additionally render them biocompatible since they are immediately coated with the respective template. In addition to the cavity, the surface of particles can also be addressed with even more possibilities than the inner cavity since the surface can be partly covered with discrete-sized preformed nanostructures on the surface with high accuracy, and also the surface can be completely covered with a homogeneous inorganic shell (Figure 2.6). For these approaches, one can rely on the native available protein structures, peptide sequences and amino acids, but all these can be tailored *via* genetic modifications, providing even more opportunities to create complex and novel functional hybrid bionanostructures. Below we discuss in brief both globular and anisotropic protein templates with, in particular, ferritin and TMV, respectively.

2.3.1 Globular Protein Cages for Particle Synthesis

2.3.1.1 *Cavity Confined*

Ferritin, with its unique cavity, is able to house many different inorganic materials. The highly functional protein surface in combination with technologically interesting inorganic cores yields valuable composite bioinorganic

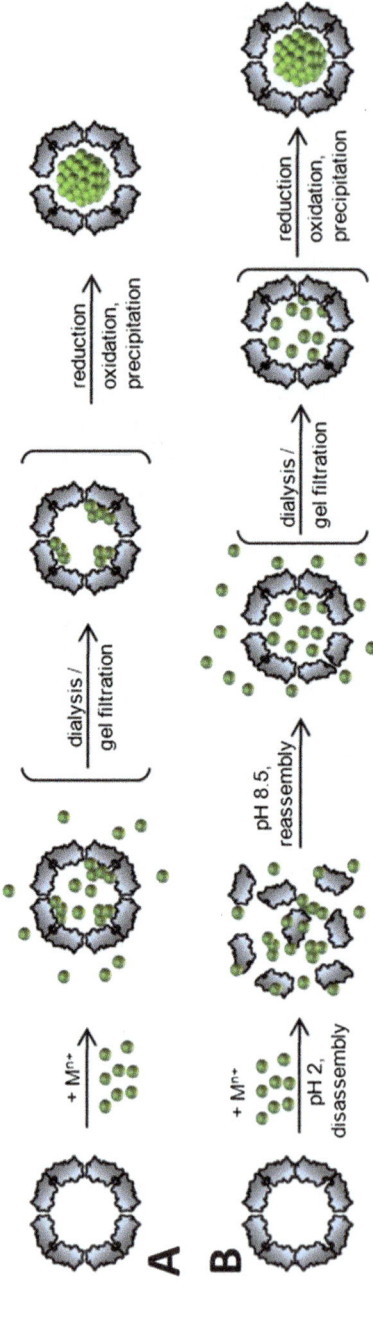

Figure 2.5 Schematic representations of different approaches for the formation of the inner core of ferritin. The two main approaches consist of using the pores inside the ferritin shell to transport ions, which are dissolved from the core or condensed upon particle formation or completely disassemble the ferritin shell and reassemble it in the presence of the desired species to be encapsulated. Reproduced from ref. 4 with permission from Elsevier.

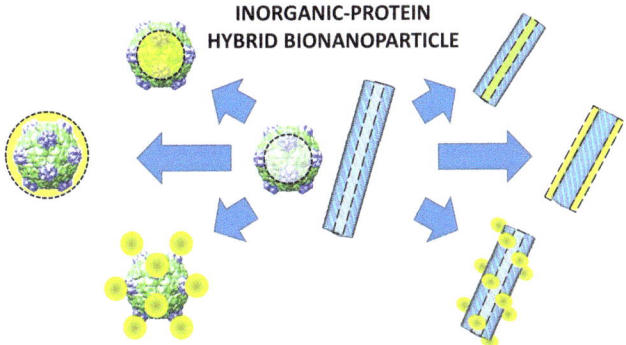

Figure 2.6 Possibilities for the addition of inorganic materials to globular and tubular virus particles. Inorganic materials can be confined to the inner voids or be confined to the surface either as a homogeneous shell or as discrete-size highly localized structures.

building blocks. Their monodisperse nature facilitates the formation of regular arrays and assemblies. Since the electronic, optical and magnetic properties of nanoparticles depend strongly on particle size and shape, the control of these aspects is important.[40] The apoferritin cage provides a viable template for size-controlled synthesis and advantageously renders the cores water soluble and biocompatible.[10,41,42] Biocompatible, functional nanoparticles are expected to have many applications in medicine and diagnostics, in chemistry and in electronics, *e.g.* as fluorescent markers, drug delivery systems and catalysts, and in data storage and quantum electronics.[43,44]

Meldrum, Mann and co-workers were the first to make use of the cage structure of ferritin as a size-constrained reaction vessel.[45,46] Although the binding of metal ions to ferritin has already been explored earlier, the interest is now in the use of ferritin as a bionanoreactor for the production and encapsulation of new inorganic nano-sized structures of different materials. Today, a large number of minerals, among them technologically interesting semiconductor nanoparticles and metal and metal alloy nanoparticles, have been formed within the apoferritin cavity.[10,47]

Two approaches have mainly been followed: either reassembly of the protein cage from the subunits in the presence of the material to be encapsulated or incubation of the empty and intact apoferritin with the respective metal ions (Figure 2.5).[48,49] Both approaches have been extensively employed for the synthesis of ferritin-encapsulated materials. The latter approach is restricted to molecules and ions that are able to penetrate the protein shell, *i.e.* that are sterically and electronically able to pass one of the channels in the protein shell (Figure 2.5).

The nucleation on the internal surface of the protein cavity is triggered by the metal ion binding capacity of negatively charged amino acids. The catalytic intra-cavity oxidation to highly oxidized ions increases the local supersaturation further and triggers the precipitation of inorganic cores. The formation of native ferrihydrite cores is a combination of the protein-facilitated redox activity and inorganic nucleation. This principle also works well with other transition metal materials. Other routes include the reduction of metal ions and complexes to metals and metal alloys, non-redox hydrolysis reactions to form hydrated oxides and hydroxides and slow coprecipitation routes. These routes with non-natural mechanisms are most often compromised by non-specific bulk precipitation. Careful control of reaction rates through reagent concentrations, pH, specific binding ligands or triggers is necessary.

Directed mineralization inside the intact capsid is a prerequisite for the controlled synthesis of nanoscale materials. It was assumed that the complementary negative electrostatic potential inside the protein shell produced by acidic amino acid residues attracts the metal ions and induces nuclei formation. After a nucleus has formed, materials may grow autocatalytically in size. The size-dependent autocatalysis on the mineral surface leads to a growth of the large particles at the expense of the smaller ones for subsequent incremental additions of ions, eventually leading to particle sizes exceeding the cavity volume.[50,51] The low catalytic activity of the protein on non-native inorganic substrates also explains the less uniform inorganic cores. While the outer surface of the protein inhibits inorganic particle formation, the inner cavity allows for precipitation and fosters uninhibited growth of inorganic matter. The metal:protein ratio often determines the nanoparticle size. However, just as important as the size of the nanoparticle are their crystalline domain sizes, which have an influence on, for example, the magnetic response.[52]

Ferritin has been used as a nanobioreactor for many different materials, including various minerals, oxides, metals and even alloys. The first non-native inorganic cores that were synthesized inside ferritin were iron sulfide, manganese oxide and uranyl oxide.[53] Iron sulfide was synthesized by chemical conversion of the native iron oxide with H_2S or Na_2S. The final cores had a slightly smaller average size than the starting ferritin cores and were amorphous.[53,54] Other mineral cores that have been synthesized include manganese oxide cores, which were formed by reconstitution of the mineral core inside demineralized ferritin in $MnCl_2$ solutions at pH 8.0–9.2.[45,53] Uranyl oxide from uranyl acetate, commonly used for staining protein assemblies and virus particles, was also found to be able to penetrate the protein shell.[53,55] The uranium-loaded ferritin particles are anticipated to provide a novel system for neutron capture therapy.

For the synthesis of nanoparticles inside protein-based cage structures, nanomaterials with useful properties, such as magnetism or luminescence, are regarded as highly promising for new applications in biomedical applications.[56–59] Magnetic iron oxide cores were synthesized in the protein cavity (magneto-ferritin) if the reconstitution experiments were performed under conditions favourable for the formation of Fe_3O_4.[46,60,61] The magnetic stability of the cores was very sensitive with respect to their size in such a way that the magnetic response changed by three orders of magnitude whereas the core volume changed by only one order of magnitude.[62] Many synthetic procedures have been developed over the years to enhance core formation and control the core properties better also for many other metal oxide combinations, such as with arsenate, molybdate and vanadate.[63–67]

The alkaline earth metal ions calcium, strontium and barium have low binding affinities to apoferritin.[48] Thus, although there is no active mechanism of particle formation inside the cavity, successful mineralization can be achieved by actively inhibiting the precipitation in bulk solution. Mann and co-workers added water-soluble polyelectrolytes, namely poly(methacrylic acid) or sodium polyphosphate, to the mineralization solution.[68] Such polymers act as potent inhibitors; however, since they cannot enter the apoferritin cavity because of their large size, their action is restricted to the bulk solution and mineral precipitation can occur only within the protein shell. $BaCO_3$, $SrCO_3$ and hydroxyapatite cores could be synthesized by the same approach, yielding particles with an average size of 5.1–5.3 nm. Other approaches and more efficient synthetic methods have also been developed by others, not only for carbonates but also other oxide and hydroxide variations combined with Fe, Eu and Ti.[69]

Metal oxides and other mineral cores have been discussed above, but pure metal nanoparticles also have many potential applications in, *e.g.*, catalysis. This approach is preferred when particles are monodisperse in size since many attributes are connected to this such as catalytic power and plasmon resonance, and also physiologically, *e.g.* interactions with organisms. For this reason, synthesis of metal nanoparticles inside the ferritin nanocage has been regarded as a convenient approach to achieve monodisperse particle sizes in addition to making them directly suspendable in water due to the protein coat.[9,10]

Pd(0) nanoclusters inside the ferritin capsid were used by Ueno *et al.* for the catalytic hydrogenation of olefins in aqueous medium.[70] The synthesis was carried out by first incubation of the apoferritin with Pd(II) salt under slightly alkaline conditions and then reduction to Pd(0) clusters by $NaBH_4$. Although no black precipitates were formed as in protein-free control experiments, it was shown by UV/Vis spectroscopy, fast protein liquid chromatography (FPLC) and native polyacrylamide gel electrophoresis (PAGE) that small Pd clusters were present inside the ferritin cavity. Similarly, Pd–Au alloy nanoparticles and bimetallic Pd@Au or Au@Pd core–shell particles were prepared by a co-reduction or sequential reduction.[71] The different catalytic behaviours of these particles in the hydrogenation of acrylamide discriminated the different nanoparticle morphologies.

For creating metallic platinum particles inside the ferritin template, Deng *et al.* incubated apoferritin with K_2PtCl_6 and obtained Pt–ferritin hybrid particles after reduction with $NaBH_4$.[72] The size distribution of the nanoparticles obtained was fairly broad with an average size of 4.7 ± 0.9 nm. The effect of Pt-loaded ferritin on the reduction of oxidative stress and cytotoxicity on cells was explored by Knez and co-workers.[73] Many approaches using ferritin for metallic nanoparticle synthesis have been developed, including Co, Ni, Cu, Ag and Au and even alloys (see the references cited in ref. 10).

In addition to metallic nanoparticles, semiconductor materials can also be incorporated into the ferritin biotemplate, which expands the scope of ferritin use for controlled biohybridization and templated synthesis containers. Yamashita *et al.* used a slow reaction system for the synthesis of CdSe nanoparticles.[74] They stabilized the Cd(II) ions through complexation with ammonia as $[Cd(NH_3)_4]^{2+}$. Further, they used selenourea, which slowly degrades in aqueous solution and provides a continuous source of small amounts of selenium ions. Very similar in approach was the synthesis by Xing *et al.* using ethylenediaminetetraacetic acid (EDTA) as Cd(II) complexing ligand and NaHSe as selenium source.[75] The quantum dots obtained with a cubic zinc blende structure showed no cytotoxicity in a simple test with HeLa and HepG2 cells.

Many semiconductor nanoparticles have been synthesized inside the ferritin nanocage under various conditions and in different yields, indicating that there is not a common route that can be used and that for each particle optimum conditions need to be established. In various studies, particles such as CdS, ZnSe, CuS, Au_2S and PbS have been synthesized within the cavity of ferritin (see the references cited in ref. 10).

2.3.1.2   Surface Confined

When discussing protein nanocages, using them as a bioreactor due to the isolated inner compartment seems obvious. Above, ferritin was discussed in combination with inorganic materials. Although globular virus particles are less frequently used in that manner, they are more often used for encapsulation with organic components and even other protein structures or combined with predefined inorganic templates as discussed in Section 2.2.[11,15,17–20,76–78] More frequently, viral nanoparticles are used as a template for adherence or condensation of inorganic materials on the virus particle surface. Owing to the presence of surface reactivity, charge or introduced *via* genetic modifications, inorganic materials can be selectively deposited.[11,15,17–20,76–78]

Since the protein subunits display distinctive areas of reactivity, it is possible to make presynthesized nanostructures adhere to exact positions on the surface in addition to covering the complete surface homogeneously with a shell coating. Peptide sequences present on the surface can be targeted using immunolabelling (Figure 2.7). Complementary labels can be attached to any kind of particle, such as gold, silver or semiconductor nanoparticles. This allows for very exact positioning since these labels adhere only to specific domains. This approach allows the surface still to be accessible for other modifications.

Additionally, since the surface has many functional groups displayed, coordination of certain inorganic material precursors is also possible. Upon coordination of *e.g.* various metal ions such as Ca^{2+}, Ag^+ and Pt^{2+}, with subsequent precipitation or reduction, an inorganic shell is formed around the protein cage.[76–80] It is possible to use the native reactivity and coordination capabilities of the virus structure, but also *via* introducing additional peptide

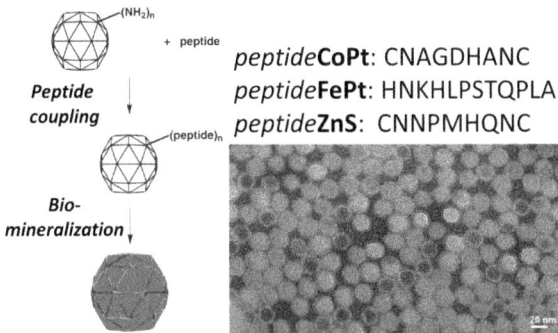

*peptide***CoPt**: CNAGDHANC
*peptide***FePt**: HNKHLPSTQPLA
*peptide***ZnS**: CNNPMHQNC

Figure 2.7 Schematic representation of the addition of peptide structures to the surface *via* coupling to the available lysine residues. Each peptide is able to coordinate specific ions and, after reduction to colloidal metal, a shell as shown in the TEM image is formed around the virus particle. Adapted from ref. 79 with permission from the Royal Society of Chemistry.

sequences either chemically or genetically, such that more complex strategies can be developed.[80] Many peptide sequences are known that are able to perform various chemistries such as condensation of tetraethyl orthosilicate (TEOS) into silicon dioxide, use of organotitanates as precursors for titanium dioxide and formation of hydroxyapatite *via* biomineralization processes either with or without additional reducing reagents. Such a general approach has been reported by Evans and co-workers[23] and others. Evans and co-workers added chemically to the surface of cowpea mosaic virus (CPMV) different peptides or polymers, which were able to coordinate specific ions and ion combinations very well and, by reducing the metal ions using NaBH$_4$, various surface-confined inorganic coatings were formed around the CPMV (Figure 2.7).[23] The protein structure not only provides the reactivity in those cases but also functions as a template to control the dimensions of the inorganic–protein hybrid nanostructures.

So far either internal particle formation or surface-confined inorganic material deposition has been targeted and, although one can make use of an alloy-type material as discussed above, it will still provide only a limited number of applications. However, since a cage protein such as ferritin and globular virus particles both have an addressable interior in addition to an exterior, both can be used to confine different materials with different properties (Figure 2.8). This allows for more functions to be incorporated selectively with close control over the morphology, size and position of materials. One such approach was described by Wang and co-workers, who used Simian virus 40 (SV40), which is 24 nm in diameter, is composed of 12 pentamers of the major capsid protein VP1 and has a cage structure with $T = 1$ icosahedral symmetry as a multifunctional scaffold.[81] A semiconductor nanoparticle was trapped inside the virus structure *via* a disassembly/reassembly procedure in the presence of the respective quantum dots. After the quantum dot encapsulation,

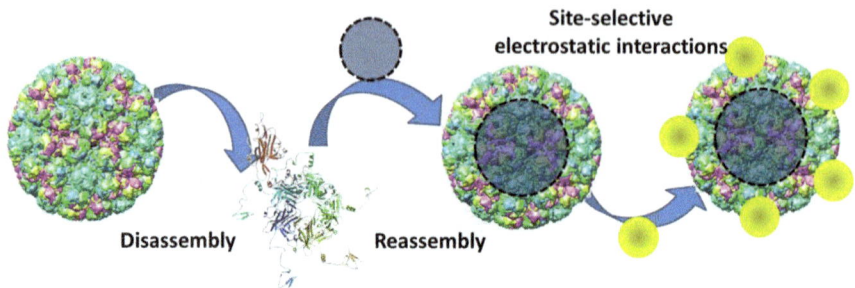

Figure 2.8   Schematic representation of the use of SV40 for dual targeting. Incorporation of nanoparticles into the inner compartment *via* disassembly/reassembly followed by electrostatic deposition of negatively charged nanoparticles site selectively on the surface of the virus particle.

gold nanoparticles were introduced *via* electrostatic interactions using bis(*p*-sulfonatophenyl)phenylphosphine (BSPP)-passivated gold nanoparticles. This approach displays the advantage of using both the inner compartments and the cage outer surface for the introduction of inorganic materials.

Various combinations of properties, such as fluorescence, plasmonic absorption, reactivity and electronic properties, can be introduced for use as novel nanostructures or nanodevices for theranostics, energy harvesting, biosensing and catalysis.

2.3.2   Anisotropic Protein Structures for Particle Synthesis

So far, globular protein cage structures have been discussed; however, the use of anisotropic structures will broaden the scope of materials, particle structures and hence properties. It is easily imaginable that long, thin particles would act as wires or needles and have a completely altered surface-to-volume ratio, and can target different length scales since one dimension can be in the low-nanometre regime whereas the second one can be much larger, *e.g.* TMV with 18 nm diameter and 300 nm length, PVX (potato virus X) with 13.5 nm diameter and 515 nm length and M13 bacteriophage with 6–7 nm diameter and 900 nm length. Although the general approach for targeting the surface is very similar to that for the globular proteins, the inner cavity is more problematic since this extends, as mentioned above, to sizes much larger than conventionally used with globular protein cage particles. Encapsulation *via* disassembly/reassembly requires a presynthesized particle of dimensions slightly smaller than those of the cavity, which is synthetically very challenging.

2.3.2.1   *Cavity Confined*

For the incorporation of inorganic materials into an elongated cavity, a rigid structure is most convenient since flexibility during the formation of the inorganic materials will most likely result in damaged structures due to the

ANISOTROPIC CAVITY TEMPLATING FOR NANOWIRE PRODUCTION

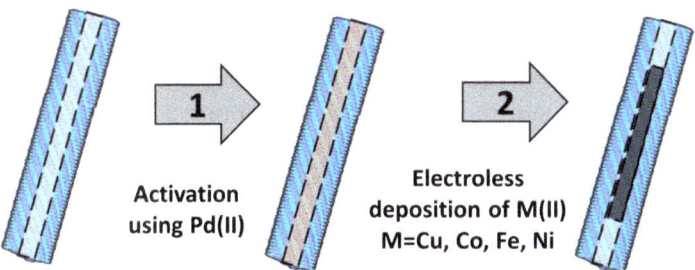

Figure 2.9   Schematic approach for nanowire formation inside the cavity of TMV, which extends throughout the length of the virus particle structure. The electroless deposition occurs first *via* activation with Pd(II) (1) and subsequent treatment with Cu(II), Co(II), Fe(II), Ni(II) or combinations (2) provides the formation of the nanowire.

extremely small diameter of those channels. TMV is most suited for this purpose since it is not only highly robust but also very rigid.[82,83] There have only been very few successful attempts at utilizing the channel for the incorporation of inorganic materials. On other occasions, small molecular components have been incorporated since this is much easier and here one can use the disassembly/reassembly process.[84] Only few studies have resulted in the successful formation of nanowires inside a tubular virus structure. Bittner and co-workers began such work in 2003 and are currently able to implement Cu, CoFe and CoFe(Ni) nanowires into the extended cavity of TMV, which was performed *via* electroless deposition (ELD) using Pd(II) activation prior to the deposition procedure.[85–87] This resulted in metallic nanowires 3–4 nm in width and 150 nm in length which were not across the entire length of the cavity (Figure 2.9).

Yamashita and co-workers also successfully implemented inorganic materials inside a tubular virus assembly, namely tomato mosaic virus (ToMV). They genetically engineered the particle to have more coordinating amino acid residues on the inside of the virus and, upon incubation with metal ions and subsequent reduction, both nanowires and nanoparticles were formed inside the structure.[88] Using this approach, they easily achieved alignment of magnetic nanoparticles, which is interesting for various electronic applications and devices.

2.3.2.2   Surface Confined

Surface confinement of inorganic materials on anisotropic virus particles such as TMV, PVX and M13 can be addressed in a similar fashion to that for globular virus particles. One can rely on the natural interactions such as electrostatic interactions or it is possible to introduce chemically or genetically groups or peptide sequences that facilitate coordination or mineralization.

More interesting for long anisotropic structures with respect to the smaller globular structures is that the structure can become electrically conductive and span across a larger distance than the globular virus structures. This is very promising for the production of nanowires and electronic arrays. Many groups have studied and reported on the development of synthetic procedures for producing such structures with either native virus particles or those genetically modified using a variety of virus structures such as TMV, ToMV, PVX, M13 and others.[17,78,89,90]

The formation of a closed shell-covered virus particle or a particle-covered virus particle depends on the application or the property that is targeted. A closed shell composed of a metal is conductive across the entire length of the virus particle whereas a metal nanoparticle-covered virus structure displays gaps between the conductive metal structure that could impair conductivity (Figure 2.10). On the other hand, with a fully covered closed shell, the virus surface is no longer accessible for further modifications whereas this would still be possible with a nanoparticle-covered surface. Rigidity increases when a surface is completely covered with a closed shell, which potentially would not allow for any applications where flexibility would be an asset, such as interactions with biological systems, but does improve mechanical stability when the particles need further processing for device manufacture. The systems developed are numerous and a few are highlighted below.

Several approaches for the formation of a closed shell have been investigated. One of the most straightforward methods was provided by Harris and co-workers, who were able to coat a wild-type TMV particle with silicon dioxide from TEOS in 50 : 50 v/v methanol–water containing 1 mM ammonia.[93,94] This resulted in slow condensation on the surface of the TMV particle, creating a homogeneous shell around it. There was increased tolerance towards methanol after adding the silica coating; however, the particles displayed mechanical instability towards mechanical perturbations such as resuspending the particles after centrifugation and washing. In addition to the wild-type, they also designed a new TMV particle, not to improve the interaction of the surface towards TEOS condensation but to elongate the

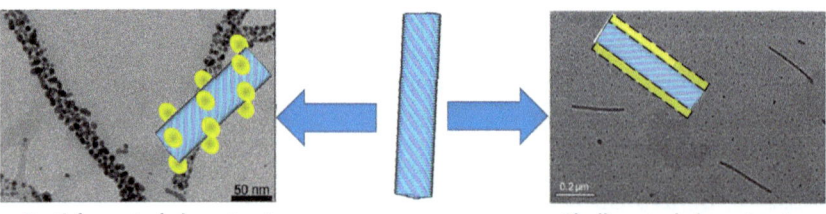

Particle coated virus structure   **Shell coated virus structure**

Figure 2.10   Schematic structure of an anisotropic virus structure covered either with nanoparticles (left, TMV with gold nanoparticles) (adapted from ref. 91 with permission from the Royal Society of Chemistry) or with a closed shell (right, TMV with a closed CdS shell) (adapted from ref. 92 with permission from the Royal Society of Chemistry).

virus particle in order to achieve longer wire-like structures. The so-called E50Q mutant had a length of almost 1000 nm and could be subjected to the same coating process.

Coating such elongated structures is interesting since also in Nature such approaches and structures can be identified in the process of forming certain types of tissues. Bone mineralization relies on anisotropic protein-based structures, namely aggregated collagen, which is mineralized, thereby forming the structurally stiff matrix found in bone.[95] Nam and co-workers used the filamentous bacteriophage fd phage as a template and modified it genetically to display surface-exposed peptide sequences that favour coordination towards either Co(II) or Ca(II).[96] The fd phage is extremely long (1–2 μm) in comparison with its diameter (6 nm). Co(II) was reduced using $NaBH_4$ and the Ca(II) was subjected to mineralization conditions for the formation of hydroxyapatite. The biomineralized fd phage hydroxyapatite structures formed bundles of fibres and could very easily be mistaken for mineralized collagen fibres, indicating that such mineralizing particles could be used in such types of applications. Also, it has been shown that TMV particles bearing mineralizing peptide sequences also stimulate the mineralization process of bone tissue.[97]

Earlier, one of the few approaches that resulted in the inclusion of inorganic materials inside the virus structure *via* ELD was discussed. Interestingly, this approach can also be used to coat the surface of anisotropic virus particles. Soto and co-workers used the ELD approach in combination with TMV, fd and M13 bacteriophages, which were combined with Pd and Cu.[98] The main difference between the inclusion experiments performed by Bittner and co-workers was that Pd particles were initially adsorbed on the surface rather than activating the structures directly with Na_2PdCl_4.

For particle formation on the surface of anisotropic virus structures, different approaches can be used. Although some approaches are very similar to the formation of homogeneous coatings *via* either mineralization or reduction using chemical agents, in some cases they result in the formation of particles rather than a shell coating. Which process will occur is difficult to predict but will probably depend on the amount that initially adheres to the surface and the capability of structures to grow together. Mann and co-workers used TMV combined with $HAuCl_4$ and $NaBH_4$ in small consecutive steps to form particles on the surface that slowly grew with each repetition of adding $HAuCl_4$ and $NaBH_4$.[91] The virus particles displayed a high density of gold nanoparticles and from UV/Vis measurements the increase in particle size could be followed, where both the intensity and the wavelength increased on performing the addition–reduction procedure five times (Figure 2.11). This synthetic strategy allows control over the properties of the particles that adhere to the surface. The approach presented was considered to be an improvement on their earlier work in which Pt and Ag nanoparticles were adhered using a one-step approach under much more acidic pH.[99]

As in some cases reducing agents are needed to create the desired particle structure, a preferred method would be biomineralization since it often

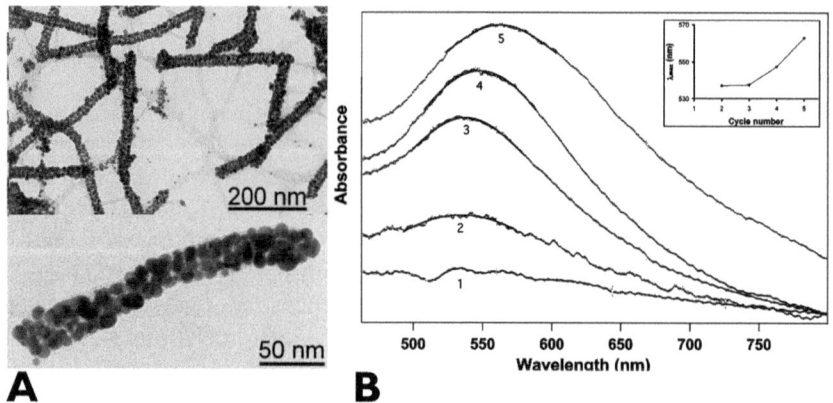

Figure 2.11 (A) TEM of TMV with gold nanoparticles grown on the surface and (B) analysis of the surface plasmon resonance of the gold nanoparticles with each growth-step. Adapted from ref. 91 with permission from the Royal Society of Chemistry.

proceeds under milder conditions, which potentially cause less damage to the virus structure. However, biomineralization will not occur in all cases. For wild-type TMV (TMV-wt), as shown earlier, simple mixing would give a homogeneous shell coating.[93] However, when PVX was used under the same conditions, no coating could be observed.[100] Therefore, an often used peptide sequence for inducing condensation of TEOS was genetically engineered on the surface of PVX. It would be expected that a similar coating to that shown for TMV would be formed, but this was not the case. Instead of a homogeneous shell coating, individual particles were formed on the surface.[100] This could be caused by the more flexible nature of the PVX compared with TMV; however, M13 and fd bacteriophages did not display this behaviour, as was shown earlier. The main difference is that the peptides engineered on the surface are not displayed on every coat protein of the virus. Therefore, not every virus coat protein initiates mineralization, hence the growth is confined to just the subunit on which the mineralization process commenced. Not having a completely homogeneously coated surface has the advantage of some virus particle surface still being available and accessible for targeting. This was subsequently performed after mineralization *via* immuno-gold labelling, which targets specific domains on the surface, and adherence occurs *via* simple mixing of the nanoparticle with the appropriate immuno-label on the surface together with the respective virus structure. This double approach, both using extremely mild conditions, resulted in double-coated PVX particles (Figure 2.12).[100]

One of the aspects that is vital when using biotemplates such as assembled protein structures is that they cannot tolerate extremely harsh conditions. When inorganic structures are used as templates, then high temperatures, extreme pH and the presence of reactive chemical entities are often not a problem. Protein structures such as virus particles and other protein

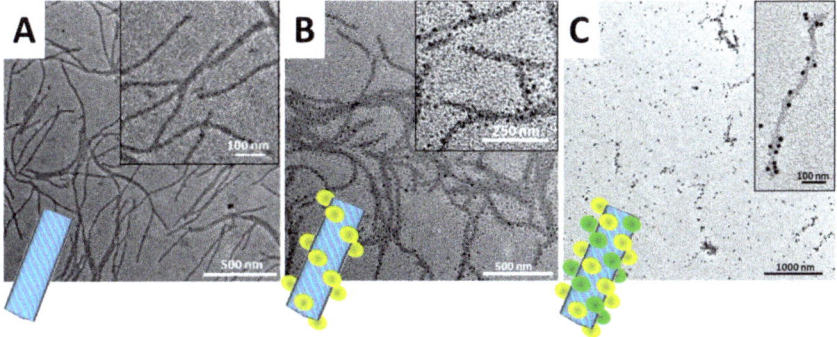

Figure 2.12 Schematic representation and TEM images for the double-coating approach going from genetically engineered PVX bearing a peptide for inducing biomineralization (A) to the biomineralization step producing SiO$_2$ nanoparticles (B) and subsequently targeted *via* immuno-gold labelling (C). Adapted from ref. 100 with permission from Wiley.

assemblies will disassemble and the protein subunits will denature. Therefore, it is always advantageous when mild and general approaches can be developed for the further use these highly versatile and promising templates. Kahn *et al.* showed such an easy and mild reaction using wild-type TMV (TMV-wt). Negatively charged gold nanoparticles (capped with citrate) and iron oxide were connected irreversibly to TMV-wt simply by mixing at pH 3, which is around the p*I* of TMV.[101] The TMV protonates to a greater extent than the nanoparticles and van der Waals forces ensure high-affinity binding. The binding becomes so strong that changing the pH to 7 or using urea does not detach the particles from the TMV surface. Although this method is easy and does not require genetically modified TMV particles, it cannot be regarded as general owing to the operating pH of 3. TMV is known for its high stability, but other structures might not be as tolerant towards these conditions.

The above examples demonstrate that different approaches under very mild conditions can be used to create various complex protein–inorganic hybrid bionanoparticles. Although no real applications have been identified for most systems, certainly more development of novel systems, combinations and synthetic conditions will provide more opportunities to achieve truly applicable systems.

2.4 Morphology Control of Inorganic–Protein Hybrid Structures

Achieving control over the position where inorganic components are integrated into protein assemblies is important for tuning the properties of each component. So far, various methods have been discussed, with nearly every combination concerning surface coatings: particle coating or dense shells and inclusion of inorganic materials in the inner voids of the proteins structures

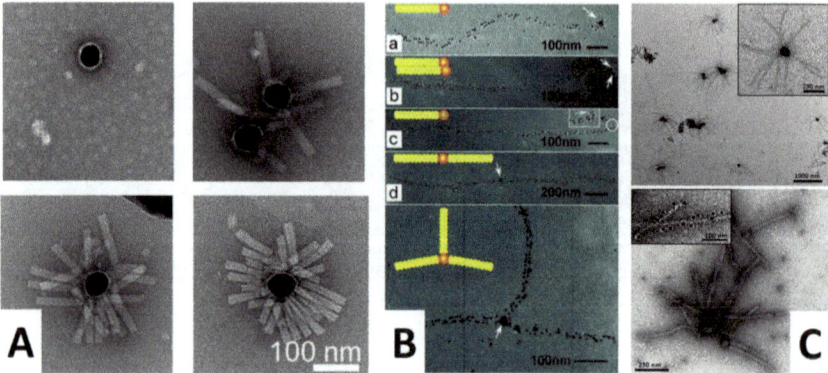

Figure 2.13   TEM images of complex inorganic–virus hybrid structures with novel morphologies. (A) Core coated with different density RNA strands, which increases the density of assembled TMV particles on the surface. Adapted from ref. 32 with permission from Wiley. (B) TEM of M13 bacteriophage that has adhesion peptides towards streptavidin on one end of the virus particle, thereby controlling the attachment of the particles to a central streptavidin-labelled inorganic core. Adapted from ref. 102 with permission from the American Chemical Society. (C) TEM images of genetically modified PVX displaying biomineralization peptides on the surface, which induce SiO$_2$-core formation, thereby incorporating PVX particles while the PVX surface is still accessible for immuno-gold labelling. Adapted from ref. 100 with permission from Wiley.

in both globular and anisotropic structures. However, an interesting feature associated with such endeavours is that similar adhesion and coordination approaches can also be used to change and control the overall morphology, thereby creating more complex nano-, micro- and mesostructures.

One example was briefly mentioned in Section 2.2, but can be considered as an alternative approach for complex structure formation. A nanoparticle is coated with different densities of RNA, which is recognized by the TMV capsid proteins, which are able to self-assemble around the individual RNA strands, producing close to native TMV structures [Figure 2.13(A)].[32] Both the density and length of the assembled virus structure can be controlled, making this a very elegant approach for complex nanohybrid structures.

Belcher and co-workers took a slightly different approach and made use of the nowadays well-established phage display method. The M13 bacteriophage has one end that is different in composition to the rest of the virus. Specific binding peptides can be expressed at this end, *e.g.* peptides that bind to streptavidin. This means that for any structure displaying streptavidin moieties on the surface, the M13 will bind to it only with one end.[102] The remainder of the surface is then still available for other modifications having other binding motifs displayed. This system has the opportunity to bind complete structures very selectively in a highly controlled fashion, forming linear single-substituted, linear double-substituted and triangular triple-substituted structures around a single inorganic core [Figure 2.13(B)].

The system shown in Figure 2.13(C) has a similar morphology to that in Figure 2.13(A) but closer to the dimensions shown in Figure 2.13(B). van Rijn *et al.* used a genetically modified PVX particle displaying silicification peptides on the surface able to convert TEOS into SiO_2.[100] Whereas the combination with TEOS gave isolated nanoparticles on the surface of the PVX, mixing TEOS and 3-aminopropyltrimethoxysilane in a 1:1 ratio induced a novel star-shaped mesostructure in the centre of which a mesoporous silicon dioxide core was found with PVX particles protruding from it that were attached at one end. Virus particles may display different reactivities at the two ends, but *via* immuno-gold end-labelling it was verified that there was not a specific particle end connected to the mesoporous core. The three-dimensional nature of the star-shaped particles was determined using cryo-TEM tomography. It should be noted that PVX-wt did not display any structure formation, indicating that the biomineralization process is crucial for structure formation. It was shown that after the mineralization process, the PVX surface was still accessible for immuno-gold labelling, thereby adding additional features to the already complex architecture.

The last approach is less controlled than in the other two example where very hierarchically built structures were formed. However, the last approach was based on simple mixing without any special designs other than the genetic modifications of the PVX concerning the surface-displayed mineralization peptides. There is always a trade-off between morphology control and system complexity, and the final target depends on the final application in mind.

2.5 Discussion and Conclusion

Many combinations of protein structures and inorganic materials, ranging through metallic, mineral and semiconducting structures, are possible. Often cage proteins such as ferritin, but also virus particles, are very attractive protein assemblies that can be combined with inorganic materials also because either the interior or the exterior can be targeted, or even both. The double address-ability and the many different combinations make it unimaginable that this would not lead to new or improved applications. However, so far no real applications have emerged from the new structures and combinations discussed in this chapter. It always takes a while to obtain a really commercially viable product or to implement something new in known systems, but some of the structures shown with ferritin were found in the early 1990s. Hence the benefits of structure control and the apparently limitless number of combinations between inorganic and protein materials are probably not sufficient to achieve a viable technology that can be is implemented in commercial products.

It could very well be that upscaling is one of the major drawbacks. Genetic modifications show great potential for introducing new features, but upscaling from milligram levels in the laboratory to grams and maybe even kilograms or more poses a problem. For this one would need a bacterial expressed system and even preferably yeast. This is not always possible, and when it is then mostly only protein subunits can be expressed and genetic material for viruses cannot. Therefore, whenever synthetic approaches and methodologies are more easily accessible, the true development of bionanoparticles and their hybrids will most likely be hampered. However, also in the area of bulk protein expression, there are and will be new developments, thereby achieving competitive processes allowing the assembled protein structures to reach their true potential.

References

1. N. Hassan, A. Soltero, D. Pozzo, P. V. Messina and J. M. Ruso, *Soft Matter*, 2012, 9553–9562.
2. T. Davids, M. Schmidt, D. Böttcher and U. T. Bornscheuer, *Curr. Opin. Chem. Biol.*, 2013, **17**, 215–220.
3. A. Schulz, H. Wang, P. van Rijn and A. Böker, *J. Mater. Chem.*, 2011, **21**, 18903.
4. G. Jutz and A. Böker, *Polymer*, 2011, **52**, 211–232.
5. K. M. L. Taylor-Pashow, J. Della Rocca, R. C. Huxford and W. Lin, *Chem. Commun.*, 2010, **46**, 5832–5849.
6. J. A. Barreto, W. O'Malley, M. Kubeil, B. Graham, H. Stephan and L. Spiccia, *Adv. Mater.*, 2011, **23**, H18–H40.
7. B. Willner, E. Katz and I. Willner, *Curr. Opin. Biotechnol.*, 2006, **17**, 589–596.
8. K. M. L. Taylor-Pashow, J. Della Rocca, R. C. Huxford and W. Lin, *Chem. Commun.*, 2010, **46**, 5832–5849.
9. P. van Rijn and A. Böker, *J. Mater. Chem.*, 2011, **21**, 16735.
10. G. Jutz, P. van Rijn, B. Santos Miranda and A. Böker, *Chem. Rev.*, 2015, **115**, 1653–1701.
11. S.-Y. Lee, J.-S. Lim and M. T. Harris, *Biotechnol. Bioeng.*, 2012, **109**, 16–30.
12. T. Douglas and M. Young, *Science*, 2006, **312**, 873–875.
13. A. de la Escosura, R. J. M. Nolte and J. J. L. M. Cornelissen, *J. Mater. Chem.*, 2009, **19**, 2274.
14. T. Douglas and M. Young, *Adv. Mater.*, 1999, **11**, 679–681.
15. M. Mahmoudi, I. Lynch, M. R. Ejtehadi, M. P. Monopoli, F. B. Bombelli and S. Laurent, *Chem. Rev.*, 2011, **111**, 5610–5637.
16. D. F. Moyano and V. M. Rotello, *Langmuir*, 2011, **27**, 10376–10385.
17. L. M. Bronstein, *Small*, 2011, 7, 1609–1618.
18. F. Li and Q. Wang, *Small*, 2014, **10**, 230–245.
19. M. Young, D. Willits, M. Uchida and T. Douglas, *Annu. Rev. Phytopathol.*, 2008, **46**, 361–384.
20. K. E. Sapsford, W. R. Algar, L. Berti, K. B. Gemmill, B. J. Casey, E. Oh, M. H. Stewart and I. L. Medintz, *Chem. Rev.*, 2013, **113**, 1904–2074.
21. S. Mann, *Nat. Mater.*, 2009, **8**, 781–792.
22. J. Sun, C. Dufort, M. Daniel, A. Murali, C. Chen, K. Gopinath, B. Stein, M. De, V. M. Rotello, A. Holzenburg, C. C. Kao and B. Dragnea, *Proc. Natl. Acad. Sci. U. S. A.*, 2007, **104**, 1354–1359.

23. A. A. A. Aljabali, G. P. Lomonosso and D. J. Evans, *Biomacromolecules*, 2011, 2723–2728.
24. P. Aggarwal, J. B. Hall, C. B. McLeland, M. A. Dobrovolskaia and S. E. McNeil, *Adv. Drug Delivery Rev.*, 2009, **61**, 428–437.
25. S. E. Aniagyei, C. Dufort, C. C. Kao and B. Dragnea, *J. Mater. Chem.*, 2008, **18**, 3763–3774.
26. C. Chen, M. Daniel, Z. T. Quinkert, M. De, B. Stein, V. D. Bowman, P. R. Chipman, V. M. Rotello, C. C. Kao and B. Dragnea, *Nano Lett.*, 2006, **6**, 611–615.
27. M. De, C. You, S. Srivastava and V. M. Rotello, *J. Am. Chem. Soc.*, 2007, **129**, 10747–10753.
28. I. Tsvetkova, C. Chen, S. Rana, C. C. Kao, V. M. Rotello and B. Dragnea, *Soft Matter*, 2012, **8**, 4571.
29. B. Berger, P. W. Shor, L. Tucker-Kellogg and J. King, *Proc. Natl. Acad. Sci. U. S. A.*, 1994, **91**, 7732–7736.
30. R. Díaz-Avalos and D. L. Caspar, *Biophys. J.*, 1998, **74**, 595–603.
31. C. B. Chang, C. M. Knobler, W. M. Gelbart and T. G. Mason, *ACS Nano*, 2008, **2**, 281–286.
32. F. J. Eber, S. Eiben, H. Jeske and C. Wege, *Angew. Chem., Int. Ed.*, 2013, **52**, 7203–7207.
33. A. Mueller, F. J. Eber, C. Azucena, A. Petershans, A. M. Bittner, H. Gliemann, H. Jeske and C. Wege, *ACS Nano*, 2011, **5**, 4512–4520.
34. S. Balci, D. M. Leinberger, M. Knez, A. M. Bittner, F. Boes, A. Kadri, C. Wege, H. Jeske and K. Kern, *Adv. Mater.*, 2008, **20**, 2195–2200.
35. C. Azucena, F. J. Eber, V. Trouillet, M. Hirtz, S. Heissler, M. Franzreb, H. Fuchs, C. Wege and H. Gliemann, *Langmuir*, 2012, **28**, 14867–14877.
36. S. K. Dixit, N. L. Goicochea, M.-C. Daniel, A. Murali, L. Bronstein, M. De, B. Stein, V. M. Rotello, C. C. Kao and B. Dragnea, *Nano Lett.*, 2006, **6**, 1993–1999.
37. M. Kim, Y. Rho, K. S. Jin, B. Ahn, S. Jung, H. Kim and M. Ree, *Biomacromolecules*, 2011, **12**, 1629–1640.
38. J. C. Cheung-Lau, D. Liu, K. W. Pulsipher, W. Liu and I. J. Dmochowski, *J. Inorg. Biochem.*, 2014, **130**, 59–68.
39. T. Peng, D. Paramelle, B. Sana, C. F. Lee and S. Lim, *Small*, 2014, **10**, 3131–3138.
40. E. Roduner, *Chem. Soc. Rev.*, 2006, **35**, 583–592.
41. K.-I. Sano, K. Ajima, K. Iwahori, M. Yudasaka, S. Iijima, I. Yamashita and K. Shiba, *Small*, 2005, **1**, 826–832.
42. B. Zheng, I. Yamashita, M. Uenuma, K. Iwahori, M. Kobayashi and Y. Uraoka, *Nanotechnology*, 2010, **21**, 045305.
43. B. Le Droumaguet, J. Nicolas, D. Brambilla, S. Mura, A. Maksimenko, L. De Kimpe, E. Salvati, C. Zona, C. Airoldi, M. Canovi, M. Gobbi, N. Magali, B. La Ferla, F. Nicotra, W. Scheper, O. Flores, M. Masserini, K. Andrieux and P. Couvreur, *ACS Nano*, 2012, **6**, 5866–5879.
44. M. P. Monopoli, C. Aberg, A. Salvati and K. A. Dawson, *Nat. Nanotechnol.*, 2012, **7**, 779–786.

45. F. C. Meldrum, T. Douglas, S. Levi, P. Arosio and S. Mann, *J. Inorg. Biochem.*, 1995, **58**, 59–68.
46. F. C. Meldrum, B. R. Heywood and S. Mann, *Science*, 1992, **257**, 522–523.
47. H. Yoshimura, *Colloids Surf., A*, 2006, **282–283**, 464–470.
48. S. Pead, E. Durrant, B. Webb, C. Larsen, D. Heaton, J. Johnson and G. D. Watt, *J. Inorg. Biochem.*, 1995, **59**, 15–27.
49. I. G. Macara, T. G. Hoy and P. M. Harrison, *Biochem. J.*, 1972, **126**, 151–162.
50. K. K. W. Wong, T. Douglas, S. Gider, D. D. Awschalom and S. Mann, *Chem. Mater.*, 1998, **10**, 279–285.
51. K. K. W. Wong and S. Mann, *Adv. Mater.*, 1996, **8**, 928–932.
52. C. C. Jolley, M. Uchida, C. Reichhardt, R. Harrington, S. Kang, M. T. Klem, J. B. Parise and T. Douglas, *Chem. Mater.*, 2010, **22**, 4612–4618.
53. F. C. Meldrum, V. J. Wade, D. L. Nimmo, B. R. Heywood and S. Mann, *Nature*, 1991, **349**, 684–687.
54. T. G. St. Pierre, W. Chua-anusorn, P. Sipos, I. Kron and J. Webb, *Inorg. Chem.*, 1993, **32**, 4480–4482.
55. J. F. Hainfeld, *Proc. Natl. Acad. Sci. U. S. A.*, 1992, **89**, 11064–11068.
56. M. Young, D. Willits, M. Uchida and T. Douglas, *Annu. Rev. Phytopathol.*, 2008, **46**, 361–384.
57. T. Douglas and M. Young, *Nature*, 1998, **393**, 152–155.
58. C. C. Jolley, M. Uchida, C. Reichhardt, R. Harrington, S. Kang, M. T. Klem, J. B. Parise and T. Douglas, *Chem. Mater.*, 2010, **22**, 4612–4618.
59. M. T. Klem, M. Young and T. Douglas, *J. Mater. Chem.*, 2008, **18**, 3821.
60. H.-A. Hosein, D. R. Strongin, M. Allen and T. Douglas, *Langmuir*, 2004, **20**, 10283–10287.
61. C. Jolley, V. Pool, Y. Idzerda and T. Douglas, *Chem. Mater.*, 2011, **23**, 3921–3929.
62. V. Pool, M. Klem, C. Jolley, E. A. Arenholz, T. Douglas, M. Young and Y. U. Idzerda, *J. Appl. Phys.*, 2010, **107**, 09B517.
63. M. Uchida, M. Terashima, C. H. Cunningham, Y. Suzuki, D. A. Willits, A. F. Willis, P. C. Yang, P. S. Tsao, M. V. McConnell, M. J. Young and T. Douglas, *Magn. Reson. Med.*, 2008, **60**, 1073–1081.
64. T. Prozorov, S. K. Mallapragada, B. Narasimhan, L. Wang, P. Palo, M. Nilsen-Hamilton, T. J. Williams, D. A. Bazylinski, R. Prozorov and P. C. Canfield, *Adv. Funct. Mater.*, 2007, **17**, 951–957.
65. J. Polanams, A. D. Ray and R. K. Watt, *Inorg. Chem.*, 2005, **44**, 3203–3209.
66. T. Douglas and V. T. Stark, *Inorg. Chem.*, 2000, **39**, 1828–1830.
67. B. Zhang, J. N. Harb, R. C. Davis, J.-W. Kim, S.-H. Chu, S. Choi, T. Miller and G. D. Watt, *Inorg. Chem.*, 2005, **44**, 3738–3745.
68. M. Li, C. Viravaidya and S. Mann, *Small*, 2007, **3**, 1477–1481.
69. M. T. Klem, J. Mosolf, M. Young and T. Douglas, *Inorg. Chem.*, 2008, **47**, 2237–2239.
70. T. Ueno, M. Suzuki, T. Goto, T. Matsumoto, K. Nagayama and Y. Watanabe, *Angew. Chem., Int. Ed.*, 2004, **43**, 2527–2530.

71. M. Suzuki, M. Abe, T. Ueno, S. Abe, T. Goto, Y. Toda, T. Akita, Y. Yamada and Y. Watanabe, *Chem. Commun.*, 2009, 4871–4873.
72. Q. Y. Deng, B. Yang, J. F. Wang, C. G. Whiteley and X. N. Wang, *Biotechnol. Lett.*, 2009, **31**, 1505–1509.
73. L. Zhang, L. Laug, W. Münchgesang, E. Pippel, U. Gösele, M. Brandsch and M. Knez, *Nano Lett.*, 2010, **10**, 219–223.
74. I. Yamashita, J. Hayashi and M. Hara, *Chem. Lett.*, 2004, **33**, 1158–1159.
75. R. Xing, X. Wang, L. Yan, C. Zhang, Z. Yang, X. Wang and Z. Guo, *Dalton Trans.*, 2009, 1710–1713.
76. M. Fischlechner and E. Donath, *Angew. Chem., Int. Ed.*, 2007, **46**, 3184–3193.
77. N. F. Steinmetz and D. J. Evans, *Org. Biomol. Chem.*, 2007, **5**, 2891–2902.
78. C. M. Soto and B. R. Ratna, *Curr. Opin. Biotechnol.*, 2010, **21**, 426–438.
79. A. A. A. Aljabali, S. N. Shah, R. Evans-Gowing, G. P. Lomonossoff and D. J. Evans, *Integr. Biol.*, 2011, **3**, 119–125.
80. J. M. Galloway and S. S. Staniland, *J. Mater. Chem.*, 2012, **22**, 12423.
81. F. Li, H. Chen, Y. Zhang, Z. Chen, Z.-P. Zhang, X.-E. Zhang and Q. Wang, *Small*, 2012, **8**, 3832–3838.
82. A. Klug, *Philos. Trans. R. Soc. London, Ser. B*, 1999, **354**, 531–535.
83. J. Atabekov, N. Nikitin, M. Arkhipenko, S. Chirkov and O. Karpova, *J. Gen. Virol.*, 2011, **92**, 453–456.
84. T. L. Schlick, Z. Ding, E. W. Kovacs and M. B. Francis, *J. Am. Chem. Soc.*, 2005, **127**, 3718–3723.
85. M. Knez, A. M. Bittner, F. Boes, C. Wege, H. Jeske, E. Mai and K. Kern, *Nano Lett.*, 2003, **3**, 1079.
86. S. Balci, A. M. Bittner, K. Hahn, C. Scheu, M. Knez, A. Kadri, C. Wege, H. Jeske and K. Kern, *Electrochim. Acta*, 2006, **51**, 6251–6257.
87. S. Balci, K. Hahn, P. Kopold, A. Kadri, C. Wege, K. Kern and A. M. Bittner, *Nanotechnology*, 2012, **23**, 045603.
88. M. Kobayashi, M. Seki, H. Tabata, Y. Watanabe and I. Yamashita, *Nano Lett.*, 2010, **10**, 773–776.
89. A. J. Love, V. Makarov, I. Yaminsky, N. O. Kalinina and M. E. Taliansky, *Virology*, 2014, **449**, 133–139.
90. M. Young, D. Willits, M. Uchida and T. Douglas, *Annu. Rev. Phytopathol.*, 2008, **46**, 361–384.
91. K. M. Bromley, A. J. Patil, A. W. Perriman, G. Stubbs and S. Mann, *J. Mater. Chem.*, 2008, **18**, 4796.
92. D. Ma, Y. Xie, J. Zhang, D. Ouyang, L. Yi and Z. Xi, *Chem. Commun.*, 2014, **50**, 15581–15584.
93. E. Royston, S.-Y. Lee, J. N. Culver and M. T. Harris, *J. Colloid Interface Sci.*, 2006, **298**, 706–712.
94. E. S. Royston, A. D. Brown, M. T. Harris and J. N. Culver, *J. Colloid Interface Sci.*, 2009, **332**, 402–407.
95. Y. Liu, Y.-K. Kim, L. Dai, N. Li, S. O. Khan, D. H. Pashley and F. R. Tay, *Biomaterials*, 2011, **32**, 1291–1300.
96. N. Korkmaz, Y. J. Kim and C. H. Nam, *Macromol. Biosci.*, 2013, **13**, 376–387.

97. P. Sitasuwan, L. A. Lee, K. Li, H. G. Nguyen and Q. Wang, *Front. Chem.*, 2014, **2**, 31.

98. J. C. Zhou, C. M. Soto, M.-S. Chen, M. A. Bruckman, M. H. Moore, E. Barry, B. R. Ratna, P. E. Pehrsson, B. R. Spies and T. S. Confer, *J. Nanobiotechnol.*, 2012, **10**, 18.

99. E. Dujardin, C. Peet, G. Stubbs, J. N. Culver and S. Mann, *Nano Lett.*, 2003, **3**, 413–417.

100. P. van Rijn, L. S. van Bezouwen, R. Fischer, E. J. Boekema, A. Böker and U. Commandeur, *Part. Part. Syst. Charact.*, 2015, **32**, 43–47.

101. A. A. Khan, E. K. Fox, M. L. Górzny, E. Nikulina, D. F. Brougham, C. Wege and A. M. Bittner, *Langmuir*, 2013, **29**, 2094–2098.

102. Y. Huang, C. Chiang, S. K. Lee, Y. Gao, E. L. Hu, J. De Yoreo and A. M. Belcher, *Nano Lett.*, 2005, **5**, 1429–1434.

CHAPTER 3

Channel Protein FhuA as a Promising Biomolecular Scaffold for Bioconjugates

LEILEI ZHU[a], MARCUS ARLT[a], HAIFENG LIU[a], MARCO BOCOLA[a], DANIEL F. SAUER[b], STEVE GOTZEN[b], JUN OKUDA[b], AND ULRICH SCHWANEBERG*[a]

[a]Lehrstuhl für Biotechnologie, RWTH Aachen University, Worringerweg 3, 52074 Aachen, Germany; [b]Institut für Anorganische Chemie, RWTH Aachen University, Landoltweg 1, 52056 Aachen, Germany
*E-mail: u.schwaneberg@biotec.rwth-aachen.de

3.1 Introduction

Ferric hydroxamate uptake protein component A (FhuA) is a monomeric β-barrel protein from the outer membrane of *Escherichia coli*,[1] and was the second pure protein isolated from the *E. coli* cell envelope.[2] FhuA has attracted great attention since the beginning of phage genetics and molecular biology. FhuA-containing proteoliposomes were used to study the mechanism of phage infection.[3] With the crystal structures of FhuA solved,[4–6] its function and properties characterized and the FhuA channel re-engineered, interest in the applications of FhuA expanded to its use as a nanopore integrated in liposome/polymersome membranes for the translocation of

RSC Smart Materials No. 16
Bio-Synthetic Hybrid Materials and Bionanoparticles: A Biological Chemical Approach Towards Material Science
Edited by Alexander Böker and Patrick van Rijn
© The Royal Society of Chemistry 2015
Published by the Royal Society of Chemistry, www.rsc.org

compounds in/out of liposomes and polymersomes.[7–9] Various redesigned FhuA variants were investigated, including deletions of the large extracellular loops, deletion of the cork domain, elongation of the hydrophobic membrane region and enlargement of its diameter.[10–12] Molecular understanding of sterically controlled compound release from the FhuA channel provided structural information to achieve triggerable compound release through the FhuA channel.[13,14] Because of its remarkable resistance towards organic solvents, high temperatures and alkaline pH[15] and its robustness in genetic modification, FhuA has become an attractive scaffold for hybrid catalysts (artificial metalloenzymes) to accommodate metal/organic catalysts for improving enantioselectivity.[16] The amino acid side-chain in the interior of the FhuA channel provides a chirally defined second coordination sphere and improves the enantioselectivity of hybrid catalysts. By substituting the amino acid residues surrounding the catalyst complex in the interior of the FhuA channel, one can also optimize the accessibility of the coupling site and the enantioselectivity of the hybrid catalyst.

3.2 FhuA

FhuA is a large monomeric transmembrane protein (Figure 3.1) in the *E. coli* outer membrane. It consists of 714 amino acids folded into 22 antiparallel β-strands (barrel domain) and a globular cork domain. FhuA is an energy-coupled transporter and receptor in the outer membrane of *E. coli* cells. It functions as a receptor for ferrichrome (ferric siderophores >700 Da) and the structurally closely related antibiotic albomycin, for several bacteriophages and for the bacterial toxin colicin M. The barrel domain of FhuA folds into a roughly cylindrical channel with the hydrophilic side-chains oriented inside the channel and the hydrophobic residues exposed to the phospholipid bilayer. The barrel domain of FhuA has an elliptically shaped cross-sectional

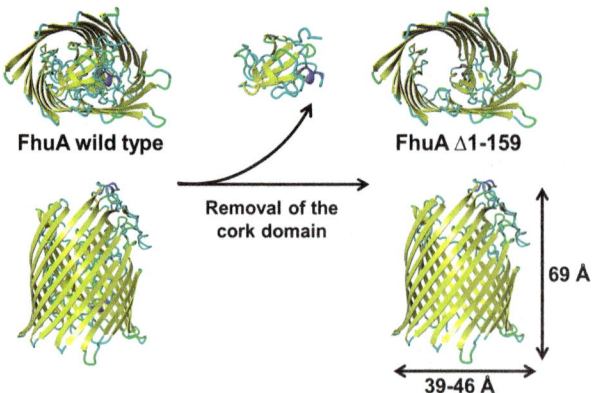

Figure 3.1   Structures of wild-type FhuA and FhuA Δ1–160 in which the cork domain was removed.

(39–46 Å), and the β-strands run antiparallel to one another, conferring exceptional robustness. The hydrogen bond network of the backbones between the neighbouring β-strands provides an extraordinary stiffness to the FhuA channel. Therefore, FhuA is stable under a broad range of experimental and application conditions (temperature, pH and unconventional solvents). The FhuA channel is tolerant to redesign in various ways, including direct genetic engineering[10,11] and covalent modifications.[17–19] By removing the capping globular domain (deletion of amino acids 1–160, Figure 3.1), FhuA became a large passive diffusion channel (FhuA Δ1–160) with an inner diameter of about 2.0 nm.

3.3 The Pore-Forming Ability of FhuA Variants in Black Lipid Membranes Investigated by Single-Channel Electrical Recording

Single-channel electrical recording of FhuA variants embedded in black lipid membranes (BLMs) is an important approach for investigating the pore-orming ability, permeability and estimation of the pore size. BLMs are known to consist of three components: lipid bilayers, a thicker annulus that forms at the interface between the supporting substrate and the bilayer, and microlenses (pockets of decane). BLMs are a reasonably close semblance of an actual cell membrane because they permit simultaneous access to the solution and electrical control of both sides of the bilayer, allowing for the mimicking of physiological conditions.[20] Single-channel electrical recording of FhuA variants in BLMs was performed to investigate the permeability of FhuA variants for ions.[21,22] Deletion of residues 322–355 (FhuA Δ322–355) converted FhuA into a TonB independent and non-specific diffusion channel that permits diffusion of ferric siderophores through FhuA. FhuA deletion variant FhuA Δ322–355 and FhuA ΔC/Δ4L[12] were reconstituted in BLMs and formed stable channels that were confirmed by conductance measurements.[21] After removing the cork domain and a single extracellular loop (335–355), FhuA variants FhuA Δ1–160/Δ335–355 and FhuA Δ1–160/Δ322–355 formed stable channels in BLMs and showed similar conductance values, 3.0 ± 0.5 and 3.0 ± 1.5 nS, respectively.[23] After removing the cork domain and four extracellular loops, FhuA ΔC/Δ4L also produces a stable channel embedded in the BLM with a higher conductance value (4.8 ± 1.3 nS).[23] Furthermore, engineered FhuA ΔC/Δ4L protein nanopore maintained its stable open state in the BLM for long periods at temperatures up to 65 °C and acidic pH (pH 3.5). It was found that FhuA ΔC/Δ4L is a cation-selective channel, probably because it is rich in negatively charged amino acid residues in the interior of the channel.[24]

Single-channel electrical recording also allows the evaluation of the diameter of the FhuA channel in polymer exclusion experiments. When FhuA ΔC/Δ4L was reconstituted into the BLM, estimation of the channel access resistance using impermeable dextran polymers indicated an average internal diameter of ~2.4 nm. Polymer exclusion experiments with poly(ethylene glycols) suggested that FhuA ΔC/Δ4L has an approximately conical internal geometry, in accord with the asymmetric nature of the crystal structure of the wild-type FhuA.[12]

3.4 FhuA as a Channel for Molecule Translocation in/out of Liposomes and Polymersomes

The combination of natural channel proteins with polymersomes allows the formation of a novel nanocompartment, synthosome.[7] Synthosomes combine the mechanical and chemical stability of the amphiphilic block copolymer with the functional specificity of channel proteins. Synthosomes[7] (Figure 3.2) are nanocontainers (50–1000 nm) formed with a polymeric membrane that harbours the channel protein, *e.g.* FhuA Δ1–160, OmpF and Tsx. In the synthosome system, the transmembrane channel proteins control the compound flux. Protein engineering allows the design of functionalized protein channels in order to control the compound release and flux. Protein channels that can function as an on/off switch offer opportunities for the design of functional nanocompartments with potential applications in medicine (drug release) and industrial biotechnology (chiral nanoreactors). So far, only a few channel proteins (*e.g.* OmpF, FhuA, Tsx and AqpZ)[25,26] have been incorporated functionally in polymer membranes of the synthosomes. FhuA Δ1–160 offers a faster release than any other reported channel proteins and represents by far the largest channel protein (39–46 Å in elliptical cross-section) employed in synthosomes (Tsx is the second largest, 20 × 30 Å in cross-section[27]).

Selective product recovery and enzymatic conversions have been achieved in synthosomes. In order to recover DNA using synthosomes,[8] nanophosphor particles were entrapped in the synthosomes formed with an ABA triblock

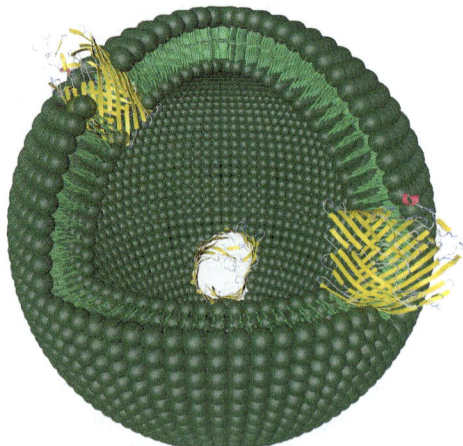

Figure 3.2   Schematic representation of synthosome composed of polymersome and reconstituted FhuA channel.

copolymer and small DNA fragments with a TAMRA (5-carboxytetramethyl-rhodamine) fluorescent dye hybridized at the 3'-end was supplemented in the synthosome solution. After translocation through the channel protein, the DNA fragment hybridized to the complementary sequence of the DNA-labelled nanophosphor particles. The hybridization event allowed an energy transfer (fluorescence resonance energy transfer, FRET) from the nanophosphors to the TAMRA dye upon excitation. The emission half-life of the nanophosphors was measured to monitor the hybridization of single-stranded DNA. Horseradish peroxidase (HRP) and 3,3',5,5'-tetramethylbenzidine (TMB) were used as a model system for the investigation of bioconversion in synthosomes. The FhuA Δ1–160 channel allows the translocation of TMB molecules, which are then oxidized by the HRP entrapped inside the synthosomes.[7] However, the encapsulation efficiency of HRP enzymes was around 15% and needs to be improved.

3.4.1 Triggered Compound Release Through FhuA in Liposomes and Polymersomes

FhuA Δ1–160 has been used as a diffusion channel in liposomes and polymersomes for the development of the triggered release of model compounds (chromophore or fluorophore[9,28]) by reducing reagents and UV radiation. In the FhuA-based reducing reagent-triggered release system, the amino group labelling agents 3-(2-pyridyldithio)propionic acid *N*-hydroxysuccinimide ester (pyridyl label) and [2-(biotinamido)ethylamido]-3,3'-dithiodipropionic acid *N*-hydroxysuccinimide ester (biotinyl label) were employed to label the lysine residues in the FhuA channel interior and acted as steric blocks for the channel. Pyridyl and biotinyl labelling of the FhuA channel prior to reconstitution in the polymersome membrane turned FhuA into a reduction-responsive channel. The six chemically (biotin and pyridyl) labelled lysine residues in the channel interior of FhuA Δ1–160 blocked the channel and prevented the translocation of calcein molecules out of loaded synthosomes. Upon reduction of the disulfide bond of the pyridyl or biotinyl label, fast release of calcein was recorded, whereas there was no detectable calcein release from polymersome without FhuA.[9]

In order to identify the contribution of each lysine residue upon labelling to restriction of the compound fluxes, six lysine residues (Figure 3.3) were systematically analysed by generating FhuA Δ1–160 variants in two subsets: subset A, six FhuA Δ1–160 variants in which one of the six lysines in the interior of the FhuA channel was replaced with to alanine; subset B, six FhuA Δ1–160 variants in which only one lysine inside the barrel was not changed to alanine.[14] HRP/TMB assay was employed to quantify the steric hindrance of pyridyl-labelled lysines on the TMB substrate. Pyridyl labelling of the FhuA Δ1–160 variant that contains only the K556 lysine in the channel interior reduces TMB translocation to reach nearly background levels in liposomes. Residue K556 was identified as the key residue for the steric control of compound fluxes through the channel of FhuA Δ1–160 reconstituted in the liposome membrane. A B-factor analysis based on molecular dynamics

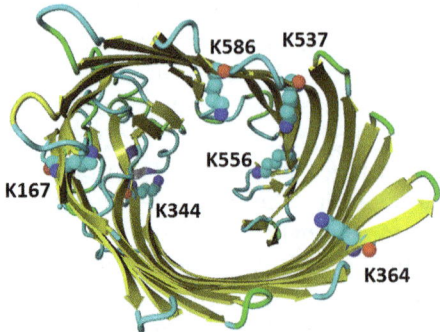

Figure 3.3   Bottom view of FhuA Δ1–160 with the six lysine residues (Lys167, Lys344, Lys364, Lys537, Lys556 and Lys586) in the interior of the FhuA channel.

simulations confirmed that position K556 is the least fluctuating lysine among the six in the channel interior and therefore can efficiently control the compound flux through steric hindrance.

In the FhuA-based UV radiation-triggered release system,[28] a photo-cleavable (366 nm) label, 6-nitroveratryloxycarbonyl chloride (NVOC-Cl), was used to label Lys556 in the interior of the FhuA channel, which was reconstituted in liposome. Kinetic studies on liposome inserted with FhuA variants, using TMB/HRP as the detection system, showed that the single labelled amino acid position, Lys556, acted sufficiently as a gate and controlled TMB translocation through the FhuA channel. There are six lysine residues in the interior of the FhuA barrel. Position Lys556 was identified as the gate keeper by replacing all of the other five lysines in the barrel interior with alanine so that the Lys556 remained the only free amino group in the FhuA Δ1–160 barrel interior.[28] Various liposomes/polymersomes have been studied as a container for drug delivery and release. Drug release from liposomes/polymersomes is governed by the diffusion of the drug through the membrane. The driving force is a concentration gradient of the drug between liposomes/polymersomes and the surrounding medium. A controlled drug release is more favourable because it releases a drug over an extended period at a controlled rate and in a predesigned manner to achieve more effective therapies while eliminating the potential for both under- and overdosing. The developed triggered (by UV radiation[28] and a reducing agent[9]) compound release from synthosomes (with FhuA integrated as channel) offers a great chance to design a controlled drug release system responding to external stimuli, *e.g.* near-infrared radiation.

3.4.2   Re-engineering of FhuA for Different Geometries

Many efforts have been made to obtain FhuA variants with different geometries (length and diameter) by rational design. To increase the length of the hydrophobic area in FhuA Δ1–159, the last five amino acids of each of the 22 β-sheets (prior to the periplasmatic β-turns) were doubled, leading to an extended FhuA channel (FhuA Δ1–159 Ext) by 1 nm. FhuA Δ1–159 Ext was successfully expressed, extracted and embedded in functional form in triblock copolymer vesicles.[11] The increased hydrophobic transmembrane region (by 1 nm) led to a predicted lower hydrophobic mismatch between the FhuA channel and polymer membrane, minimizing the insertion energy penalty. FhuA Δ1–159 Ext is the first FhuA variant that can be efficiently inserted cost-effectively in polymeric nanocompartments. The strategy of adding amino acids to the FhuA hydrophobic part can be further extended to increase the hydrophobicity of the FhuA channel by using more hydrophobic amino acids, promoting more efficient embedding into more hydrophobic block copolymer membranes. A FhuA channel with increased diameter will increase the translocation efficiency and facilitate the translocation of bulky molecules (*e.g.* double-stranded DNA and proteins) for biosensing applications. Motivated by this possibility, a FhuA variant with expanded diameter (by 0.4 nm), FhuA Δ1–159 Exp, was also successfully generated by doubling the amino acid sequence of the first two N-terminal β-strands. The total number of β-strands increased from 22 to 24 and the channel surface area was expected to increase by ~16%. FhuA Δ1–159 Exp was successfully reconstituted in liposomes and showed 17% faster diffusion kinetics than FhuA Δ1–159.

In general, using a non-ionic detergent to extract FhuA from an *E. coli* membrane is a laborious procedure with unsatisfactory yields. Therefore, expression of FhuA as inclusion body followed by a refolding procedure is a useful alternative approach for FhuA production. A cost-effective inclusion body production technology for FhuA was developed by deleting the N-terminal signal sequence.[29] The produced inclusion bodies of FhuA Δ1–159 Δsignal were refolded with an efficient and simple refolding protocol using a commercially available diblock copolymer, PE–PEG [polyethylene–poly(ethylene glycol) with an M_n of 2250] as an alternative to detergents such as octyl-POE (*n*-octylpolyoxyethylene). Deconvolution of the recorded circular dichroism (CD) spectra of the refolded FhuA Δ1–159 Δsignal and comparison with CD spectra obtained for FhuA Δ1–159 extracted from the membrane suggested that the FhuA Δ1–159 Δsignal β-sheet structure was successfully recovered.

3.5   FhuA as a Protein Scaffold for Hybrid Catalysts

Hybrid catalysts, also called artificial metalloenzymes, combine the structural diversity of a biomolecular scaffold with the various types of metal catalyst/organocatalyst.[30] Importantly, the hybrid catalyst offers great opportunities for genetic optimization. Introduction of point mutations on the host protein allows both the activity and the selectivity of the hybrid catalysts to be fine-tuned.

One of the important factors for developing a successful hybrid catalyst is the selection of a suitable biomolecular scaffold, typically a protein. Important considerations in the selection of proteins as biomolecular scaffolds are the size of the cavity for accommodating the catalyst complex and substrate and their temperature stability, organic solvent resistance and pH stability. The protein scaffold must be stable under the reaction conditions for incorporating the catalyst complex and the reactions catalysed by the hybrid catalyst. Up to now, the choice of protein scaffold has been limited; only proteins such as avidin (Av),[31] streptavidin (Sav),[32] bovine serum albumin (BSA),[33] nitrobindin[34,35] and apomyoglobin (apo-Mb),[36] which were used to bind the catalyst, additionally offered enough space for accepting substrates to achieve enantioselective catalysis.

Streptavidin has an exceptionally strong affinity for biotin as a linker between the protein and metal complex. It has been used in the development of asymmetric reactions, Pd-mediated allylic alkylation[37] and ruthenium-catalysed transfer hydrogenation of ketones.[38] FhuA Δ1–159, an open channel with an inner diameter of ~2 nm and length of ~7 nm, provides sufficient space to harbour the catalyst complex and substrates. Furthermore, the hydrogen bonding-interaction in the β-barrel structure of the FhuA channel makes it very stable. The amino acid side-chains in the interior of the FhuA channel provide a chirally defined second coordination sphere and introduce enantioselectivity to the metal catalyst. FhuA Δ1–159 shows superior resistance to organic solvents, *e.g.* THF (up to 45 vol%), alkaline pH (6–10) and thermal unfolding (T_m value: up to 60–65 °C). Precipitated FhuA Δ1–159 in the presence of higher concentrations of organic solvent or at higher temperature can be refolded simply by dialysis against a buffer including suitable stabilizing detergents. Therefore, coupling of a metal complex to unfolded FhuA Δ1–159 and then refolding of the FhuA Δ1–159–metal complex to obtain the folded FhuA channel harbouring the metal catalyst is an alternative strategy for protein–catalyst coupling. In addition, the large-scale production of FhuA Δ1–159 and its variants was achieved by expression as inclusion bodies in the well-understood *E. coli* expression system[39] and extraction with green organic solvents.

Twelve crystal structures of FhuA have been solved. The availability of these FhuA crystal structures allows the prediction of binding sites of FhuA for catalyst complexes and optimization of the surrounding second sphere for improvement of enatioselectivity. FhuA is very robust towards the introduction of a single mutation,[28] insertion of DNA sequence fragments[11,40] and deletion of loop regions.[21] It is assumed that engineering of the size of the protein pores (channel) can be used to discriminate between substrates based on their shape and size. Varying lengths and diameters of FhuA Δ1–159 have been successfully achieved by protein engineering.[11,40] This will expand the application of the FhuA channel as a scaffold for hybrid catalysts accepting substrates of different sizes and shapes.

3.5.1 Re-engineering of FhuA to Accommodate Metal Catalysts

FhuA Δ1–159 is an attractive candidate as a biological scaffold for hybrid catalyst development. Several genetic optimization steps have been performed to optimize FhuA Δ1–159 further for accommodating catalyst complexes and for analytical purposes. The FhuA variant FhuA ΔCVF^{TEV} was generated and used to prepare a hybrid catalyst by covalently anchoring a Grubbs–Hoveyda-type olefin metathesis catalyst at a single accessible cysteine in the barrel interior (Figure 3.4).

Prerequisites for stereoselective chemical reactions (such as polymerization, metathesis or hydrogenation) using hybrid catalysts are a rigid protein structure, a defined environment and a conformationally stable binding position. The amino acid position 545 (equivalent to position Lys556 in the FhuA variant with a His tag) located in the interior of the FhuA channel was identified to control the compound flux through the FhuA Δ1–159 channel protein upon labelling.[28] For covalent binding with chemical catalysts, the lysine at position 545 was replaced with a cysteine. The amino acids around Lys545Cys of the FhuA were rationally re-engineered to ensure solvent accessibility of the thiol group of Cys545. Based on the X-ray crystal structure of FhuA (PDB code: 1BY3),[6,41] the solvent accessibility of the mutated Lys545Cys using the YASARA[42] software package was determined. The results for the minimized model showed Asn seems to be oriented towards the position Cys545. Thus the thiol group is not freely accessible for covalent coupling of a linker [see Figure 3.5(a)]. A systematic computational mutation analysis using FoldX[43] was performed to identify promising amino acid exchanges at position 548. Preconditions were substitutions that do not destabilize the protein and allow steric access to Cys545. The best predicted substitution was valine, which has a comparable size to asparagine, but is oriented away from Cys545 [see Figure 3.5(b)]. The determination of surface exposure showed a more accessible Cys545 environment.[16]

Furthermore, residue Glu501 was identified as a potential complexation site for a ruthenium catalyst that would lead to inactivation due to the strong coordination of ruthenium complexes to negative charges. The negatively

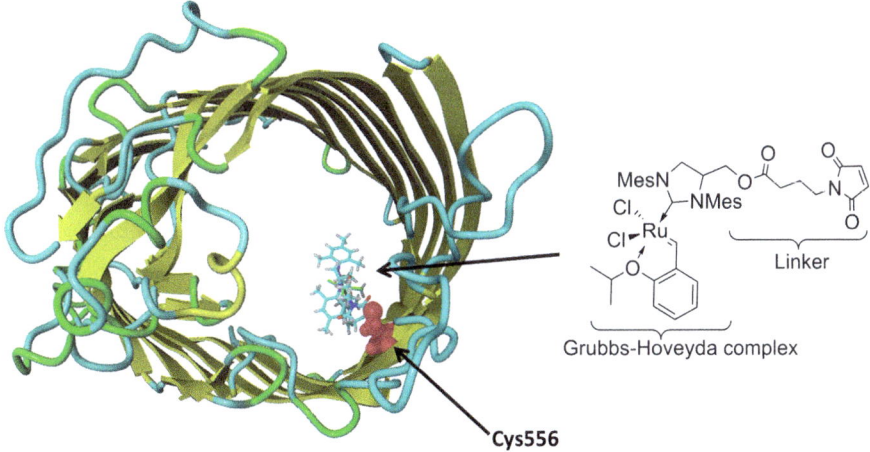

Figure 3.4   Hybrid catalyst composed of FhuA ΔCVF^{TEV} protein and a Grubbs–Hoveyda-type catalyst with a maleimide linking unit.

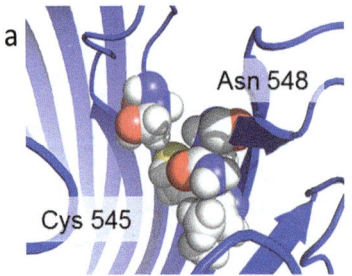 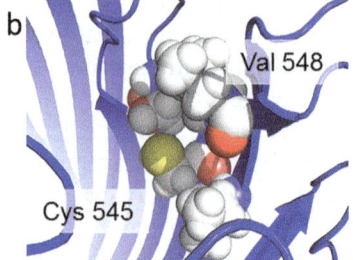

Figure 3.5 (a) Cys545 partially blocked by Asn548. (b) Cys545 with full solvent accessibility when Asn is replaced with Val at position 548.

charged Glu501 forms a salt bridge to Lys545 in wild-type FhuA that is not present in the FhuA Δ1–159 C545 variant and therefore tends to undergo charge balancing. FoldX analysis at position 501 predicted phenylalanine as the preferred amino acid substituent, and the structural model of the hybrid catalyst variant Glu501Phe suggested beneficial interactions with the aromatic ligands of the ruthenium catalyst. The designed FhuA ΔCVF[TEV] variant for hybrid catalyst synthesis therefore includes the additional substitutions Asn548Val and Glu501Phe.[16]

The use of ruthenium catalysts in chemical[44–47] and biochemical[48] transformations for the preparation of natural products[49,50] has attracted great attention. Water-soluble Grubbs and Grubbs–Hoveyda catalysts are known to catalyse the ring-opening metathesis polymerization (ROMP) of several 7-oxanorbornenes with preferential *trans* selectivity.[51–53] Thus, Grubbs–Hoveyda catalysts embedded in a protein surrounding are promising for the ROMP of strained cycloolefins.

The coupling of a Grubbs–Hoveyda-type catalyst to FhuA ΔCVF[TEV] was performed *via* cysteine at position 545 with reactive linkers attached to a hydroxyl group at the NHC ligand of the Grubbs–Hoveyda catalyst.[16]

3.5.2 Detection of the Covalent Modifications of Cys545 using MALDI-TOF-MS

For the determination of organometallic complexes[54,55] and membrane proteins[56,57] of about 100 kDa, matrix-assisted laser desorption/ionization time-of-flight mass spectrometry (MALDI-TOF-MS) has demonstrated its strengths of showing high throughput, high sensitivity and extraordinary measurement accuracy. Its enhanced tolerance towards several detergents and also salts[58] is particularly important for protein samples (improved co-crystallization and ionization). Typical matrix compounds include sinapinic acid, α-cyano-4-hydroxycinnamic acid (HCCA) and 2,5-dihydroxybenzoic acid (DHB) co-crystallized with the sample. The laser ionization of the matrix leads to very mild ionization of the sample, allowing the investigation of complete proteins or digestion products without fragmentation.

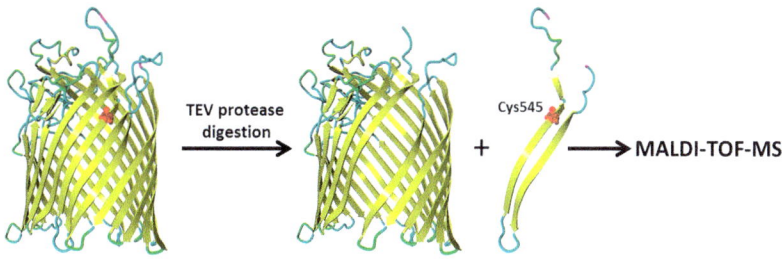

Figure 3.6   TEV protease digestion of FhuA ΔCVF^{TEV} to obtain a 5.9 kDa fragment for MALDI-TOF-MS analysis.

However, the prominent hydrophobic core of integral membrane proteins (such as FhuA) tends to self-aggregate. Subsequent co-crystallization of the protein sample with the matrix is of low quality, hence the ionization efficiency is very limited. Furthermore, detergents or chaotropic reagents that are necessary to maintain FhuA's structure lead to significant signal suppression. Enzymatic digestion of FhuA (*e.g.* using trypsin, AspN, GluC and LysC) allows the investigation of fragments of FhuA. However, the cleaving efficiency for hydrophobic regions of FhuA is very limited owing to the shielding of detergent. Therefore, a tobacco etch virus (TEV)-recognition site (ENLYFQ|G)[59] was introduced at the extracellular loops S7 and S8 connecting two β-strands harbouring the Cys545 (Figure 3.6). Suitable positions for the TEV recognition site within the extracellular loops (S7 and S8) not covered by detergents were identified by molecular modelling using the YASARA software package. By extension incorporation, the loops were elongated with the TEV recognition sites, leading to elongated loops. Maximized solvent accessibility was achieved by elongating the loops with the TEV recognition sequence. The TEV protease cleavage of FhuA ΔCVF^{TEV} resulted in peptide fragments 5.9, 40.3 and 17.5 kDa in size.

TEV protease cleavage of FhuA ΔCVF^{TEV} resulted in a 5.9 kDa fragment carrying the Cys545 (Figures 3.6 and 3.7). MALDI-TOF-MS analysis of the modified fragment showed typical fragmentation of the covalent attachment of linker and spacer to Cys545 (Figure 3.7). The fragment of 6093.1 Da (calc. 6093.3 Da, dashed line in Figure 3.7) corresponds to an ester cleavage at the NHC ligand, an additional carboxyl reduction to the aldehyde results in the fragment 6077.1 Da (calc. 6077.1 Da), and the fragment at 6109.6 Da (calc. 6111.3 Da) results from the NHC ester cleavage and hydrolysis of the maleimide ring. All these results indicate the successful coupling of the Grubbs–Hoveyda-type catalysts to FhuA ΔCVF^{TEV}.

3.5.3   Detection of the Covalent Modifications of Cys545 Using the Cysteine Titration Method

Since there is a single free cysteine in FhuA ΔCVF^{TEV}, cysteine titration of FhuA ΔCVF^{TEV} in microtitre plate format is a fast and simple method to detect free Cys545 and estimate the efficiency of covalent modifications of Cys545 by a maleimidic linker at a Grubbs–Hoveyda catalyst. Among fluorescent thiol

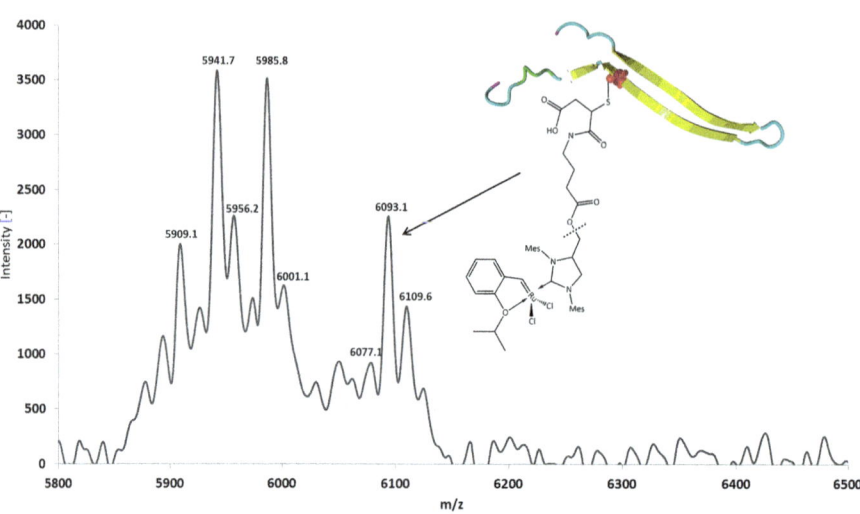

Figure 3.7   MALDI-TOF-MS analysis of TEV-cleaved FhuA ΔCVF^{TEV} coupled with Grubbs–Hoveyda catalyst.

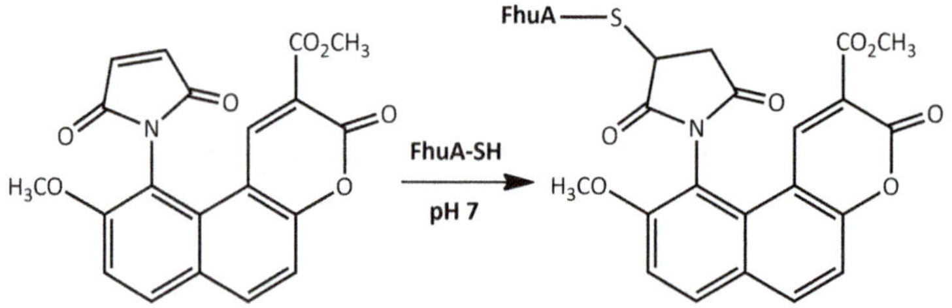

Figure 3.8   Reaction mechanism of cysteine titration using MMBC (ThioGlo1).

reagents (*e.g.* fluorescein-5-maleimide, BodiPy and MMBC), MMBC [methyl-10-(2,5-dioxo-2,5-dihydro-1*H*-pyrrol-1-yl)-9-methoxy-3-oxo-3*H*-benzo[*f*] chromene-2-carboxylate; ThioGlo1] was found to be the most suitable owing to the high signal-to-noise ratio and small influence of buffers and detergents.[60] MMBC (ThioGlo1) binds selectively to cysteine *via* a nucleophilic attack of the thiol group on the maleimidic function (Figure 3.8). Upon binding of MMBC to the free Cys545 in FhuA ΔCVF^{TEV}, a strong fluorescence signal (λ_{ex} = 379 nm, λ_{em} = 510 nm) was measured for the free cysteine. In comparison, a significantly lower fluorescence signal (93% lower) compared with that of the free Cys545 indicated the successful coupling of the Grubbs–Hoveyda-type catalyst to position Cys545.

3.5.4   Evaluation of the Secondary Structure of the Grubbs–Hoveyda–FhuA ΔCVF^{TEV} Hybrid Catalyst by CD

It is crucial that FhuA ΔCVF^{TEV} maintains its secondary structure under the reaction conditions of the Grubbs–Hoveyda–FhuA ΔCVF^{TEV} hybrid catalyst to ensure a defined second ligand sphere. Therefore, CD measurements

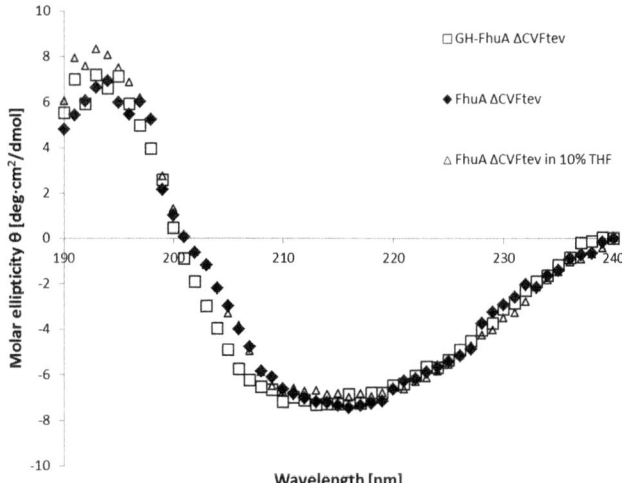

Figure 3.9   CD spectrum of FhuA ΔCVF^{TEV} and Grubbs–Hoveyda–FhuA ΔCVF^{TEV}.

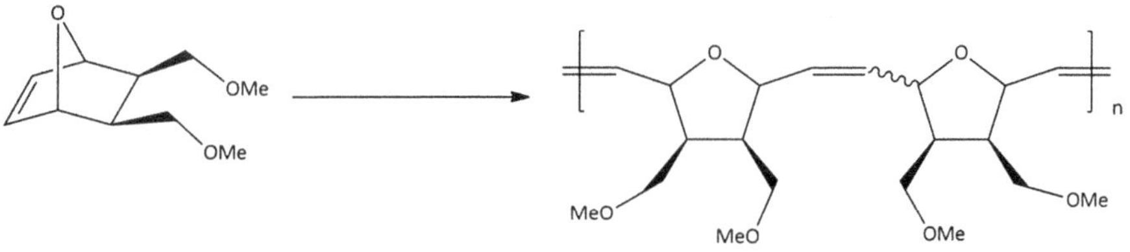

Scheme 3.1   Ring-opening metathesis polymerization of 2,3-bis(methoxymethyl)-*exo*-7-oxabicyclo[2.2.1]hept-5-ene catalyzed by Grubbs–Hoveyda catalyst.

were performed on this hybrid catalyst. Figure 3.9 shows that FhuA ΔCVF^{TEV} displays the characteristic minimum at 220 nm and maximum at 195 nm, which are typical for proteins with a high β-sheet content. The attachment of the Grubbs–Hoveyda catalyst did not result in structural alterations of FhuA ΔCVF^{TEV}.

3.5.5   Catalytic Properties of the Grubbs–Hoveyda–FhuA ΔCVF^{TEV} Hybrid Catalyst

A ROMP reaction catalysed by the Grubbs–Hoveyda–FhuA ΔCVF^{TEV} hybrid catalyst (Scheme 3.1) was carried out with water-soluble 7-oxanorbornene as substrate. This substrate has the advantage, in addition to its solubility features, of not affecting the secondary structure of FhuA ΔCVF^{TEV} under the reaction conditions. The Grubbs–Hoveyda–FhuA ΔCVF^{TEV} hybrid catalyst showed significant ROMP activity (37%). Control experiments with FhuA ΔCVF^{TEV} and FhuA ΔCVF^{TEV} with Grubbs–Hoveyda catalyst lacking the linking structure showed no catalytic activity. With the hybrid catalyst, a slight change in the *cis–trans* selectivity of polyoxanorbonene was observed compared with a protein-free control experiment using the catalyst under identical reaction conditions.

3.6   Prospects and Challenges for the Application of FhuA

FhuA, as a robust protein scaffold, has general applicability in liposome/polymersome systems for compound translocation. Re-engineering of FhuA by directed evolution and rational design provide the possibility of obtaining tailor-made channels for different application purposes. Synthosomes are very promising systems for drug delivery systems. However, the insertion efficiency of FhuA in synthosomes should be improved and quantified if required. The immunogenicity of synthosomes must be determined for medical applications. The structural properties (diameter, length and shape of the channel) and the superior resistance of the FhuA channel towards organic solvent and high temperatures make it a very attractive protein scaffold for hybrid catalyst development. Protein engineering of the FhuA channel allows effective improvements of the activity and selectivity of the hybrid catalyst. However, the use of detergents to stabilize the FhuA channel in aqueous solution complicates the analytics and applications. Using whole cells with the FhuA channel on the outer membrane might be an alternative for hybrid catalyst development instead of using detergent-stabilized FhuA.

Acknowledgements

The authors gratefully acknowledge financial support from the Cluster of Excellence RWTH Aachen 'Tailor-Made Fuels from Biomass' and the Deutsche Forschungsgemeinschaft through the International Research Training Group 'Selectivity in Chemo- and Biocatalysis.'

References

1. H. Hoffmann, E. Fischer, H. Kraut and V. Braun, *J. Bacteriol.*, 1986, **166**, 404–411.
2. V. Braun, *J. Bacteriol.*, 2009, **191**, 3431–3436.
3. J. Bohm, O. Lambert, A. S. Frangakis, L. Letellier, W. Baumeister and J. L. Rigaud, *Curr. Biol.*, 2001, **11**, 1168–1175.
4. A. D. Ferguson, V. Braun, H. P. Fiedler, J. W. Coulton, K. Diederichs and W. Welte, *Protein Sci.*, 2000, **9**, 956–963.
5. V. Braun, M. Braun and H. Killmann, *Adv. Exp. Med. Biol.*, 2000, **485**, 33–43.
6. A. D. Ferguson, E. Hofmann, J. W. Coulton, K. Diederichs and W. Welte, *Science*, 1998, **282**, 2215–2220.

7. M. Nallani, S. Benito, O. Onaca, A. Graff, M. Lindemann, M. Winterhalter, W. Meier and U. Schwaneberg, *J. Biotechnol.*, 2006, **123**, 50–59.
8. M. Nallani, O. Onaca, N. Gera, K. Hildenbrand, W. Hoheisel and U. Schwaneberg, *Biotechnol. J.*, 2006, **1**, 828–834.
9. O. Onaca, P. Sarkar, D. Roccatano, T. Friedrich, B. Hauer, M. Grzelakowski, A. Guven, M. Fioroni and U. Schwaneberg, *Angew. Chem., Int. Ed.*, 2008, **47**, 7029–7031.
10. M. Krewinkel, T. Dworeck and M. Fioroni, *J. Nanobiotechnol.*, 2011, **9**, 33.
11. N. Muhammad, T. Dworeck, M. Fioroni and U. Schwaneberg, *J. Nanobiotechnol.*, 2011, **9**, 8.
12. D. J. Niedzwiecki, M. M. Mohammad and L. Movileanu, *Biophys. J.*, 2012, **103**, 2115–2124.
13. A. Guven, T. Dworeck, M. Fioroni and U. Schwaneberg, *Adv. Eng. Mater.*, 2011, **13**, B324–B329.
14. A. Guven, M. Fioroni, B. Hauer and U. Schwaneberg, *J. Nanobiotechnol.*, 2010, **8**, 14.
15. S. J. Tenne and U. Schwaneberg, *Int. J. Mol. Sci.*, 2012, **13**, 2459–2471.
16. F. Philippart, M. Arlt, S. Gotzen, S. J. Tenne, M. Bocola, H. H. Chen, L. Zhu, U. Schwaneberg and J. Okuda, *Chem.–Eur. J.*, 2013, **19**, 13865–13871.
17. H. Bayley and P. S. Cremer, *Nature*, 2001, **413**, 226–230.
18. J. Sanchez-Quesada, A. Saghatelian, S. Cheley, H. Bayley and M. R. Ghadiri, *Angew. Chem., Int. Ed.*, 2004, **43**, 3063–3067.
19. L. Movileanu, *Soft Matter*, 2008, **4**, 925–931.
20. R. S. Ries, H. Choi, R. Blunck, F. Bezanilla and J. R. Heath, *J. Phys. Chem. B*, 2004, **108**, 16040–16049.
21. H. Killmann, R. Benz and V. Braun, *EMBO J.*, 1993, **12**, 3007–3016.
22. H. Killmann, R. Benz and V. Braun, *J. Bacteriol.*, 1996, **178**, 6913–6920.
23. M. M. Mohammad, K. R. Howard and L. Movileanu, *J. Biol. Chem.*, 2011, **286**, 8000–8013.
24. M. M. Mohammad, R. Iyer, K. R. Howard, M. P. McPike, P. N. Borer and L. Movileanu, *J. Am. Chem. Soc.*, 2012, **134**, 9521–9531.
25. O. Onaca, M. Nallani, S. Ihle, A. Schenk and U. Schwaneberg, *Biotechnol. J.*, 2006, **1**, 795–805.
26. P. Baumann, P. Tanner, O. Onaca and C. G. Palivan, *Polymers*, 2011, **3**, 173–192.
27. J. Q. Ye and B. van den Berg, *EMBO J.*, 2004, **23**, 3187–3195.
28. A. Guven, T. Dworeck, M. Fioroni and U. Schwaneberg, *Adv. Eng. Mater.*, 2011, **13**, B324–B329.
29. T. Dworeck, A. K. Petri, N. Muhammad, M. Fioroni and U. Schwaneberg, *Protein Expression Purif.*, 2011, **77**, 75–79.
30. V. Kohler, J. C. Mao, T. Heinisch, A. Pordea, A. Sardo, Y. M. Wilson, L. Knorr, M. Creus, J. C. Prost, T. Schirmer and T. R. Ward, *Angew. Chem., Int. Ed.*, 2011, **50**, 10863–10866.
31. M. E. Wilson and G. M. Whitesides, *J. Am. Chem. Soc.*, 1978, **100**, 306–307.
32. J. Collot, J. Gradinaru, N. Humbert, M. Skander, A. Zocchi and T. R. Ward, *J. Am. Chem. Soc.*, 2003, **125**, 9030–9031.

33. T. Kokubo, T. Sugimoto, T. Uchida, S. Tanimoto and M. Okano, *J. Chem. Soc., Chem. Commun.*, 1983, 769–770.
34. A. Onoda, K. Fukumoto, M. Arlt, M. Bocola, U. Schwaneberg and T. Hayashi, *Chem. Commun.*, 2012, **48**, 9756–9758.
35. K. Fukumoto, A. Onoda, E. Mizohata, M. Bocola, T. Inoue, U. Schwaneberg and T. Hayashi, *ChemCatChem*, 2014, **6**, 1229–1235.
36. M. Ohashi, T. Koshiyama, T. Ueno, M. Yanase, H. Fujii and Y. Watanabe, *Angew. Chem., Int. Ed.*, 2003, **42**, 1005–1008.
37. J. Pierron, C. Malan, M. Creus, J. Gradinaru, I. Hafner, A. Ivanova, A. Sardo and T. R. Ward, *Angew. Chem., Int. Ed.*, 2008, **47**, 701–705.
38. M. Creus, A. Pordea, T. Rossel, A. Sardo, C. Letondor, A. Ivanova, I. Le Trong, R. E. Stenkamp and T. R. Ward, *Angew. Chem., Int. Ed.*, 2008, **47**, 1400–1404.
39. T. Dworeck, A. K. Petri, N. Muhammad, M. Fioroni and U. Schwaneberg, *Protein Expression Purif.*, 2011, **77**, 75–79.
40. M. Krewinkel, T. Dworeck and M. Fioroni, *J. Nanobiotechnol.*, 2011, **9**, 33.
41. A. D. Ferguson, V. Braun, H. P. Fiedler, J. W. Coulton, K. Diederichs and W. Welte, *Protein Sci.*, 2000, **9**, 956–963.
42. E. Krieger, T. Darden, S. B. Nabuurs, A. Finkelstein and G. Vriend, *Proteins*, 2004, **57**, 678–683.
43. J. Schymkowitz, J. Borg, F. Stricher, R. Nys, F. Rousseau and L. Serrano, *Nucleic Acids Res.*, 2005, **33**, W382–W388.
44. H. Fuhrmann, T. Dwars and G. Oehme, *Chem. Unserer Zeit*, 2003, **37**, 40–50.
45. S. Mecking, A. Held and F. M. Bauers, *Angew. Chem., Int. Ed.*, 2002, **41**, 544–561.
46. N. Pinault and D. W. Bruce, *Coord. Chem. Rev.*, 2003, **241**, 1–25.
47. K. H. Shaughnessy, *Chem. Rev.*, 2009, **109**, 643–710.
48. J. B. Binder and R. T. Raines, *Curr. Opin. Chem. Biol.*, 2008, **12**, 767–773.
49. A. H. Hoveyda and A. R. Zhugralin, *Nature*, 2007, **450**, 243–251.
50. T. M. Trnka and R. H. Grubbs, *Acc. Chem. Res.*, 2001, **34**, 18–29.
51. J. P. Gallivan, J. P. Jordan and R. H. Grubbs, *Tetrahedron Lett.*, 2005, **46**, 2577–2580.
52. J. P. Jordan and R. H. Grubbs, *Angew. Chem., Int. Ed.*, 2007, **46**, 5152–5155.
53. D. M. Lynn, B. Mohr, R. H. Grubbs, L. M. Henling and M. W. Day, *J. Am. Chem. Soc.*, 2000, **122**, 6601–6609.
54. N. Srinivasan, C. A. Haney, J. S. Lindsey, W. Zhang and B. T. Chait, *J. Porphyrins Phthalocyanines*, 1999, **3**, 283–291.
55. M. F. Wyatt, *J. Mass Spectrom.*, 2011, **46**, 712–719.
56. B. Rosinke, K. Strupat, F. Hillenkamp, J. Rosenbusch, N. Dencher, U. Krüger and H.-J. Galla, *J. Mass Spectrom.*, 1995, **30**, 1462–1468.
57. A. Poetsch, D. Schlusener, C. Florizone, L. Eltis, C. Menzel, M. Rogner, K. Steinert and U. Roth, *J. Biomol. Tech.*, 2008, **19**, 129–138.
58. F. Hillenkamp, M. Karas, R. C. Beavis and B. T. Chait, *Anal. Chem.*, 1991, **63**, 1193A–1203A.
59. R. Zenobi and R. Knochenmuss, *Mass Spectrom. Rev.*, 1998, **17**, 337–366.
60. S. K. Wright and R. E. Viola, *Anal. Biochem.*, 1998, **265**, 8–14.

CHAPTER 4

Altering the Function and Properties of Plant Viral Assemblies via Genetic Modification

KERSTIN UHDE-HOLZEM[a], RAINER FISCHER[a], AND ULRICH COMMANDEUR*[a]

[a]Institut für Molekulare Biotechnologie, RWTH Aachen University, Worringerweg 1, 52074 Aachen, Germany
*E-mail: commandeur@molbiotech.rwth-aachen.de

4.1 Introduction

Plant virus capsids are made up of many copies of one or a few types of protein subunits. They assemble to either icosahedral or helical symmetry and are usually arranged around a single-stranded RNA genome. Viral particles can be easily produced in large quantities, either *via* the infection of plants or by heterologous expression of the subunits using various expression systems, such as *Escherichia coli*, yeast or insect cells.[1] The high stability of the capsids in combination with their simplicity and the high production yields in plants or heterologous expression systems make plant viruses or virus-like particles (VLPs) an interesting tool for application in bionanotechnology. Since naturally occurring viral particles rarely feature the functional groups

RSC Smart Materials No. 16
Bio-Synthetic Hybrid Materials and Bionanoparticles: A Biological Chemical Approach Towards Material Science
Edited by Alexander Böker and Patrick van Rijn
Published by the Royal Society of Chemistry, www.rsc.org

desired for chemical modification, they first need to be subjected to genetic and chemical modification of the coat proteins. Subsequently, modified particles can be used *e.g.* for the encapsulation of foreign material or for incorporation into supramolecular structures.

Genetic modification of coat proteins was first described in the 1990s,[2,3] with the aim of expressing antigenic peptides for the production of potential novel subunit vaccines.[4] Following this, the chemical modification of plant virus particles was investigated. When chemical modification is planned, information about the numbers and types of potential addressable groups and their accessibility is required. Therefore, information about the topology of the coat protein in the assembled virions is advantageous. If appropriate groups are lacking, they can be inserted into surface-exposed positions by genetic modification. To maintain intact viral particles, mild reaction conditions are preferred for subsequent chemical modifications.

The broad category of bionanoparticles (BNPs) or protein-based nanoparticles comprises nanosized, protein-based systems, such as virus and VLPs, ferritins, the heat shock protein (Hsp) family, enzyme complexes, cellular micro-compartments and amyloid and other similar biological templates.[5–8] The key definition of a BNP is the organization of multiple proteins through non-covalent interactions, leading to a highly organized and uniform nanoparticle, which is less than 300 nm in at least one of the dimensions. This nano-scale, highly repetitive organization leads to unique advantageous characteristics: if the protein is genetically modified, a single type of surface-exposed reactive group can be presented all over the particles with multiple copies displayed in a defined geometry, giving rise to a powerful correlation between modification sites and spatial resolution. However, to be able to genetically engineer viral particles, infectious cDNA clones of the virus of interest need to be available. In the early stages of plant virus conditioning, infectious clones of RNA plant viruses were generated *in vitro* from RNA transcripts by *E. coli* or T7 polymerases.[9] Only after the introduction of clones incorporating the 35S promoter of the cauliflower mosaic virus (CaMV) a significant improvement was achieved.[10–13] The CaMV 35S promoter made inoculation of plants with plasmid DNA containing cDNA of the plant virus genome possible. Later, the plant virus infection efficiency was further increased, creating clones suitable for agroinfection.[14–16] Using these methods, after T-DNA delivery into the plant cell nucleus, viral RNA transcripts are generated *in situ*, by the host RNA polymerase. Consequently, high yields of wild-type (wt) or genetically engineered plant virus particles can be garnered from infected leaves and easily purified for subsequent chemical modification.

The most common chemical groups desired for coat protein placement can be added *via* mutation of relevant amino acids in the coat protein gene sequence. The desired groups include those which can be used for chemical conjugation with diazonium salts, *i.e.* the amino groups of lysines, the thiol group of cysteines, the carboxyl group of aspartic/glutamic acids and the phenolic group of tyrosine residues. Functional ligands that can be tethered to BNPs using such groups include fluorescent tags, imaging agents, carbohydrates and drugs.[17-19] In addition to specific codon modification by site-directed mutagenesis, there are two other common genetic modifications that are suitable for the generation of specific properties on nanoparticles: (1) the introduction of small "functional" peptides, which have either a high binding affinity to specific materials, such as metals, semiconducting oxides and other technological compounds,[20] or a targeting function,[21,22] and (2) the incorporation of proteins with enzymatic or catalytic functions.[23,24]

High-affinity peptides can be inserted either at the C- or N-terminus, or at an internal loop, provided that these residues are present on the surface of the virus or VLPs. The selection of the short peptide(s) and also the insertion site need to be made so as not to interfere with particle assembly. The number of useful binding peptides available is continuously growing. For the selection of new, suitable peptides combinatorial biology techniques are used:[25] large bacteriophage libraries, presenting random peptides with a specific length on their surface, are used for successive rounds of "panning" on suitable binding material, *e.g.* inorganic surfaces. Only those peptides which bind strongly to the surface are selected.[25-27] Using this phage display technology, peptide sequences were identified that stimulate the binding of inorganic materials. Viral nanoparticles (VNPs) were genetically modified for the presentation of such peptides and subsequently shown to bind a variety of synthetic hybrid structures. These modified particles have great potential, *e.g.* for applications in electronics (Table 4.1).[27-29] Films and arrays can be produced by the assembly of these metalized particles into higher order structures.

Viral protein cages are a very interesting tool for the packaging of foreign nanomaterials. Encapsidation of the foreign molecules into the capsid can take place either during capsid assembly or afterwards when the capsids are already preassembled into particles. Icosahedral viruses are the most common tool used for the creation of viral protein cages. However, the internal

Table 4.1   Peptide sequences with interesting binding properties.

Material	Peptide aa sequence	Ref.
Silica (SiO_2)	YSDQPTQSSQRP	30 and 167
FePt	HNKHLPSTQPLA	31–33
ZnS	CNNPMHQNC or VISNHAESSRRL	28,29 and 31
CdS	SLTPLTTSHLRS	29 and 31
CoPt	CNAGDHANC	31 and 32
Co^{2+}	EPGHDAVP	34
Gold	VSGSSPDS	35–38
Silver	EEEE	39
Co_3O_4	EEEE	36,37 and 39
$FePO_4$	EEEE	40
Silver	NPSSLFRYLPSD (AG4)	41
Metal-binding peptide	DKDGDGWLEFEE	42

channel of rod-shaped TMV particles, with their two open ends, has also been trialed.[1,43–45]

The encapsulation of synthetic nanoparticles, such as fluorescent quantum dots and other metallic structures, is performed to enable imaging applications. The major advantages of nanoparticle encapsulation include biocompatibility, prevention of aggregation and the possibility of achieving specific tissue targeting through either the bioconjugation of functional ligands (targeting molecules) or the presentation of targeting peptides. For the *in vitro* self-assembly of VLPs and the encapsulation of synthetic nanoparticles, two significant principles exist: origin-of-assembly site (OAS) templating and polymer templating. For OAS templating, an artificial OAS is presented on a nanoparticle of interest (*e.g.* gold nanoparticles, magnetic nanoparticles or quantum dots) and initiates the self-assembly process of coat protein monomers (*e.g.* RCNMV).[46] For polymer templating, the synthetic nanoparticle core is coated with negatively charged polymers, mimicking the negative charge of the natural nucleic acid cargo (*e.g.* BMV) and stimulating the templating process.

VNPs are especially interesting for the fabrication of new materials.[47] Here the focus lies on viral capsids. Their main function is the protection of the genome, *i.e.* the nucleic acid. For this reason, the viral capsids are highly robust, which is a major advantage for use as a tool for bionanotechnological applications. This quality offers conditions for a wide range of modifications and incorporates a high tolerance to temperature, the integrity of the particles over a wide range of pH values and tolerance of the particles to organic solvent–water mixtures. Further beneficial features are the various shapes available, the ease of production and that high particle yields, *i.e.* in milligram quantities, can be produced.

Many different plant viruses have been extensively studied and various different-shaped viral capsids are available for nanotech applications. The sizes and shapes range from nanorods of about 13 nm diameter and length up to the micrometer scale and spherical particles with diameters ranging from about 18 nm (*Nanoviridae*) to 70 nm (*Reoviridae*) (http://www.dpvweb.net). Consequently, for each bionanotechnological application, a suitable nanoparticle can be chosen from this range of nanoparticles with discrete shapes and sizes (Figure 4.1).[1,45,48]

In the following sections, five icosahedral plant VNPs and two rod-shaped plant VNPs are introduced, with their potential applications.

4.2   Icosahedral VNPs

4.2.1   Cowpea Chlorotic Mottle Virus (CCMV)

CCMV belongs to the family *Bromoviridae* and is a tripartite ssRNA virus. The icosahedral capsid ($T = 3$) is composed of 180 identical copies of the coat protein (CP) (20 kDa, 190 aa) and is approximately 28 nm in diameter. The RNA is packed inside the central cavity, which is surrounded by a 2–4 mm

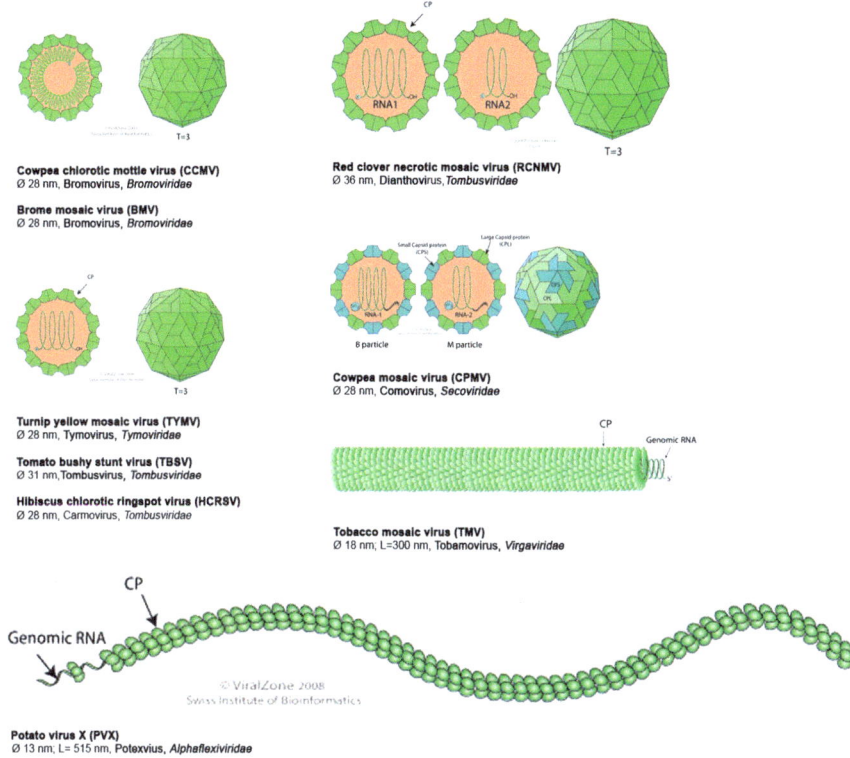

Figure 4.1   Illustrations of plant viral nanoparticles demonstrating the various shapes and sizes. Icosahedral virus particles: CCMV, cowpea chlorotic mottle virus; BMV, brome mosaic virus; TYMV, turnip yellow mosaic virus; TBSV, tomato bushy stunt virus; HCRSV, hibiscus chlorotic ringspot virus; RCNMV, red clover necrotic mosaic virus; CPMV, cowpea mosaic virus. Rod shaped particle: TMV, tobacco mosaic virus. Flexible rod-shaped particle: PVX, potato virus X. Source: ViralZone, http://www. expasy.org/viralzone. Copyright Swiss Institute of Bioinformatics.

thick protein cage. Particles can be obtained from the natural host *Vigna unguiculata*, where particles accumulate at 1–2 mg viral particles per gram of infected leaf material. Alternatively, it can be produced recombinantly in yeast expression systems; for example, a *Pichia pastoris* expression system yields up to 0.5 mg/g of wet cell mass.[49] These yields are comparable to those obtained from infected plants. A major advantage of a heterologous expression system is the possibility of producing wt or genetically modified capsids, with or without specific nucleic acids of interest. This also allows the production of VLPs, which cannot assemble or accumulate in the natural host cells, *e.g.* particles with mutations on the interior surface of the viral capsid, or with mutations that alter the CP subunit interface.

CCMV has been consistently used as a model system for virus assembly studies for the last 40 years[50] and it is the most extensively used plant virus for internal capsid modifications. CP monomers from this system assemble into intact particles *in vitro*[51] and the loading and release mechanism is based on the structural transitions and swelling behavior that occur in response to calcium ions. The swelling mechanism is induced by altering the pH to above 6.5 in the absence of divalent cations. Under these conditions, the particle diameter undergoes a reversible structural transition, which results in a 10% increase in size and the formation of about 60 pores of ~20 Å each.[52,53] Consequently, CCMV particles are best suited for use as a carrier for small molecules. During the opening and closing process, a mineralization reaction with negatively charged materials is also possible. This is driven by electrostatic attraction and occurs on the positively charged interior surface of the particles. The interior surface of wt CCMV particles carries a positive charge, due to the presence of nine basic amino acid (aa) residues (six Arg, three Lys), which are located in the amino-terminal region of each subunit of the coat protein. In general, these positively charged residues interact with the negatively charged viral RNA, but they can also be responsible for the mineralization of the interior surface of the particle, *e.g.* by the addition of anionic polyoxometallate salts or titania, for the production of defined inorganic nanoparticles.[54,55] It is also possible to alter the interior charge by genetic modification, replacing the basic aa residues with acidic residues. Genetically modified CCMV particles were shown to bind ferrous and ferric ions successfully, and could be used for magnetic resonance imaging (MRI) contrasting.[43] A clinical MRI contrast agent was also produced when a metal-binding peptide was genetically fused to the N-terminus of the CCMV subunit. CCMV particles presenting this metal-binding peptide bound Gd^{3+} with higher affinity than wt particles.[42] However, in this case a second chemical modification procedure was shown to be more promising for binding Gd^{3+}.

The controlled disassembly and assembly process allows also for the production of multifunctionalized CCMV particles. Disassembly and purification of the two types of particles allowed them to be differentially labeled and then reassembled in a controlled fashion using varying ratios of the two types of subunits.[56]

It has also been shown that enzymes and polymers can be enclosed in CCMV capsids during particle self-assembly. The entrapment of a single horseradish peroxidase (HRP) molecule in one assembled CCMV VLP has already been demonstrated.[57]

When deletions were introduced into the N-terminus of the CCMV coat protein, it was able to assemble into structures distinct from the natural virions.[58] In contrast to the natural particles with 180 subunits, the modified particles contain only 60 or 120 subunits. Tube-like structures could also be generated that were composed of CCMV coat protein dimers.[59] Furthermore, CCMV can be genetically modified to insert two solvent-exposed cysteines per CP for the introduction of thiol-selective dyes, in which circumstance it was observed that one-third of the thiols introduced are solvent accessible.[60]

Klem *et al.*[61] bound genetically and chemically altered CCMV particles to a gold surface for directed array formation. Using site-directed mutagenesis, the alanine residue at position 163 was replaced with a cysteine residue. After expression and purification from the heterologous yeast expression system, the CCMV particles presented 180 sulfhydryl groups on the particle surface. To be certain that the particles did not aggregate through disulfide bond formation, a reducing agent such as 2-mercaptoethanol was used. Because of the high symmetry of CP subunits, the integrated cysteine residues are evenly distributed on the entire viral particle surface. To avoid the uncontrolled aggregation that may occur due to this multivalent presentation, the symmetry of the particle surface needs to be disrupted. This was achieved through binding of the A163C mutant to a Sepharose resin with an activated thiol. After removal of free virus particles and inactivation of free thiols with iodoacetic acid, bound CCMV A163C was released using 2-mercaptoethanol. Subsequently, the particles obtained could be used for directional derivatization and formed the initial stage of a monolayer array on an gold substrate for magnetic semiconductors.[61]

By nanoindentation measurements with an atomic force microscope, Michel *et al.*[5] analyzed the elastic properties of CCMV capsids at pH 4.8. Surprisingly, a single amino acid substitution, of Lys42 to Arg, increased the capsid stiffness. This mutant had previously been shown to be a salt-stable mutant, which remained stable at pH 7.5 and showed no dissociation under high ionic strength.[62] All capsids tested were found to be very elastic: wt capsids, both empty and RNA genome containing full capsids of the salt-stable mutant and empty capsids of the sub E mutant, whose nine basic residues at the N-terminus have been exchanged by glutamic acid in addition to the Lys42 substitution.[43] It was notable that full capsids resisted indentation more than empty capsids.

4.2.2   Brome Mosaic Virus (BMV)

BMV belongs to the family *Bromoviridae* and is a positive-sense tripartite ssRNA virus. The icosahedral capsid is composed of 180 identical copies of the CP and is about 28 nm in diameter. The spherical shape originates from a $T = 3$ icosahedral symmetry. In addition to production in plants, particles have been expressed in the heterologous expression system *Saccharomyces cerevisiae.* Furthermore, BMV CP monomers assemble *in vitro* into intact VLPs and, analogously to CCMV, particles have a swelling mechanism that is pH and ion dependent.[63,64] Although these structural features are clearly able to be improved by genetic modifications, only wt BMV particles have been used so far for nanotechnological applications. Wt BMV particles were primarily used for the encapsidation of gold nanoparticles, whereby the diameter of the gold NPs was found to influence the size of the resultant capsid.[65] In this reaction, the encapsidation efficiency was improved by coating BMV particles with carboxyl-terminated poly(ethylene glycol).[66] Encapsidation was thereby achieved using polymer templating, where the synthetic nanoparticle core was coated with negatively charged polymers that mimic the negative charge of a natural nucleic acid cargo. In addition to gold nanoparticles, magnetic iron oxide nanoparticles and CdSe/ZnS quantum dots have also been successfully encapsulated by BMV particles.[65-68] Furthermore, the production of superparamagnetic particles, which could be used for magnetic imaging and biosensing, was accomplished by assembling BMV capsids around iron oxide nanotemplates.[68]

4.2.3   Red Clover Necrotic Mosaic Virus (RCNMV)

RCNMV is a bipartite ssRNA virus and belongs to the *Tombusviridae* family. The icosahedral particles are composed of 180 identical copies of the CP (37 kDa) and are about 36 nm in diameter ($T = 3$). The assembly process is initiated by the recognition of the origin of the assembly site on the viral RNA by the CP and the particles are stabilized by the internal protein/RNA cage.[46]

RCNMV has a swelling mechanism similar to those of BMV and CCMV, which is dependent on the calcium ion concentration.[69] This advantageous feature can be exploited for the encapsulation of dyes and chemotherapeutics, such as the cytotoxic drug doxorubicin,[70] and RCNMV particles can thus be applied as a targeted therapeutic device.[71] Similarly to BMV, a range of gold nanoparticles were encapsulated into RCNMV particles, with the hybrid metal-containing particles able to be used for biosensing purposes or as building blocks for the construction of new nanostructured materials.[72] In contrast to the polymer templating of BMV, the *in vitro* self-assembly process of RCNMV coat protein monomers is initiated by presenting an artificial OAS on a nanoparticle of interest (*e.g.* gold nanoparticles, magnetic nanoparticles or quantum dots).[46] Consequently, coat protein monomers of RCNMV assemble around artificial template cores of <17 nm diameter, in accordance with the inner diameter of RCNMV particles.[69,70,73]

To date, only wt particles have been used for nanotech applications, so genetic modification of RCNMV could further enhance its application diversity.

4.2.4   Hibiscus Chlorotic Ringspot Virus (HCRSV)

HCRSV, similarly to RCNMV, belongs to the family *Tombusviridae* and is an ssRNA virus. The icosahedral capsid is composed of 180 identical copies of the CP and is about 28 nm in diameter. The spherical shape originates from a $T = 3$ icosahedral symmetry. During the natural self-assembly process, the coat protein monomers encapsulate negatively charged nucleic acids. By taking advantage of this natural process, artificial nucleic acids or other negatively charged polymers can be trapped inside the HCRSV particles. Purified coat proteins combined with negatively charged polymers have been shown to form hybrid VLPs containing the polymer cargo.[54,74-76] This property has been exploited for targeted drug-delivery, enclosing complexes of anionic polymers, such as poly(styrenesulfonic acid) and poly(acrylic acid)[75] and cytotoxic drugs such as doxorubicin into empty particles.[77] However, neutrally charged dextran molecules could not be encapsidated. For better targeting of the drug-containing particles to cancerous cells, folic acid was conjugated to lysine residues on the outer virus surface.[77] Consequently, the uptake and cytoxicity of doxorubicin to ovarian cancer cells were improved and such modified plant virus particles are of potential use for targeted drug delivery in cancer chemotherapy. For further improvements, genetic modifications with suitable targeting peptides are necessary.

4.2.5　Cowpea Mosaic Virus (CPMV)

CPMV belongs to the family *Secoviridae* and is a bipartite ssRNA virus. Virus particles assemble from 60 copies of each of the two structural proteins, the small (S) and the large (L) subunit, which together build up an icosahedral particle of 28 nm with T = pseudo-3 symmetry. Particles can be obtained from the natural host *Vigna unguiculata*. CPMV is one of the most commonly used plant viruses and it is a particularly versatile plant virus that can be used for various purposes. Infectious cDNA clones of both RNAs have been available for more than 20 years, and allow site-directed genetic manipulation.[9,78] In addition to genetic modification, CPMV particles possess excellent properties for chemical modification: the three-dimensional structure is known down to atomic resolution, which is advantageous for directed chemical modifications,[79] purification from infected tissue is possible in large quantities and particles are remarkably robust, showing stability for up to 4 hours at 60 °C, across a pH range of 4–10, and also in organic solvent–water mixtures.[80] Furthermore, surface-exposed insertion sites have been identified, which are suitable for the successful presentation of peptides without affecting particle assembly.[81,82]

Naturally occurring CPMV particles contain a number of residues that are exposed and thus accessible for chemical modification: five lysines and eight or nine exposed carboxylates (aspartic and glutamic acid residues) per asymmetric unit (one copy each of the L and S protein build up an asymmetric unit). Therefore, a complete particle presents about 300 accessible amine groups and 480–540 accessible carboxyl groups. A number of tyrosines are also present, but no cysteines. All these groups were investigated for their suitability for chemical modification purposes. In this context, the most frequently utilized group has been the ε-amino group of the surface-exposed lysines. Of these, four out of five accessible groups are modified under forcing conditions (see Figure 4.2),[83,84] resulting in the attachment of 240 dye molecules per particle. One of the first examples of lysine modification was to permit the addition of poly(ethylene glycol) (PEG).[85] PEG has a beneficial effect when particles are to be used as potential vaccines, since it acts to hide the particles partially from the immune system. However, significant investigations were carried out and showed that efficient shielding of particles from the immune response is only achieved if long PEG chains are added.[86]

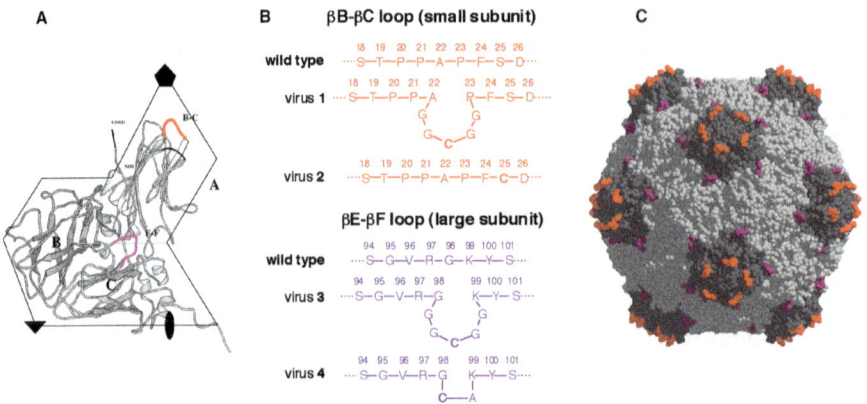

Figure 4.2   Cysteine-added mutants of CPMV. (A) The atomic structure of the CPMV coat protein, with the sites of mutational insertion highlighted in red (βB–βC loop) and purple (βE–βF loop). (B) Amino acid sequences corresponding to native and mutant CPMVs 1–4. (C) A model structure of the entire particle shows the addition of a five-residue insert (GGCGG) at the two positions of interest in the wt CPMV structure. The resulting mutant viruses correspond to viruses 1 and 3 in (B). Note that the BC loop resides further "up" on the protruding cap than the EF loop at each fivefold axis of the icosahedral structure. Reproduced with permission from ref. 83.

Many applications require the modification of all lysine residues; however, if only specific lysines need to be modified, lysines need to be eliminated selectively. This has been accomplished by genetic modification, creating a series of Lys-minus CPMV mutants where the lysine residues were replaced with arginine residues.[87] It was demonstrated that all of the native surface amines in CPMV particles are reactive, but to varying extents.

The introduction of cysteine residues at appropriate positions on the surface of CPMV particles has been used for the generation of, for example, voltage-controlled switchable networks, for use in memory nanocircuits. These residues allowed the binding of gold particles, which were subsequently linked by molecular wires. Thereby, a three-dimensional conducting network was generated on the surface of each virus particle. The interconnection of these particles was accomplished using thiol-terminated organic molecules.[88,89]

The coat protein of CPMV has previously been genetically modified to express antigenic peptides for the production of potential novel subunit vaccines.[2] More recently, the surface expression of appropriate peptides on CPMV particles was also exploited for subsequent introduction of chemical modifications. For example, the introduction of several hexahistidine residues on the virus surface allowed the binding of the modified virus particles on glass slides coated with nickel nitrilotriacetic acid (Ni-NTA).[90] Subsequently, a bifunctional Ni-NTA–biotin reagent was bound to the particles and permitted the fixation of quantum dots *via* a biotin–avidin interaction.

Peptides can also mediate the specific mineralization of virus particles. Virus chimeras presenting peptides, facilitating the sedimentation of silica, are shown in Table 4.2.

CPMV particles have been genetically modified to present 60 peptides specific for SiO_2 or FePt (Table 4.1). Subsequently, these modified CPMV particles were used for the synthesis of silica and amorphous FePt nanoparticles.[30,33] The silicate–CPMV particles, with a diameter of ~32 nm (2 nm silica coating), were visible in unstained electron microscopy images.[30] Similarly, the addition of an iron–platinum (FePt) composition on the virion surface (βB–βC loop of the S protein) resulted in mineralized, hollow particles ~30 nm in diameter, which were highly monodisperse.[33] These particles have applications in magnetic devices and in catalysis.

By controlling chemical dye attachment to multivalent CPMV sites, a signal amplification was obtained and the signal of fluorescently labeled CPMV could be increased threefold.[91] Therefore CPMV can be used as a tracer for significant signal amplification.[92]

Surprisingly, CPMV specifically interacts with the mammalian vascular endothelium. This characteristic was used to create an ideal tool for intravital and tumor neovascular imaging in mouse and chicken, by modifying CPMV to incorporate surface-attached fluorophores. Using this system, it was also possible to distinguish arterial from venous vasculature.[93]

4.2.6   Tomato Bushy Stunt Virus (TBSV)

TBSV is a monopartite, single-stranded (ss), positive-sense RNA virus and the type member of the *Tombusviridae*. The icosahedral capsid is composed of 180 identical copies of the CP (41 kDa)[94] that self-assembles into particles about 31 nm in diameter. The spherical shape originates from a $T = 3$ icosahedral symmetry and the ternary structure of the CP subunit is subdivided into three separate domains: the shell domain (S), the C-terminal protruding (P) domain and the RNA binding domain (R), which is formed by the N-terminal residues. Regarding the quaternary arrangement, the C-terminal P domain is located on the outer surface of the viral particles, whereas the R domain is located in the interior.[95] Therefore, genetic fusion of peptides to the C-terminus of the CP TBSV coat protein results in presentation on the viral surface.

Cargo can be loaded inside TBSV VNPs by a switchable pH-dependent gating mechanism. Under slightly alkaline conditions combined with chelation of divalent ions, particles start to swell[96] and will encapsulate foreign material.

In the primary structure of the TBSV CP, 13 lysine residues are present. However, in the quaternary structure of the assembled virions, these lysines are not accessible on the viral surface. Genetic modification with a Flag tag (DYKDDDDK) fusion has been used to introduce two lysine residues, which were subsequently accessible at the viral surface and were used for bioconjugation chemistry.[97]

Table 4.2 Various BNPs with identified modification sites.

BNP	Genetic modification	Description	Function/used for
CCMV	SubE mutant[43]	Nine basic residues at the N-terminus were substituted with glutamic acid (subE mutant) Selective reactivity at additional cysteine residues	CCMV particles bound ferrous and ferric ions and could be used for MRI contrasting
	Cys[60,61]		Formation of self-assembled monolayers (SAMs) on Au surfaces of genetically modified and symmetry-broken capsids
	DKDGDGWLEFEE[42]	Metal-binding peptide with high affinity for Gd^{3+}	MRI contrasting
	Amino acid substitution of Lys42 to arginine[5]	Salt-stable mutant Increased stiffness compared with the wild-type	
BMV	—		
RCNMV	—		
HCRSV	—		
CPMV	Peptide antigens[157–159]		Used for the production of potential novel subunit vaccines
	Cysteine[88,89,160]	See above	Binding of metallic gold, polymer chains, carbohydrate derivatives, oligonucleotides and other small molecules; generation of electrically conductive molecular networks
	Lysine elimination[87]	Lys-minus CPMV mutants replacing the lysine residues with arginine residues	
	His6tag[90,161,162]	His-tag presenting particles	Particles were bound to glass slides coated with NiNTA for subsequent binding of a bifunctional NiNTA–biotin reagent to the particles and the fixation of quantum dots *via* the biotin–avidin interaction
	RGD[21]	RGD (arginine–glycine–aspartic acid) sequence: human adenovirus type 2 (HAdV-2)-derived cell adhesion integrin binding motif oligopeptide sequence	Binding and internalization into cancer cells, adds an extra positive charge
	SiO_2 or FePt specific peptides (see also Table 4.1)[30,33]	Synthesis of silica and amorphous FePt nanoparticles	Can be applied for magnetic devices and in catalysis

TBSV	Presentation of peptides: Flag tag (DYKDDDDDK)[97] DDDDDHHHH[98]	The flag tag introduces two lysine residues This peptide adds two different charged amino acids at the carboxy terminus	Used for bioconjugation chemistry Genetically modified particles formed mono-layers with different structures
TYMV	RGD[99]	See above	See above
TMV	Antigenic peptides[113,163]	Presentation of immunogenic peptides	Generation of vaccines candidates based on TMV
	RGD peptides[164,165]	See above	See above
	Cysteine[126]	See above	Attachment of thiol-reactive Cy5- and Cy3-maleimide-linked dyes
	Lysine[116]	library-based approach for lysine display at the C-terminus of the TMV U1 CP	Addition of amino acid side chains targeted for modification on the virion surface
	His6tag[127,166]	C-terminally engineered His-tag	His–TMV CP created new forms of hierarchical assembly and extended the range of solution conditions suitable for TMV scaffolded nanostructures
PVX	HIV epitope[132]	Presentation of the highly conserved, neutralizing epitope ELDKWA from human immunodeficiency virus type 1 (HIV-1) glycoprotein (gp) 41 on PVX particles	Potential HIV vaccine
	HCV R9 "mimotope"[133]	Presentation of the R9 epitope, a consensus sequence derived from diverse variants of the hypervariable region 1 from the hepatitis C virus (HCV) envelope protein E2	Potential HCV vaccine
	SiO$_2$ specific peptides (see also Table 4.1),[167]	Synthesis of silica nanoparticles	Mineralized PVX-chimaeras may be exploited in nanotechnological applications such as biomedicine and catalysis
	Protein A domain B (Uhde-Holzem *et al.*, in preparation)	Presentation of the B domain of protein A composed of 60 aa in three different ways on the surface of PVX particles: as direct fusion with the PVX coat protein, or as a fusion *via* the FMDV 2A sequence or a flexible 15 aa linker	Binding of protein A binding antibodies could be used for targeting

Lüders *et al.*[98] genetically modified TBSV by point mutations to present two different charged amino acids at the carboxy-terminus. This led to coat proteins that possess six aspartic acid (negative charge, 6× Asp) and four histidine (positive charge, 4× His) residues. They demonstrated that TBSV is able to self-assemble into monolayers compromising large ordered areas on native and chemically modified mica. When they compared the layers formed by wt and genetically modified virus particles, the latter led to enhanced surface coverage, even when low virus concentrations were used. Lüders *et al.* concluded that such genetic modification is advantageous for the production of homogeneously coated large surface areas.[98]

4.2.7   Turnip Yellow Mosaic Virus (TYMV)

TYMV is a monopartite, single-stranded, positive-sense RNA virus and the type member of the *Tymoviridae* family. The icosahedral capsid is composed of 180 identical copies of the coat protein, which self-assembles into particles approximately 28 nm in diameter. The spherical shape originates from a $T = 3$ icosahedral symmetry. The host range is confined almost entirely to the *Cruciferae* and virus particles can be isolated in gram quantities from turnips or Chinese cabbage.

The chemical reactivity of TYMV was exploited by applying carbodiimide activation and "click chemistry" Copper-catalyzed azide–alkyne cycloaddition (CuAAC) reaction. Oligo(ethylene glycol) (OEG) short chains, Coumarintriazole and RGD-containing peptides have been coupled to the surface of TYMV.[99,100] Particles presenting the RGD motif were applied for the production of composite films, composed of TYMV–RGD44 and poly(allylamine hydrochloride) (PAH). These TYMV/RGD films showed enhanced bone marrow stromal cell (BMSC) adhesion in comparison with wt TYMV.[101] TYMV is also a suitable candidate for bioactive peptide display for use in cell culture studies.

Modified plant virus particles are suitable materials for tissue engineering, drug delivery and biosensing.

4.3   Rod-Shaped Viruses

4.3.1   Tobacco Mosaic Virus (TMV)

TMV is the type member of the genus *Tobamovirus* of the family *Virgaviridae*. Its (ss) linear RNA genome is 6395 nucleotides long and of plus-sense polarity. The coat protein gene is located at the 3′ end of the genome and encodes a 17.6 kDa protein of 158 amino acids.[102] Ridged rod-shaped TMV particles are helically assembled from 2130 identical protein subunits around the genomic RNA.[103] TMV particles are 300 nm long and 18 nm wide and the diameter of the hollow core is 4 nm.[102] TMV virions are remarkably stable, remaining intact at temperatures up to 90 °C and at pH values between 3.5 and 9.[104,105] Coat protein subunits of TMV have the ability to self-assemble *in vitro* with or without RNA, with the latter particles sometimes extending even longer than the normal wt length.[106]

Since high-resolution X-ray diffraction data of TMV particles and the coat protein subunits were determined long ago,[107] possible surface-located insertion points for foreign peptide sequence integration could be readily identified.[4] Takamatsu *et al.*[3] used a direct CP fusion with a small pentapeptide at the exposed C-terminus. The recombinant virus was unable to spread systemically, indicating that particle formation was impaired. To overcome this negative impact on virion assembly, Hamamoto *et al.*[108] utilized an indirect fusion by termination codon suppression of the CP ORF and subsequent read through.[109] By this means, both wt CP and CP–peptide fusion proteins are generated and incorporated into virus particles. Other insertions have been tried successfully between positions 154 and 155, close to the C-terminus,[110,111] and between positions 63 and 66.[112] Small peptides might have a greater chance to be tolerated in the assembly process than larger proteins. However, even these minor changes might lead to host responses that would ultimately reduce the yield of virions. Bendahmane *et al.*[113] showed that a isoelectric point (pI) of the fusion peptide higher than that of wt CP led to the induction of necrosis on inoculated leaves of tobacco plant and cell death of infected protoplasts. The above-mentioned CP small-peptide fusions served as effective alternative antigens for the development of vaccines.[114] However, the display of larger proteins and the creation of vaccine scaffolds utilizing recombinant TMV particles could be achieved by insertion of a heptapeptide with affinity to streptavidin near the C-terminus of the CP. Plant protein extracts containing streptavidin were incubated with recombinant TMV virions, and complexes of viral particles with the protein of interest were collected by centrifugation. Recombinant TMV–streptavidin complex could be dissociated under mild conditions.[115] In the opposite case, coupling of biotin to the TMV CP was achieved by the engineered insertion of a lysine-containing, surface-exposed peptide. Transient coexpression of streptavidin–fusion proteins resulted in decoration of TMV virions which could be purified from plants.[116]

The naturally charged amino acids of TMV capsids were used to deposit a range of inorganic materials along their sides.[117] Semiconducting nanocrystals (PbS, CdS) were created by nucleation of precursor salts on TMV. Iron oxide mineralization and the sol–gel deposition of SiO_2 on the TMV exterior surfaces were also reported, demonstrating the versatility of TMV as the template.[118]

Surface reactive sites for bioconjugation are not abundant on wt TMV particles, as amine and thiol groups are absent or not accessible. Glutamic acid residues 97 and 106 could only be modified by carbodiimide coupling reactions attaching amines on the inner surface after reassembly from modified monomers.[119] The tyrosine at position 139 on the outside of the virus particle became addressable for chemical ligation using an electrophilic substitution reaction at the *ortho* position of the phenol ring with diazonium salts[119] and by CuAAC reactions.[120] Among others, these surface modifications have led to applications in nanoelectronics, nanomaterials and biomedicine.[48,121,122]

The limited addressability of the TMV particle surface called for a genetic engineering approach. Viral mutants displaying reactive cysteine or lysine residues on the solvent-exposed exterior have been made, allowing decoration *via* thiol- or amine-selective chemistry.[123–126]

A genetically engineered TMV coat protein with a C-terminal His-tag self-assembled into disks, hexagonally packed arrays of disk structures, stacked disks, helical rods, fibers and elongated rafts. The attachment of the His-tag created an additional chemical handle for the control of the structures formed by His–TMV CP, enabling new forms of hierarchical assembly to be developed and extending the range of solution conditions suitable for TMV scaffolded nanostructures.[127]

Various surfaces, such as gold, mica, glass and silicon wafers, have been investigated for their potential for the adsorption of TMV,[128] and a technique for the rapid and large-scale assembly of thin-film coatings and ordered fibers, consisting of aligned TMV particles, has also been reported.[129] Partially disassembled TMV virions expose the RNA at the 5′ end of the particle, allowing an oriented arrangement by specific hybridization on to solid supports.[125,126] The RNA-guided self-assembly process of two selectively addressable TMV CP mutants resulted in distinct molecule species grouped and ordered along the longitudinal axis. This technology paves the way towards rod-shaped scaffolds with predefined, selectively reactive barcode patterns on the nanometer scale.[130]

4.3.2 Potato Virus X (PVX)

PVX is a monopartite, positive-strand RNA virus that belongs to the family *Alphaflexiviridae*. The flexible particle is composed of 1270 identical copies of the CP (25 kDa), which self-assembles into particles of 515 nm length and 13 nm diameter.

Natural host plants are members of the family *Solanaceae*, but *Nicotiana benthamiana* plants can be mechanically inoculated with particles. Using this system, quantities of particles up to 200 mg can be obtained from 1 kg of systemically infected leaf material.[135]

Baulcombe *et al.* developed infectious cDNA clones of the PVX genome and genetic modification protocols[131] are now well established. They have been mainly used for the genetic fusion of peptides and proteins to the surface-exposed N-terminus for the presentation of various peptides and proteins, which were used *e.g.* for the generation of potential vaccines.[132,133] For successful presentation of proteins or certain peptides, which are able to impede the assembly process, partial modification of the coat protein biosynthesis is necessary. This is achieved by insertion of the foot and mouth disease virus (FMDV) 2A peptide sequence between the foreign sequence and the N-terminus of the coat protein gene sequence.[134] Utilizing the 2A peptide, Carette *et al.* expressed enzyme lipase B from *Candida antarctica* on the viral surface. The virus-anchored enzyme molecules retained catalytic activity and could act as a novel anchored biocatalyst.[24]

The presentation of foreign epitopes on recombinant PVX particles led to a number of point mutations and deletions after several passages on

N. benthamiana plants, predominantly affecting positively charged residues.[135] Therefore, successful and stable epitope display on PVX particles is possible only if the epitope pI : charge value and its impact on the overall pI is advantageous. To predict whether a peptide of interest would be successfully presented on the viral surface, Lico *et al.* investigated N-terminal PVX CP fusions of many peptides varying in length and amino acid composition.[136] They established that successful expression of the chimeric constructs depends on the pI and the tryptophan content of the foreign peptide. This knowledge can be applied to the design of promising peptides for display on PVX particles.[137]

To investigate the suitability of PVX for chemical bioconjugation protocols, Steinmetz *et al.* examined the reactivity of functional groups on the viral surface in detail.[138] The PVX coat protein contains 11 lysine, 10 aspartic acid and 10 glutamic acid residues, which, if presented on the viral surface, are potential targets for bioconjugation strategies. However, only some of the lysine residues have been found to be surface exposed, with an average of just over one lysine residue per subunit modified when bioconjugation was performed.[138] The combination of chemical functionalization and genetic engineering of PVX coat protein would further improve its application in this field. For example, PVX particles presenting a cell-targeting peptide could be chemically modified for imaging.

4.4   Bioconjugation to VNPs Using Natural Amino Acids

Of the 20 natural canonical amino acids, five possess reactive side-chain moieties, which are particularly suited for the formation of biocompatible covalent bonds: lysine, cysteine, glutamic acid, aspartic acid and tyrosine (Figure 4.3). Among these, lysine with 6.3% frequency is the most common amino acid.[139] Furthermore, lysine is hydrophilic and consequently predominantly surface exposed, allowing for high-density modification, but hampering specific targeting. Nevertheless, lysine residues are the most commonly used amino acid. The primary amines (RNH_2) are accessible to *N*-hydroxysuccinimidyl (NHS) chemistry. In contrast, Michael addition to maleimides is used for targeting cysteine residues, while aspartic and glutamic acids exhibit carboxylic acids for carbodiimide activation.[18] The frequency of tyrosines at 3.3% is very low and therefore the reaction is rather specific. The phenol groups of tyrosine react with benzenediazonium derivatives, giving rise to unique chemical bonds. All these efficient bioconjugation chemistries have been used to attach epitopes, nucleic acids, polymers and small molecules to the surfaces of viral capsids. CuAAC is the most common example of so-called "click chemistry".[140,141] Compared with the aforementioned commonly used bioconjugation chemistries, "click chemistry" is a powerful new approach, since it is bioorthogonal and proceeds faster and with higher fidelity than corresponding NHS or maleimide reactions.

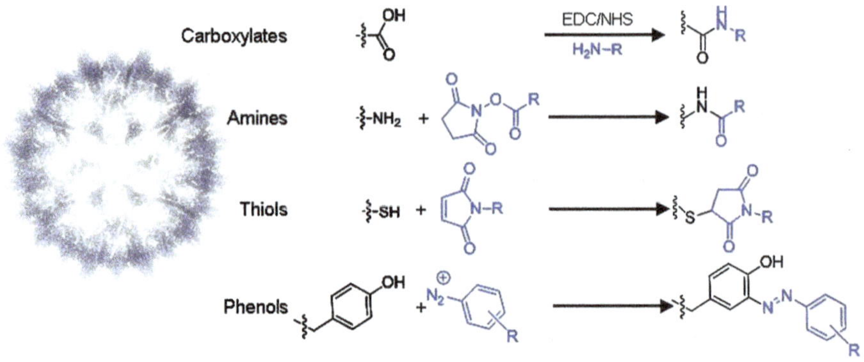

Figure 4.3 Conventional conjugations for chemically introduced functional ligands on BNPs. Fluorescent tags, imaging agents, carbohydrates and drugs can be tethered to BNPs *via* the amino group of lysines, the thiol of cysteines and the carboxyl group of aspartic/glutamic acids, and also on the phenolic group of tyrosine residues with diazonium salts. EDC, 1-ethyl-3-(ε-dimethylaminopropyl)carbodiimide; NHS, *N*-hydroxysuc-cinimide. Reproduced with permission from ref. 6.

The reaction takes place at low reagent concentrations between azide and alkyne moieties and an appropriate ligand in the presence of a Cu^I source, to form a substituted triazole.

4.5 Future Prospects: Bioconjugation to VNPs Using Unnatural Amino Acids

Azide or alkyne side-chains were incorporated into the capsids of RNA phage Qβ and Hepatitis B virus several years ago.[142] However, the strategy of incorporating unnatural amino acids into protein scaffolds has not yet been applied to plant viruses. The use of unnatural amino acids has great potential in this field, especially for the targeted chemical modification of plant viruses.

To be able to introduce homopropargylglycine or azidohomoalanine residues to replace methionine residues, a methionine-auxotroph *E. coli* strain is deprived of methionine, and instead the desired unnatural amino acid is introduced. In this way, the natural thioether is compensated with a "click-able" group. Since this method can only be applied with proteins expressed in *E. coli*, not all plant viruses are suitable for this approach. Furthermore, a suitable methionine residue must be present or needs to be introduced for substitution. However, once these conditions are present, assuming they do not hinder the downstream application, it is an extremely power-ful tool for chemical modification. The incorporation of such unnatural amino acids was successful in Qβ and HBV and after incorporation nearly all available unnatural amino acids were conjugated by a "click" protocol.[142] The ability to include unnatural and natural amino acid side-chains offers multiple opportunities for the design of complex VNPs for the development of next-generation devices and therapeutics. Therefore, the combination of molecular biology and bioconjugation chemistry offers scope for the incorporation of multiple functional modalities such as therapeutic moieties, targeting ligands and imaging molecules.

4.6 Future Prospects: Virus Engineering – Functionalization and Stabilization

For a wide variety of biotechnological applications, it may be very important to have a virus particle that does not lose its integrity under harsh environmental conditions. Further benefits include stabilization of the virus against dissociation into subunits or against a conformational rearrangement. To date, there has been only one report on the engineering of plant virus particles for improved physical stability. Culver *et al.*[143] mutated a number of Asp and Glu residues in TMV, which resulted in increased resistance to dissociation at alkaline pH. The carboxyl groups of these residues were found to be involved in increased electrostatic repulsions. The capsids of several viruses are stabilized by naturally occurring disulfide bonds, and therefore it should be possible to modify the thermal stability of virus particles by rational engineering of intersubunit disulfide bonds. A natural CCMV mutant demonstrates this, where an evolved Cys mutation allowed the formation of intersubunit disulfide bonds, leading to particles that were more resistant to dissociation.[144] Therefore, the conformational stabilization of viral particles could be achieved by engineering of disulfide bonds between capsid subunits. This could be an attractive tool for the improvement of particles for a variety of applications.[145]

4.7 Summary

BNPs are an emerging field, with increasing interest in them as a preferred option over inorganic particles composed of gold, silver, silicon oxide and iron.[146–148] Biological nanosized materials are primarily used as building blocks for the production of nanomaterials, which have a wide range of applications such as in biosensors and for electronic nanodevices, such as batteries and data storage devices.[37,39,40,149] Additionally, BNPs have great potential for use in various biomedical applications,[150,151] such as tumor targeting,[152] tissue engineering (development of VNP films and arrays),[153,154] tissue-specific imaging,[33] therapeutic and pharmaceutical uses,[155] fabrication of metalized nanoparticles using mineralization techniques,[117] *e.g.* for nuclear magnetic resonance imaging,[156] and as advanced vaccine carrier vehicles and multifunctional drug/gene delivery agents. The major advantage of virus particles as protein nanosystems is their intrinsic genetic programmability. Just a specific single codon modification can lead to an amino acid residue change with major effects. Site-directed mutagenesis of certain codons is now being used to add the special functional amino acids that organic chemists need for further functionalization of BNPs.

4.8   Critical Evaluation and Outlook

Plant virus capsids vary in size, stability and functionality, enabling us to select an appropriate template to suit synthetic needs for targeted applications. Capsids and thus BNPs can also be modified using chemical or genetic approaches or both to incorporate specific functionalities.

Harsh conditions such as pH extremes, high temperature or organic solvents will certainly limit the application of plant virus particles for synthetic approaches in materials science, but rational protein engineering and semi-rational combinatorial/directed evolution techniques have the potential to select for capsids that are stable under many conditions.

References

1. G. P. Lomonossoff and D. J. Evans, Applications of plant viruses in bionanotechnology, *Curr. Top. Microbiol. Immunol.*, 2014, **375**, 61–87.
2. G. P. Lomonossoff and J. E. Johnson, Use of macromolecular assemblies as expression systems for peptides and synthetic vaccines, *Curr. Opin. Struct. Biol.*, 1996, **6**(2), 176–182.
3. N. Takamatsu, Y. Watanabe, H. Yanagi, T. Meshi, T. Shiba and Y. Okada, Production of enkephalin in tobacco protoplasts using tobacco mosaic virus RNA vector, *FEBS Lett.*, 1990, **269**(1), 73–76.
4. C. Porta and G. P. Lomonossoff, Scope for using plant viruses to present epitopes from animal pathogens, *Rev. Med. Virol.*, 1998, **8**(1), 25–41.
5. J. P. Michel, I. L. Ivanovska, M. M. Gibbons, W. S. Klug, C. M. Knobler, G. J. Wuite, *et al.*, Nanoindentation studies of full and empty viral capsids and the effects of capsid protein mutations on elasticity and strength, *Proc. Natl. Acad. Sci. U. S. A.*, 2006, **103**(16), 6184–6189.
6. L. A. Lee, H. G. Nguyen and Q. Wang, Altering the landscape of viruses and bionanoparticles, *Org. Biomol. Chem.*, 2011, **9**(18), 6189–6195.
7. P. van Rijn and A. Böker, Bionanoparticles and hybrid materials: tailored structural properties, self-assembly, materials and developments in the field, *J. Mater. Chem.*, 2011, **21**, 16735–16747.
8. P. van Rijn, H. Park, K. Ozlem Nazli, N. C. Mougin and A. Böker, Self-assembly process of soft ferritin–PNIPAAm conjugate bionanoparticles at polar–apolar Interfaces, *Langmuir*, 2012, **29**(1), 276–284.
9. R. Eggen, J. Verver, J. Wellink, A. De Jong, R. Goldbach and A. van Kammen, Improvements of the infectivity of in vitro transcripts from cloned cowpea mosaic virus cDNA: impact of terminal nucleotide sequences, *Virology*, 1989, **173**(2), 447–455.
10. J. C. Boyer and A. L. Haenni, Infectious transcripts and cDNA clones of RNA viruses, *Virology*, 1994, **198**(2), 415–426.
11. U. Commandeur, W. Jarausch, Y. Li, R. Koenig and W. Burgermeister, cDNAs of beet necrotic yellow vein virus RNAs 3 and 4 are rendered biologically active in a plasmid containing the cauliflower mosaic virus 35S promoter, *Virology*, 1991, **185**(1), 493–495.

12. J. T. Dessens and G. P. Lomonossoff, Cauliflower mosaic virus 35S promoter-controlled DNA copies of cowpea mosaic virus RNAs are infectious on plants, *J. Gen. Virol.*, 1993, **74**(5), 889–892.
13. M. Mori, K. Mise, K. Kobayashi, T. Okuno and I. Furusawa, Infectivity of plasmids containing brome mosaic virus cDNA linked to the cauliflower mosaic virus 35S RNA promoter, *J. Gen. Virol.*, 1991, **72**(2), 243–246.
14. Y. Gleba, V. Klimyuk and S. Marillonnet, Viral vectors for the expression of proteins in plants, *Curr. Opin. Biotechnol.*, 2007, **18**(2), 134–141.
15. L. Liu and G. Lomonossoff, Agroinfection as a rapid method for propagating cowpea mosaic virus-based constructs, *J. Virol. Methods*, 2002, **105**(2), 343–348.
16. K. Saunders and G. P. Lomonossoff, Exploiting plant virus-derived components to achieve in planta expression and for templates for synthetic biology applications, *New Phytol.*, 2013, **200**, 16–26.
17. K. J. Koudelka and M. Manchester, Chemically modified viruses: principles and applications, *Curr. Opin. Chem. Biol.*, 2010, **14**(6), 810–817.
18. M. T. Smith, A. K. Hawes and B. C. Bundy, Reengineering viruses and virus-like particles through chemical functionalization strategies, *Curr. Opin. Biotechnol.*, 2013, **24**, 620–626.
19. N. F. Steinmetz and D. J. Evans, Utilisation of plant viruses in bionanotechnology, *Org. Biomol. Chem.*, 2007, **5**(18), 2891–2902.
20. M. Sarikaya, C. Tamerler, A. K. Jen, K. Schulten and F. Baneyx, Molecular biomimetics: nanotechnology through biology, *Nat. Mater.*, 2003, **2**(9), 577–585.
21. M. L. Hovlid, N. F. Steinmetz, B. Laufer, J. L. Lau, J. Kuzelka, Q. Wang, *et al.*, Guiding plant virus particles to integrin-displaying cells, *Nanoscale*, 2012, **4**(12), 3698–3705.
22. A. M. Wen, S. Shukla, P. Saxena, A. A. Aljabali, I. Yildiz, S. Dey, *et al.*, Interior engineering of a viral nanoparticle and its tumor homing properties, *Biomacromolecules*, 2012, **13**(12), 3990–4001.
23. D. Cardinale, N. Carette and T. Michon, Virus scaffolds as enzyme nano-carriers, *Trends Biotechnol.*, 2012, **30**(7), 369–376.
24. N. Carette, H. Engelkamp, E. Akpa, S. J. Pierre, N. R. Cameron, P. C. Christianen, *et al.*, A virus-based biocatalyst, *Nat. Nanotechnol.*, 2007, **2**(4), 226–229.
25. S. Brown, Metal-recognition by repeating polypeptides, *Nat. Biotechnol.*, 1997, **15**(3), 269–272.
26. R. R. Naik, L. L. Brott, S. J. Clarson and M. O. Stone, Silica-precipitating peptides isolated from a combinatorial phage display peptide library, *J. Nanosci. Nanotechnol.*, 2002, **2**(1), 95–100.
27. S. R. Whaley, D. S. English, E. L. Hu, P. F. Barbara and A. M. Belcher, Selection of peptides with semiconductor binding specificity for directed nanocrystal assembly, *Nature*, 2000, **405**(6787), 665–668.
28. S. W. Lee, C. Mao, C. E. Flynn and A. M. Belcher, Ordering of quantum dots using genetically engineered viruses, *Science*, 2002, **296**(5569), 892–895.

29. C. Mao, C. E. Flynn, A. Hayhurst, R. Sweeney, J. Qi, G. Georgiou, *et al.*, Viral assembly of oriented quantum dot nanowires, *Proc. Natl. Acad. Sci. U. S. A.*, 2003, **100**(12), 6946–6951.

30. N. F. Steinmetz, S. N. Shah, J. E. Barclay, G. Rallapalli, G. P. Lomonossoff and D. J. Evans, Virus-templated silica nanoparticles, *Small*, 2009, **5**(7), 813–816.

31. C. Mao, D. J. Solis, B. D. Reiss, S. T. Kottmann, R. Y. Sweeney, A. Hayhurst, *et al.*, Virus-based toolkit for the directed synthesis of magnetic and semiconducting nanowires, *Science*, 2004, **303**(5655), 213–217.

32. B. D. Reiss, C. Mao, D. J. Solis, K. S. Ryan, T. Thomson and A. M. Belcher, Biological routes to metal alloy ferromagnetic nanostructures, *Nano Lett.*, 2004, **4**(6), 1127–1132.

33. S. N. Shah, N. F. Steinmetz, A. A. Aljabali, G. P. Lomonossoff and D. J. Evans, Environmentally benign synthesis of virus-templated, mon-odisperse, iron–platinum nanoparticles, *Dalton Trans.*, 2009, (40), 8479–8480.

34. S. K. Lee, D. S. Yun and A. M. Belcher, Cobalt ion mediated self-assembly of genetically engineered bacteriophage for biomimetic Co–Pt hybrid material, *Biomacromolecules*, 2006, **7**(1), 14–17.

35. Y. Huang, C. Y. Chiang, S. K. Lee, Y. Gao, E. L. Hu, J. De Yoreo, *et al.*, Programmable assembly of nanoarchitectures using genetically engi-neered viruses, *Nano Lett.*, 2005, **5**(7), 1429–1434.

36. A. S. Khalil, J. M. Ferrer, R. R. Brau, S. T. Kottmann, C. J. Noren, M. J. Lang, *et al.*, Single M13 bacteriophage tethering and stretching, *Proc. Natl. Acad. Sci. U. S. A.*, 2007, **104**(12), 4892–4897.

37. K. T. Nam, D. W. Kim, P. J. Yoo, C. Y. Chiang, N. Meethong, P. T. Ham-mond, *et al.*, Virus-enabled synthesis and assembly of nanowires for lithium ion battery electrodes, *Science*, 2006, **312**(5775), 885–888.

38. G. R. Souza, D. R. Christianson, F. I. Staquicini, M. G. Ozawa, E. Y. Sny-der, R. L. Sidman, *et al.*, Networks of gold nanoparticles and bacterio-phage as biological sensors and cell-targeting agents, *Proc. Natl. Acad. Sci. U. S. A.*, 2006, **103**(5), 1215–1220.

39. K. T. Nam, R. Wartena, P. J. Yoo, F. W. Liau, Y. J. Lee, Y. M. Chiang, *et al.*, Stamped microbattery electrodes based on self-assembled M13 viruses, *Proc. Natl. Acad. Sci. U. S. A.*, 2008, **105**(45), 17227–17231.

40. Y. J. Lee, H. Yi, W. J. Kim, K. Kang, D. S. Yun, M. S. Strano, *et al.*, Fab-ricating genetically engineered high-power lithium-ion batteries using multiple virus genes, *Science*, 2009, **324**(5930), 1051–1055.

41. R. M. Kramer, C. Li, D. C. Carter, M. O. Stone and R. R. Naik, Engineered protein cages for nanomaterial synthesis, *J. Am. Chem. Soc.*, 2004, **126**(41), 13282–13286.

42. L. Liepold, S. Anderson, D. Willits, L. Oltrogge, J. A. Frank, T. Douglas, *et al.*, Viral capsids as MRI contrast agents, *Magn. Reson. Med.*, 2007, **58**(5), 871–879.

43. T. Douglas, E. Strable, D. Willits, A. Aitouchen, M. Libera and M. Young, Protein engineering of a viral cage for constrained material synthesis, *Adv. Mater.*, 2002, **14**, 415–418.
44. M. L. Flenniken, M. Uchida, L. O. Liepold, S. Kang, M. J. Young and T. Douglas, A library of protein cage architectures as nanomaterials, *Curr. Top. Microbiol. Immunol.*, 2009, **327**, 71–93.
45. M. Young, D. Willits, M. Uchida and T. Douglas, Plant viruses as biotemplates for materials and their use in nanotechnology, *Annu. Rev. Phytopathol.*, 2008, **46**, 361–384.
46. T. L. Sit, A. A. Vaewhongs and S. A. Lommel, RNA-mediated transactivation of transcription from a viral RNA, *Science*, 1998, **281**(5378), 829–832.
47. Z. Liu, J. Qiao, Z. Niu and Q. Wang, Natural supramolecular building blocks: from virus coat proteins to viral nanoparticles, *Chem. Soc. Rev.*, 2012, **41**(18), 6178–6194.
48. J. K. Pokorski and N. F. Steinmetz, The art of engineering viral nanoparticles, *Mol. Pharmaceutics*, 2011, **8**(1), 29–43.
49. S. Brumfield, D. Willits, L. Tang, J. E. Johnson, T. Douglas and M. Young, Heterologous expression of the modified coat protein of cowpea chlorotic mottle bromovirus results in the assembly of protein cages with altered architectures and function, *J. Gen. Virol.*, 2004, **85**(Pt 4), 1049–1053.
50. P. Ahlquist, Bromoviruses (Bromoviridae), in *Encyclopedia of Virology*, ed. A. Granoff and R. G. Webster, Academic Press, London, 1999, vol. 2, pp. 198–204.
51. L. O. Liepold, J. Revis, M. Allen, L. Oltrogge, M. Young and T. Douglas, Structural transitions in cowpea chlorotic mottle virus (CCMV), *Phys. Biol.*, 2005, **2**(4), S166–S172.
52. A. Schneemann and M. J. Young, Viral assembly using heterologous expression systems and cell extracts, *Adv. Protein Chem.*, 2003, **64**, 1–36.
53. J. A. Speir, S. Munshi, G. Wang, T. S. Baker and J. E. Johnson, Structures of the native and swollen forms of cowpea chlorotic mottle virus determined by X-ray crystallography and cryo-electron microscopy, *Structure*, 1995, **3**(1), 63–78.
54. T. Douglas and M. Young, Host–guest encapsulation of materials by assembled virus protein cages, *Nature*, 1998, **393**, 152–155.
55. M. T. Klem, M. Young and T. Douglas, Biomimetic synthesis of β-TiO_2 inside a viral capsid, *J. Mater. Chem.*, 2008, **18**, 3821–3823.
56. E. Gillitzer, P. Suci, M. Young and T. Douglas, Controlled ligand display on a symmetrical protein-cage architecture through mixed assembly, *Small*, 2006, **2**(8–9), 962–966.
57. M. Comellas-Aragones, H. Engelkamp, V. I. Claessen, N. A. Sommerdijk, A. E. Rowan, P. C. Christianen, *et al.*, A virus-based single-enzyme nanoreactor, *Nat. Nanotechnol.*, 2007, **2**(10), 635–639.

58. J. Tang, J. M. Johnson, K. A. Dryden, M. J. Young, A. Zlotnick and J. E. Johnson, The role of subunit hinges and molecular "switches" in the control of viral capsid polymorphism, *J. Struct. Biol.*, 2006, **154**(1), 59–67.

59. S. Mukherjee, C. M. Pfeifer, J. M. Johnson, J. Liu and A. Zlotnick, Redirecting the coat protein of a spherical virus to assemble into tubular nanostructures, *J. Am. Chem. Soc.*, 2006, **128**(8), 2538–2539.

60. E. Gillitzer, D. Willits, M. Young and T. Douglas, Chemical modification of a viral cage for multivalent presentation, *Chem. Commun.*, 2002, (20), 2390–2391.

61. M. T. Klem, D. Willits, M. Young and T. Douglas, 2-D array formation of genetically engineered viral cages on au surfaces and imaging by atomic force microscopy, *J. Am. Chem. Soc.*, 2003, **125**(36), 10806–10807.

62. J. M. Fox, X. Zhao, J. A. Speir and M. J. Young, Analysis of a salt stable mutant of cowpea chlorotic mottle virus, *Virology*, 1996, **222**(1), 115–122.

63. M. Cuillel, M. Zulauf and B. Jacrot, Self-assembly of brome mosaic virus protein into capsids. Initial and final states of aggregation, *J. Mol. Biol.*, 1983, **164**(4), 589–603.

64. R. W. Lucas, S. B. Larson and A. McPherson, The crystallographic structure of brome mosaic virus, *J. Mol. Biol.*, 2002, **317**(1), 95–108.

65. J. Sun, C. DuFort, M. C. Daniel, A. Murali, C. Chen, K. Gopinath, *et al.*, Core-controlled polymorphism in virus-like particles, *Proc. Natl. Acad. Sci. U. S. A.*, 2007, **104**(4), 1354–1359.

66. C. Chen, M. C. Daniel, Z. T. Quinkert, M. De, B. Stein, V. D. Bowman, *et al.*, Nanoparticle-templated assembly of viral protein cages, *Nano Lett.*, 2006, **6**(4), 611–615.

67. S. K. Dixit, N. L. Goicochea, M. C. Daniel, A. Murali, L. Bronstein, M. De, *et al.*, Quantum dot encapsulation in viral capsids, *Nano Lett.*, 2006, **6**(9), 1993–1999.

68. X. Huang, L. M. Bronstein, J. Retrum, C. Dufort, I. Tsvetkova, S. Aniagyei, *et al.*, Self-assembled virus-like particles with magnetic cores, *Nano Lett.*, 2007, **7**(8), 2407–2416.

69. M. B. Sherman, R. H. Guenther, F. Tama, T. L. Sit, C. L. Brooks, A. M. Mikhailov, *et al.*, Removal of divalent cations induces structural transitions in red clover necrotic mosaic virus, revealing a potential mechanism for RNA release, *J. Virol.*, 2006, **80**(21), 10395–10406.

70. L. Loo, R. H. Guenther, S. A. Lommel and S. Franzen, Encapsidation of nanoparticles by red clover necrotic mosaic virus, *J. Am. Chem. Soc.*, 2007, **129**(36), 11111–11117.

71. S. Franzen and S. A. Lommel, Targeting cancer with 'smart bombs': equipping plant virus nanoparticles for a 'seek and destroy' mission, *Nanomedicine*, 2009, **4**(5), 575–588.

72. D. Lockney, S. Franzen and S. Lommel, Viruses as nanomaterials for drug delivery, *Methods Mol. Biol.*, 2011, **726**, 207–221.

73. L. Loo, R. H. Guenther, V. R. Basnayake, S. A. Lommel and S. Franzen, Controlled encapsidation of gold nanoparticles by a viral protein shell, *J. Am. Chem. Soc.*, 2006, **128**(14), 4502–4503.

74. J. B. Bancroft, E. Hiebert and C. E. Bracker, The effects of various polyanions on shell formation of some spherical viruses, *Virology*, 1969, **39**(4), 924–930.

75. Y. Ren, S. M. Wong and L. Y. Lim, In vitro-reassembled plant virus-like particles for loading of polyacids, *J. Gen. Virol.*, 2006, **87**(Pt 9), 2749–2754.

76. F. D. Sikkema, M. Comellas-Aragones, R. G. Fokkink, B. J. Verduin, J. J. Cornelissen and R. J. Nolte, Monodisperse polymer–virus hybrid nanoparticles, *Org. Biomol. Chem.*, 2007, **5**(1), 54–57.

77. Y. Ren, S. M. Wong and L. Y. Lim, Folic acid-conjugated protein cages of a plant virus: a novel delivery platform for doxorubicin, *Bioconjugate Chem.*, 2007, **18**(3), 836–843.

78. C. L. Holness, G. P. Lomonossoff, D. Evans and A. J. Maule, Identification of the initiation codons for translation of cowpea mosaic virus middle component RNA using site-directed mutagenesis of an infectious cDNA clone, *Virology*, 1989, **172**(1), 311–320.

79. N. F. Steinmetz, T. Lin, G. P. Lomonossoff and J. E. Johnson, Structure-based engineering of an icosahedral virus for nanomedicine and nanotechnology, *Curr. Top. Microbiol. Immunol.*, 2009, **327**, 23–58.

80. T. Lin and J. E. Johnson, Structures of picorna-like plant viruses: implications and applications, *Adv. Virus Res.*, 2003, **62**, 167–239.

81. J. Johnson, T. Lin and G. Lomonossoff, Presentation of heterologous peptides on plant viruses: genetics, structure, and function, *Annu. Rev. Phytopathol.*, 1997, **35**, 67–86.

82. G. P. Lomonossoff and J. E. Johnson, Eukaryotic viral expression systems for polypeptides, *Semin. Virol.*, 1995, **6**, 257–267.

83. Q. Wang, T. Lin, J. E. Johnson and M. G. Finn, Natural supramolecular building blocks, Cysteine-added mutants of cowpea mosaic virus, *Chem. Biol.*, 2002, **9**(7), 813–819.

84. Q. Wang, T. Lin, L. Tang, J. E. Johnson and M. G. Finn, Icosahedral virus particles as addressable nanoscale building blocks, *Angew. Chem., Int. Ed.*, 2002, **41**(3), 459–462.

85. K. S. Raja, Q. Wang, M. J. Gonzalez, M. Manchester, J. E. Johnson and M. G. Finn, Hybrid virus-polymer materials. 1. Synthesis and properties of PEG-decorated cowpea mosaic virus, *Biomacromolecules*, 2003, **4**(3), 472–476.

86. N. F. Steinmetz and M. Manchester, PEGylated viral nanoparticles for biomedicine: the impact of PEG chain length on VNP cell interactions in vitro and ex vivo, *Biomacromolecules*, 2009, **10**(4), 784–792.

87. A. Chatterji, W. F. Ochoa, M. Paine, B. R. Ratna, J. E. Johnson and T. Lin, New addresses on an addressable virus nanoblock; uniquely reactive Lys residues on cowpea mosaic virus, *Chem. Biol.*, 2004, **11**(6), 855–863.

88. A. S. Blum, C. M. Soto, C. D. Wilson, C. Amsinck, P. Franzon and B. R. Ratna, Electronic properties of molecular memory circuits on a nanoscale scaffold, *IEEE Trans. Nanobiosci.*, 2007, **6**(4), 270–274.

89. A. S. Blum, C. M. Soto, C. D. Wilson, T. L. Brower, S. K. Pollack, T. L. Schull, *et al.*, An engineered virus as a scaffold for three-dimensional self-assembly on the nanoscale, *Small*, 2005, **1**(7), 702–706.

90. I. L. Medintz, K. E. Sapsford, J. H. Konnert, A. Chatterji, T. Lin, J. E. Johnson, *et al.*, Decoration of discretely immobilized cowpea mosaic virus with luminescent quantum dots, *Langmuir*, 2005, **21**(12), 5501–5510.

91. C. M. Soto, K. M. Blaney, M. Dar, M. Khan, B. Lin, A. P. Malanoski, *et al.*, Cowpea mosaic virus nanoscaffold as signal enhancement for DNA microarrays, *Biosens. Bioelectron.*, 2009, **25**(1), 48–54.

92. C. M. Soto, B. D. Martin, K. E. Sapsford, A. S. Blum and B. R. Ratna, Toward single molecule detection of staphylococcal enterotoxin B: mobile sandwich immunoassay on gliding microtubules, *Anal. Chem.*, 2008, **80**(14), 5433–5440.

93. J. D. Lewis, G. Destito, A. Zijlstra, M. J. Gonzalez, J. P. Quigley, M. Manchester, *et al.*, Viral nanoparticles as tools for intravital vascular imaging, *Nat. Med.*, 2006, **12**(3), 354–360.

94. K. A. White and P. D. Nagy, Advances in the molecular biology of tombusviruses: gene expression, genome replication, and recombination, *Prog. Nucleic Acid Res. Mol. Biol.*, 2004, **78**, 187–226.

95. A. J. Olson, G. Bricogne and S. C. Harrison, Structure of tomato busy stunt virus IV. The virus particle at 2.9 Å resolution, *J. Mol. Biol.*, 1983, **171**(1), 61–93.

96. R. Aramayo, C. Merigoux, E. Larquet, P. Bron, J. Perez, C. Dumas, *et al.*, Divalent ion-dependent swelling of tomato bushy stunt virus: a multi-approach study, *Biochim. Biophys. Acta*, 2005, **1724**(3), 345–354.

97. S. Grasso, C. Lico, F. Imperatori and L. Santi, A plant derived multifunctional tool for nanobiotechnology based on tomato bushy stunt virus, *Transgenic Res.*, 2013, **22**, 519–535.

98. A. Lüders, C. Müller, K. Boonrod, G. Krczal and C. Ziegler, Tomato bushy stunt viruses (TBSV) in nanotechnology investigated by scanning force and scanning electron microscopy, *Colloids Surf., B*, 2012, **91**, 154–161.

99. X. Zan, P. Sitasuwan, J. Powell, T. W. Dreher and Q. Wang, Polyvalent display of RGD motifs on turnip yellow mosaic virus for enhanced stem cell adhesion and spreading, *Acta Biomater.*, 2012, **8**(8), 2978–2985.

100. Q. Zeng, S. Saha, L. A. Lee, H. Barnhill, J. Oxsher, T. Dreher, *et al.*, Chemoselective modification of turnip yellow mosaic virus by Cu(I) catalyzed azide–alkyne 1,3-dipolar cycloaddition reaction and its application in cell binding, *Bioconjugate Chem.*, 2011, **22**(1), 58–66.

101. G. Kaur, M. T. Valarmathi, J. D. Potts and Q. Wang, The promotion of osteoblastic differentiation of rat bone marrow stromal cells by a polyvalent plant mosaic virus, *Biomaterials*, 2008, **29**(30), 4074–4081.

102. M. A. A. B. Zaitlin, Descriptions of Plant Viruses, *Tobacco mosaic virus*, 2000, 370, 1–13. http://www.dpvweb.net/dpv/showdpv.php?dpvno=370.
103. D. Zimmern, The nucleotide sequence at the origin for assembly on tobacco mosaic virus RNA, *Cell*, 1977, **11**(3), 463–482.
104. R. N. Perham and T. M. Wilson, The characterization of intermediates formed during the disassembly of tobacco mosaic virus at alkaline pH, *Virology*, 1978, **84**(2), 293–302.
105. M. Zaitlin and P. Palukaitis, Advances in understanding plant viruses and virus diseases, *Annu. Rev. Phytopathol.*, 2000, **38**, 117–143.
106. M. Knez, A. M. Bittner, F. Boes, C. Wege, H. Jeske, E. Maiss, *et al.*, Bio-template synthesis of 3 nm nickel and cobalt nanowires, *Nano Lett.*, 2003, **3**, 1079–1082.
107. A. C. Bloomer and P. J. G. Butler, Tobacco mosaic virus: structure and self-assembly, in *The Plant Viruses*, ed. M. H. V. van Regenmortel and H. Fraenkel-Conrat, Plenum Press, New York, 1986, vol. 2, pp. 19–57.
108. H. Hamamoto, Y. Sugiyama, N. Nakagawa, E. Hashida, Y. Matsunaga, S. Takemoto, *et al.*, A new tobacco mosaic virus vector and its use for the systemic production of angiotensin-I-converting enzyme inhibitor in transgenic tobacco and tomato, *Bio/Technology*, 1993, **11**(8), 930–932.
109. J. M. Skuzeski, L. M. Nichols, R. F. Gesteland and J. F. Atkins, The signal for a leaky UAG stop codon in several plant viruses includes the two downstream codons, *J. Mol. Biol.*, 1991, **218**(2), 365–373.
110. J. Fitchen, R. N. Beachy and M. B. Hein, Plant virus expressing hybrid coat protein with added murine epitope elicits autoantibody response, *Vaccine*, 1995, **13**(12), 1051–1057.
111. M. Koo, M. Bendahmane, G. A. Lettieri, A. D. Paoletti, T. E. Lane, J. H. Fitchen, *et al.*, Protective immunity against murine hepatitis virus (MHV) induced by intranasal or subcutaneous administration of hybrids of tobacco mosaic virus that carries an MHV epitope, *Proc. Natl. Acad. Sci. U. S. A.*, 1999, **96**(14), 7774–7779.
112. T. H. Turpen, S. J. Reinl, Y. Charoenvit, S. L. Hoffman, V. Fallarme and L. K. Grill, Malarial epitopes expressed on the surface of recombinant tobacco mosaic virus, *Bio/Technology*, 1995, **13**(1), 53–57.
113. M. Bendahmane, M. Koo, E. Karrer and R. N. Beachy, Display of epitopes on the surface of tobacco mosaic virus: impact of charge and isoelectric point of the epitope on virus–host interactions, *J. Mol. Biol.*, 1999, **290**(1), 9–20.
114. V. Yusibov, S. Rabindran, U. Commandeur, R. M. Twyman and R. Fischer, The potential of plant virus vectors for vaccine production, *Drugs R&D*, 2006, **7**(4), 203–217.
115. V. Negrouk, G. Eisner, S. Midha, H. I. Lee, N. Bascomb and Y. Gleba, Affinity purification of streptavidin using tobacco mosaic virus particles as purification tags, *Anal. Biochem.*, 2004, **333**(2), 230–235.
116. M. L. Smith, J. A. Lindbo, S. Dillard-Telm, P. M. Brosio, A. B. Lasnik, A. A. McCormick, *et al.*, Modified tobacco mosaic virus particles as scaffolds for display of protein antigens for vaccine applications, *Virology*, 2006, **348**(2), 475–488.

117. W. Shenton, T. Douglas, M. Young, G. Stubbs and S. Mann, Inorganic-organic nanotube composites from template mineralization of tobacco mosaic virus, *Adv. Mater.*, 1999, **11**, 253–265.

118. M. T. Dedeo, D. T. Finley and M. B. Francis, Viral capsids as self-assembling templates for new materials, *Prog. Mol. Biol. Transl. Sci.*, 2011, **103**, 353–392.

119. T. L. Schlick, Z. Ding, E. W. Kovacs and M. B. Francis, Dual-surface modification of the tobacco mosaic virus, *J. Am. Chem. Soc.*, 2005, **127**(11), 3718–3723.

120. M. A. Bruckman, G. Kaur, L. A. Lee, F. Xie, J. Sepulveda, R. Breitenkamp, *et al.*, Surface modification of tobacco mosaic virus with "click" chemistry, *ChemBioChem*, 2008, **9**(4), 519–523.

121. M. A. Bruckman, S. Hern, K. Jiang, C. A. Flask, X. Yu and N. F. Steinmetz, Tobacco mosaic virus rods and spheres as supramolecular high-relaxivity MRI contrast agents, *J. Mater. Chem. B*, 2013, **1**(10), 1482–1490.

122. S. Y. Lee, J. S. Lim and M. T. Harris, Synthesis and application of virus-based hybrid nanomaterials, *Biotechnol. Bioeng.*, 2012, **109**(1), 16–30.

123. M. Demir and M. H. B. Stowell, A chemoselective biomolecular template for assembling diverse nanotubular materials, *Nanotechnology*, 2002, **13**, 541–544.

124. W. S. Tan, C. L. Lewis, N. E. Horelik, D. C. Pregibon, P. S. Doyle and H. Yi, Hierarchical assembly of viral nanotemplates with encoded microparticles via nucleic acid hybridization, *Langmuir*, 2008, **24**(21), 12483–12488.

125. H. Yi, S. Nisar, S. Y. Lee, M. A. Powers, W. E. Bentley, G. F. Payne, *et al.*, Patterned assembly of genetically modified viral nanotemplates via nucleic acid hybridization, *Nano Lett.*, 2005, **5**(10), 1931–1936.

126. H. Yi, G. W. Rubloff and J. N. Culver, TMV microarrays: hybridization-based assembly of DNA-programmed viral nanotemplates, *Langmuir*, 2007, **23**(5), 2663–2667.

127. M. A. Bruckman, C. M. Soto, H. McDowell, J. L. Liu, B. R. Ratna, K. V. Korpany, *et al.*, Role of hexahistidine in directed nanoassemblies of tobacco mosaic virus coat protein, *ACS Nano*, 2011, **5**(3), 1606–1616.

128. M. Knez, M. P. Sumser, A. M. Bittner, C. Wege, H. Jeske, D. M. Hoffmann, *et al.*, Binding the tobacco mosaic virus to inorganic surfaces, *Langmuir*, 2004, **20**(2), 441–447.

129. D. M. Kuncicky, R. R. Naik and O. D. Velev, Rapid deposition and long-range alignment of nanocoatings and arrays of electrically conductive wires from tobacco mosaic virus, *Small*, 2006, **2**(12), 1462–1466.

130. F. C. Geiger, F. J. Eber, S. Eiben, A. Mueller, H. Jeske, J. P. Spatz, *et al.*, TMV nanorods with programmed longitudinal domains of differently addressable coat proteins, *Nanoscale*, 2013, **5**(9), 3808–3816.

131. D. C. Baulcombe, S. Chapman and S. Santa Cruz, Jellyfish green fluorescent protein as a reporter for virus infections, *Plant J.*, 1995, **7**(6), 1045–1053.

132. C. Marusic, P. Rizza, L. Lattanzi, C. Mancini, M. Spada, F. Belardelli, *et al.*, Chimeric plant virus particles as immunogens for inducing murine and human immune responses against human immunodeficiency virus type 1, *J. Virol.*, 2001, **75**(18), 8434–8439.

133. K. Uhde-Holzem, V. Schlosser, S. Viazov, R. Fischer and U. Commandeur, Immunogenic properties of chimeric potato virus X particles displaying the hepatitis C virus hypervariable region I peptide R9, *J. Virol. Methods*, 2010, **166**(1–2), 12–20.

134. S. S. Cruz, S. Chapman, A. G. Roberts, I. M. Roberts, D. A. Prior and K. J. Oparka, Assembly and movement of a plant virus carrying a green fluorescent protein overcoat, *Proc. Natl. Acad. Sci. U. S. A.*, 1996, **93**(13), 6286–6290.

135. K. Uhde-Holzem, R. Fischer and U. Commandeur, Genetic stability of recombinant potato virus X virus vectors presenting foreign epitopes, *Arch. Virol.*, 2007, **152**(4), 805–811.

136. C. Lico, F. Capuano, G. Renzone, M. Donini, C. Marusic, A. Scaloni, *et al.*, Peptide display on potato virus X: molecular features of the coat protein-fused peptide affecting cell-to-cell and phloem movement of chimeric virus particles, *J. Gen. Virol.*, 2006, **87**(Pt 10), 3103–3112.

137. C. Lico, C. Mancini, P. Italiani, C. Betti, D. Boraschi, E. Benvenuto, *et al.*, Plant-produced potato virus X chimeric particles displaying an influenza virus-derived peptide activate specific CD^{8+} T cells in mice, *Vaccine*, 2009, **27**(37), 5069–5076.

138. N. F. Steinmetz, M. E. Mertens, R. E. Taurog, J. E. Johnson, U. Commandeur, R. Fischer, *et al.*, Potato virus X as a novel platform for potential biomedical applications, *Nano Lett.*, 2010, **10**(1), 305–312.

139. D. Gilis, S. Massar, N. J. Cerf and M. Rooman, Optimality of the genetic code with respect to protein stability and amino-acid frequencies, *Genome Biol.*, 2001, **2**(11), , research0049.

140. V. V. Rostovtsev, L. G. Green, V. V. Fokin and K. B. Sharpless, A stepwise Huisgen cycloaddition process: copper(I)-catalyzed regioselective "ligation" of azides and terminal alkynes, *Angew. Chem., Int. Ed.*, 2002, **41**(14), 2596–2599.

141. C. W. Tornøe, C. Christensen and M. Meldal, Peptidotriazoles on solid phase: [1,2,3]-triazoles by regiospecific copper(I)-catalyzed 1,3-dipolar cycloadditions of terminal alkynes to azides, *J. Org. Chem.*, 2002, **67**(9), 3057–3064.

142. E. Strable, D. E. Prasuhn Jr, A. K. Udit, S. Brown, A. J. Link, J. T. Ngo, *et al.*, Unnatural amino acid incorporation into virus-like particles, *Bioconjugate Chem.*, 2008, **19**(4), 866–875.

143. J. N. Culver, W. O. Dawson, K. Plonk and G. Stubbs, Site-directed mutagenesis confirms the involvement of carboxylate groups in the disassembly of tobacco mosaic virus, *Virology*, 1995, **206**(1), 724–730.

144. J. M. Fox, F. G. Albert, J. A. Speir and M. J. Young, Characterization of a disassembly deficient mutant of cowpea chlorotic mottle virus, *Virology*, 1997, **227**(1), 229–233.

145. M. G. Mateu, Virus engineering: functionalization and stabilization, *Protein Eng., Des. Sel.*, 2011, **24**(1–2), 53–63.
146. P. V. Kamat, Photophysical, photochemical and photocatalytic aspects of metal nanoparticles, *J. Phys. Chem. B*, 2002, **106**, 7729–7744.
147. Z. Niu, J. He, T. P. Russell and Q. Wang, Synthesis of nano/micro-structures at fluid interfaces, *Angew. Chem., Int. Ed.*, 2010, **49**(52), 10052–10066.
148. Y. Sun and J. A. Rogers, Inorganic semiconductors for flexible electronics, *Adv. Mater.*, 2007, **19**, 1897–1916.
149. I. Yildiz, S. Shukla and N. F. Steinmetz, Applications of viral nanoparticles in medicine, *Curr. Opin. Biotechnol.*, 2011, **22**(6), 901–908.
150. Y. B. Lim, K. S. Moon and M. Lee, Recent advances in functional supramolecular nanostructures assembled from bioactive building blocks, *Chem. Soc. Rev.*, 2009, **38**(4), 925–934.
151. R. V. Ulijn and D. N. Woolfson, Peptide and protein based materials in 2010: from design and structure to function and application, *Chem. Soc. Rev.*, 2010, **39**(9), 3349–3350.
152. W. Chen, P. A. Jarzyna, G. A. van Tilborg, V. A. Nguyen, D. P. Cormode, A. Klink, *et al.*, RGD peptide functionalized and reconstituted high-density lipoprotein nanoparticles as a versatile and multimodal tumor targeting molecular imaging probe, *FASEB J.*, 2010, **24**(6), 1689–1699.
153. H. Cui, M. J. Webber and S. I. Stupp, Self-assembly of peptide amphiphiles: from molecules to nanostructures to biomaterials, *Biopolymers*, 2010, **94**(1), 1–18.
154. P. X. Ma, Biomimetic materials for tissue engineering, *Adv. Drug Delivery Rev.*, 2008, **60**(2), 184–198.
155. S. Grasso and L. Santi, Viral nanoparticles as macromolecular devices for new therapeutic and pharmaceutical approaches, *Int. J. Physiol., Pathophysiol. Pharmacol.*, 2010, **2**(2), 161–178.
156. D. P. Cormode, P. A. Jarzyna, W. J. Mulder and Z. A. Fayad, Modified natural nanoparticles as contrast agents for medical imaging, *Adv. Drug Delivery Rev.*, 2010, **62**(3), 329–338.
157. M. C. Canizares, L. Nicholson and G. P. Lomonossoff, Use of viral vectors for vaccine production in plants, *Immunol. Cell Biol.*, 2005, **83**(3), 263–270.
158. L. Liu, M. C. Canizares, W. Monger, Y. Perrin, E. Tsakiris, C. Porta, *et al.*, Cowpea mosaic virus-based systems for the production of antigens and antibodies in plants, *Vaccine*, 2005, **23**(15), 1788–1792.
159. L. Nicholson, M. C. Canizares and G. Lomonossoff, Production of vaccines in GM plants, in *Plant Biotechnology: Current and Future Applications of Genetically Modified Crops*, ed. H. Ng, Wiley, Chichester, 2006, pp. 1164–1192.
160. Q. Wang, T. Lin, J. E. Johnson and M. G. Finn, Natural supramolecular building blocks. Cysteine-added mutants of cowpea mosaic virus, *Chem. Biol.*, 2002, **9**(7), 813–819.
161. A. Chatterji, W. F. Ochoa, T. Ueno, T. Lin and J. E. Johnson, A virus-based nanoblock with tunable electrostatic properties, *Nano Lett.*, 2005, **5**(4), 597–602.

162. F. Sainsbury, K. Saunders, A. A. Aljabali, D. J. Evans and G. P. Lomonossoff, Peptide-controlled access to the interior surface of empty virus nanoparticles, *ChemBioChem*, 2011, **12**(16), 2435–2440.
163. M. L. Smith, W. P. Fitzmaurice, T. H. Turpen and K. E. Palmer, Display of peptides on the surface of tobacco mosaic virus particles, *Curr. Top. Microbiol. Immunol.*, 2009, **332**, 13–31.
164. A. A. McCormick, T. A. Corbo, S. Wykoff-Clary, K. E. Palmer and G. P. Pogue, Chemical conjugate TMV–peptide bivalent fusion vaccines improve cellular immunity and tumor protection, *Bioconjugate Chem.*, 2006, **17**(5), 1330–1338.
165. L. Wu, J. Zang, L. A. Lee, Z. Niu, G. C. Horvath, V. Braxton, *et al.*, Electrospinning fabrication, structural and mechanical characterization of rod-like virus-based composite nanofibers, *J. Mater. Chem.*, 2011, **21**, 8550–8557.
166. N. Liu, C. Wang, W. Zhang, Z. Luo, D. Tian, N. Zhai, *et al.*, Au nanocrystals grown on a better-defined one-dimensional tobacco mosaic virus coated protein template genetically modified by a hexahistidine tag, *Nanotechnology*, 2012, **23**(33), 335602.
167. P. van Rijn, L. S. van Bezouwen, R. Fischer, E. J. Boekema, A. Böker and U. Commandeur, Virus-SiO$_2$ and Virus-SiO$_2$-Au Hybrid Particles with Tunable Morphology, *Part. Part. Syst. Charact.*, 2015, **32**, 43–47.

CHAPTER 5

Intra- and Intermolecular Bionanoparticle Self-Assembly as Functional Systems

BARBARA SANTOS DE MIRANDA[a,b], JELMER SJOLLEMA[a,b], AND PATRICK VAN RIJN[*a,b]

[a]Department of Biomedical Engineering, FB-40, University Medical Center Groningen, University of Groningen, A. Deusinglaan 1, 9713 AV Groningen, The Netherlands; [b]W. J. Kolff Institute for Biomedical Engineering and Materials Science, FB-41, University Medical Center Groningen, University of Groningen, A. Deusinglaan 1, 9713 AV Groningen, The Netherlands
*E-mail: p.van.rijn@umcg.nl

5.1 Introduction

Bionanoparticles offer both intra- and intermolecular self-assembly. Obviously, when considering protein-based particles/structures, intramolecular self-assembly is eminently suitable to obtain the specific three-dimensional structure. This three-dimensional structure offers the possibility of forming larger assemblies because the three-dimensional structure permits specific interparticle interactions that drive the formation of large fiber-like structures as encountered both in the intracellular matrix, *e.g.* microtubules and actin microfilaments, and in the extracellular matrix, *e.g.* collagen, elastin and

RSC Smart Materials No. 16
Bio-Synthetic Hybrid Materials and Bionanoparticles: A Biological Chemical Approach Towards Material Science
Edited by Alexander Böker and Patrick van Rijn

fibronectin.[1,2] Additionally, there are protein-based structures that assemble into a defined nanosized structure, which were originally used in Nature for storage, such as ferritin, but also complex architectures such as viruses, that are a common occurrence.[3] All these structures can be utilized for functional structures in, *e.g.*, biomaterials science and bioelectronics, as drug delivery platforms and many more. Although proteins with aggregation properties are useful in the aforementioned applications, interprotein aggregation also lies at the foundation of many neurodegenerative diseases. This usually is a result of a compromised three-dimensional structure where (partial) unfolding results in aggregation (amyloid formation) and a correct and stable intramolecular assembly is required.[4]

In this chapter, we discuss intra- and intermolecular assemblies of proteins that offer functional bionanoparticle-based structures and assemblies. As important specific structures, ferritin, collagen and virus particles are discussed.

5.2   Intra- *Versus* Intermolecular Self-Assembly of Bionanoparticles

In Nature, proteins/bionanoparticles are often not confined to one place inside the cell but are able to travel to various locations, usually guided by surface charge or the distribution of hydrophobic regions, just as a different set of these properties confines them to membranes. Without a larger regulating system, the bionanoparticles still have these properties but do not have the same overall system to guide them, therefore different types of aggregation can be induced. Here we distinguish three levels of self-assembly. On the peptide-strand level; the polypeptide folds into a defined three-dimensional structure, which holds for all proteins including those which are able to assemble further into larger structures, *e.g.* particles and fibers. On the particle level; the protein subunits assemble/disassemble reversibly into a discrete structure with a fixed number of assembling units and with specific dimensions (*e.g.* cage proteins such as ferritin and viral particles such as tobacco mosaic virus). Alternatively, when protein subunits are complementary in a more head-to-tail fashion, fiber-like structures will be the dominant morphology (*e.g.* extracellular matrix components such as collagen and fibronectin).

5.2.1   Intra-Bionanoparticle Self-Assembling Properties in Solution

Intra-bionanoparticle self-assembly in solution occurs through the precise folding of a polypeptide strand of a very specific sequence that determines the exact overall folded three-dimensional structure (Figure 5.1). Upon folding, various interactions stabilize the folded structure, *e.g.* hydrogen bonding, disulfide bridges or hydrophobic interactions (Figure 5.1). While Nature

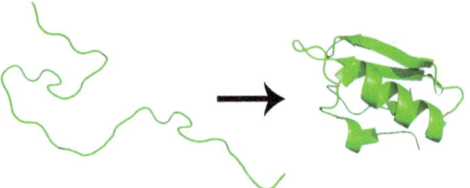

Figure 5.1   Schematic representation of a peptide strand folding into a specific
three-dimensional structure.

uses this process continuously, it is also a common approach for recombinant protein synthesis involving a defined folding step in order to obtain
the desired structure.[5,6] All bionanoparticles, when composed of several subunits as in the case of cage proteins such as ferritin (Section 5.3), Hsp (heat-
shock protein) and Dps (DNA-binding protein of starved cells), fiber-forming
structures such as collagen (Section 5.4) and well-known viral particles such
as CCMV and TMV (Section 5.5), are composed of specifically folded individual subunits, hence intramolecular self-assembly lies at the foundation
of all these structures. The folding of the initial subunits needs extremely
high accuracy since the inter-assembled structures are able to disassemble/
reassemble with high precision, which can only occur from the perfectly
reproducible subunit structure. This dynamic behavior with high-precision
reversibility allows for the removal of any unwanted natural components,
e.g. RNA or inorganic cores. While normally the structures reassemble into
their native structure again, by choosing the correct reactive additives that
can act as a template, the introduced materials, inorganic, organic or biological, can be biofunctionalized.[7,8] *Via* non-covalent interactions or due to
chemical reactions, the subunits are transferred to the templates instead of
reassembling into their native structure and newly functionalized nanoparticles are formed or new supramolecular structures evolve from the coordination.[9] Although this is an interesting approach, it is beyond the scope of
this chapter.

5.2.2   Inter-Bionanoparticle Self-Assembling
Properties in Solution

Proteins and protein subunits are able to interact with each other in certain
cases or under specific conditions due to non-covalent interactions such
as hydrophobic interactions, complementary surface charges or hydrogen
bonding motifs to form larger non-native structures. It was found that certain combinations of small peptide structures and proteins are also able to
form other structures, namely peptidic macrocycles or "proteinaceous nanostructures" as they were called by Wagner *et al.* Designing different peptide
structures increases the scope of materials that are able to be formed.[10] Taking a dimeric protein species in combination with a flexible peptide linker,
different self-assembled macrocycles in solution were obtained. The size of
the cycles was found to be dependent on the length and composition of the

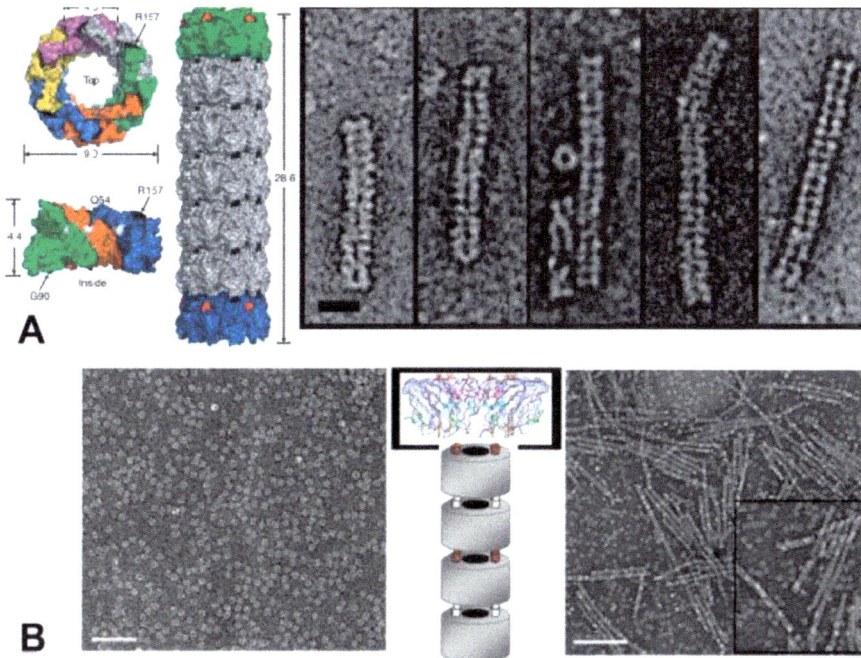

Figure 5.2   Formation of nanotubes in solution of different length scales. (A) Structure of Hcp1$_{CC}$ protein, a mutant of the Hcp1 stacked *Pseudomonas aeruginosa* with two cysteine incorporations, displaying short non-helical tubes of various lengths up to about 90 nm (scale bar: 15 nm). (B) Single structures of the wild-type protein TRAP (trp RNA-binding attenuation protein) from *Bacillus stearothermophilus* on the left and the TRAP–HCCAH mutant in the presence of DTT which induces polymerization, forming long tubular structures (scale bars: left, 50 nm; right, 100 nm). Images adapted from ref. 12 with permission from the National Academies of Science and from ref. 13 with permission from Wiley.

flexible spacer and the specific structure and interactions of the dimerizer. Another variation also produced nanorings with catalytic properties, which were found to be dependent on ring size.[11]

A ring structure immediately suggests the notion of self-assembling units forming tubular structures when stacked on top of each other. This structure is observed in the honeycomb crystal structure of stacked ring-shaped peptides, the homohexameric Hcp1 from *Pseudomonas aeruginosa*. This observation inspired Mougous and co-workers to introduce specific cysteine mutations in order to allow the formation of disulfide bonds between macrocycles in a head-to-tail fashion, creating stable tubular structures in solution (Figure 5.2).[12]

Protein-fiber structures are an important feature in biological systems, *e.g.* the cytoskeleton built from actin, which is probably why attempts are being made to recreate these structures and additionally also to prepare tubes, since an empty inner space would allow for additional functions such as storage or perhaps even transport. As in the previous example, it is most convenient to take a promising natural structure and modify it synthetically. There are two ways of addressing this. One approach is to take disk-like protein structures which are not prone to assemble into tubes but which after the introduction of a mutation are prone to do so, as we have seen previously and in other examples.[13] The other approach is to take a structure from Nature that already forms a fiber-like morphology and attempt to introduce more control over its dimensions, which was done by introducing a capping protein to the fiber system. Clark *et al.* introduced a compatible protein of a similar design which was able to associate into the fiber but prevented further growth.[14] One can imagine that control over the dimensions of such structures potentially offers different materials with various properties when different dimensions or hierarchies of self-assembly can be addressed. In some cases one does not need to rely on synthetic modifications and the native system already exhibits this control. This form of hierarchically building of structures was found for ADP-ribosyl cyclase, which assembles from the monomeric state into a defined dimer and subsequently hexamer, finally forming fibers that are crystalline, displaying nanorods and microtubes.[15] Although not every naturally occurring protein has suitable properties for the formation of these hierarchically self-assembled structures, it is being increasingly recognized that other properties can also be exploited. These include applying potential differences in order to induce polymerization of actin filaments and orient the fibers that they produce.[16] However, proteins already prone to form networks, as fibrin does in blood clots and thrombi, can also be used. By isolating the proteins from other interacting species, interesting structures were discovered depending on the concentration, as was found for fibrin, which forms thin nanosheets at low concentrations.[17]

When proteins are considered as nanoparticulate building blocks, one of the drawbacks is stability towards conditions that can denature the protein's primary structure. This instability, however, can also be used as a tool to form supramolecular structures as a response to the denaturing conditions. It was found that hen egg white lysozyme is able to form thermo-reversible gels with the use of a denaturing agent such as dithiothreitol (DTT).[18] That proteins by themselves are able to form gels due to denaturing is trivial, since this is exactly the structure of a boiled egg. However, with the knowledge that this occurs and the realization that it can be induced by different stimuli such as heat, reducing agents or even ultrasound and the fact that proteins behave like particles, interesting structures and morphologies have been discovered.[19] Combining different synthetic materials with proteins or introducing mutations genetically to alter the properties in such a fashion that new properties and self-assembling structures arise are versatile approaches and certainly an area of research where new and exciting applications can be addressed. However, it is important not to forget about the structures already provided by Nature, which already offer tremendous possibilities. In the following sections, three types of protein structures are addressed that resemble classes of self-assembling structures that provide discrete-sized particles and long fibrous morphologies.

5.3 Assembling Protein Structures for Defined Bionanoparticle Formation

Nature offers various protein-based morphologies that are well-defined in one, two or three dimensions. The protein subunit is a highly reproducible single-stranded structure, which in some cases can aggregate into fibers, *e.g.* collagen and actin microfilaments, or form defined multimeric or particle structures, *e.g.* insulin (tetramer assembly), cage proteins such as ferritin and Dps (DNA-binding protein from starved cells) (24- and 12-mer, respectively), or viral assemblies that often have a complex symmetry and range from 120 protein subunits for small virus particles such as cowpea mosaic virus up to 2130 protein subunits for larger particles such as tobacco mosaic virus. The foundation for all the protein-based structures is a perfectly reproducible folded single amino acid strand, as shown in the previous section. Ferritin, collagen and virus particles are highly investigated structures and represent defined protein-based structures of different dimensions, hence they are discussed in greater detail in the following sections.

5.3.1 Ferritin: The Benchmark Cage Protein

Ferritin is a cage protein that has been used in many studies concerning the combination of synthetic structures with proteins in order to develop new hybrid bionanoparticles.[20] Ferritin represents a family of iron storage proteins that have a ubiquitous distribution among many life forms (*e.g.* vertebrates,[21] bacterioferritin,[22] Dps[23]) and are the most abundant members of the ferritin-like superfamily. A carboxylate-bridged diiron center within the four-helix bundle can be considered as a general structure within the ferritin family. Based on sequence similarities that show the high conservation or conservative substitution of few key structural residues of the iron-binding and -chelating motifs, it was suggested that ferritins are obligate proteins for aerobic metabolism.[24] Ferritins and Dps function as dynamic iron storage, and Fe(II) sequestration not only prevents spontaneous oxidation to Fe(III) and production of toxic free radicals but also has various other functions.

The particular features of the cage structure are the basis for the extraordinary capabilities of ferritin and its numerous applications in the field of bionanomaterials science and the benchmark protein cage.

5.3.2 The Structure of the Mammalian Ferritin Protein Shell

The principle tertiary and quaternary structure of ferritin is highly conserved throughout a broad range of species, but the amino acid sequence can vary significantly.[25] Generally, 24 protein subunits self-assemble into dimers to form a dodecameric cage with an outer diameter of approximately 12 nm and an inner cavity of 7–8 nm that is filled with a ferric oxohydroxy core. Ferritin devoid of the inorganic core is called apoferritin.[26] The protein cage is temperature stable up to 85 °C and tolerates reasonably high levels of urea, guanidinium chloride and many other denaturants at neutral pH.

The ferritin shell structure has been examined by X-ray diffraction and is one of the best studied protein cage structures. The protein shell of mammalian ferritin is usually heterogeneous and consists of a mixture of two subunits of about 21 kDa, termed H for heavy chain, and of about 19 kDa, termed L for light chain.

Apoferritin can be readily disassembled and reassembled by changing the ionic strength of the aqueous medium or by reducing the pH to a value as low as 2 and increasing it above 7, respectively (Figure 5.3).[27] Because of the empty cavity inside the apoferritin and also the dynamic nature concerning the disassembly and reassembly, the protein structure is very interesting as a nanocontainer, nanoreactor, biostabilized catalyst, *etc.*[28-30] During the recombination process, the bulk solution and dissolved molecules and also nanoparticles can be entrapped in the interior, which is an efficient way to encapsulate material that is unable to enter the cavity of the intact protein shell through one of the pores. Webb *et al.* gave a rough estimation of the internal volume of 1.4×10^{-19} cm^3, which can hold solubilized species.[31] In Chapter 2 it was shown that inorganic species can be condensed inside the protein cage but also the structure can be altered genetically in a similar fashion, as described in Chapter 3 and 4, and also chemically altered *via* methods mentioned in Chapter 1. Therefore, ferritin can be regarded as an interdisciplinary particle, which has triggered the imagination of scientists in many different fields of research. Although ferritin consists of relatively simple and small protein particles, other morphologies can also be obtained from Nature that are very useful structures in various (bio)medical applications but also show promise in bionanoparticle-based electronic devices.

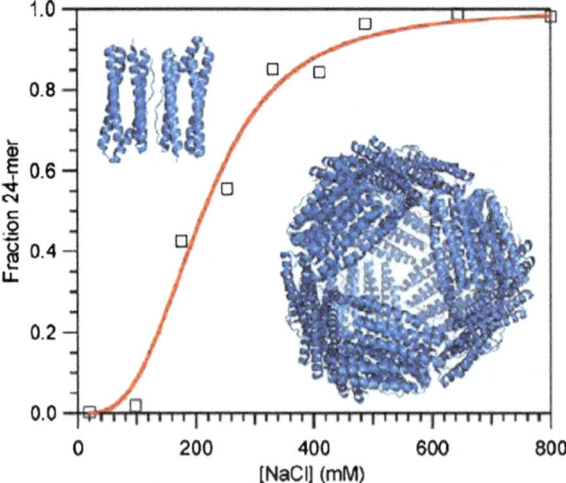

Figure 5.3   Dynamic structure of ferritin. The reversible assembly–disassembly depends on the medium with respect to pH and ionic strength. Reproduced from ref. 27 with permission from the American Chemical Society.

5.4 Collagens as Bionanoparticle-Based Biomaterials

Owing to the aging society in Europe, biomaterials are of growing importance for the restoration of body functions. However, all biomaterials have a number of disadvantages, with unwanted foreign body reactions, such as chronic inflammation, granulation tissue development, formation of foreign body giant cells, scar tissue formation (fibrosis), the release of reactive oxygen species and bacterial infection risks.[32] The reason for these side-effects is that the applied biomaterials are non-human, produced by chemical synthesis or physico-chemical processing in order to comply with mechanical and functional requirements. In all cases they are recognized as foreign bodies and prone to support bacterial biofilm formation.[33]

Because biomaterials derived from the extracellular matrix (ECM) are degradable and may reduce the typical immunological reactions, these materials are used for tissue repair, tissue engineering and regenerative medicine, and collagen is such a fibrous matrix forming protein.[34] Additionally, collagen is used in several formulations (*e.g.* sponges, gels, films, shields, tablets) to be applied as carriers for local administration of drugs, allowing the entrapment, local storage and delivery of, *e.g.*, growth factors, cytokines and antibiotics, and therefore plays important roles during organ development, wound healing and tissue repair.[35]

5.4.1 Types of Collagen and Collagen Synthesis

Collagen is one of the main constituents of the ECM and the most abundant mammalian protein, accounting for about 20–30% of total body proteins.[35] Collagen is a generic name for a protein that forms a helix-like structure comprising three polypeptide chains, identical or different, with varying composition, but characterized by glycine-rich tripeptide Gly–X–Y domains (see also Table 5.1).[36] Glycine is centered within the right-handed helix and

Table 5.1 Composition of amino acid residues in skin and bone, taken from analyses of 10 mammalian species.

Mean amino acid composition (residues per 1000)					
Amino acid	Bone	Skin	Amino acid	Bone	Skin
Glycine	330	329	Leucine	26	24
Proline	119	126	Valine	23	22
Alanine	115	109	Threonine	19	19
Hydroxyproline	95	95	Phenylalanine	14	13
Glutamic acid	75	74	Isoleucine	11	11
Arginine	49	49	Hydroxylysine	6	6
Aspartic acid	48	47	Methionine	4	6
Serine	35	36	Histidine	5	5
Lysine	26	29	Tyrosine	3	3

other amino acids are at the outer parts of the molecules. In total, 26 genetically distinct types of collagen are identified, indicated as type I–XXVI, all different in amino acid composition, structure and function,[37] the most abundant being type I, III and V collagen, found in the bone, dermis, tendons, skin, ligaments, cornea and many interstitial connective tissues, together with collagen type II, which is mainly found in cartilage. Collagen type I, II, III and V all form fibrils, highly orientated supramolecular aggregates with a characteristic supra-structure with diameters between 25 and 400 nm.

Cross-linking of the lysine and hydroxylysine amino acid in the collagen is the basis for these highly oriented structures (Figure 5.4). Fibrils may aggregate into fibers, having dimensions up to more than 100 μm, and the direction and density control the biomechanical properties of the ECM material in each of the specific tissue types. In tendons these fibers are oriented in specific directions, whereas in cartilage the fibrils are more randomly oriented. In tendons and fascia the collagen fibers provide tensile stiffness, and in bone they provide load-bearing and torsional stiffness, in particular after calcification. In addition to the fibril-forming collagens (type I, II, II, V and XI), several other subgroups of non-fibrillar collagens can be distinguished, *e.g.* collagens that

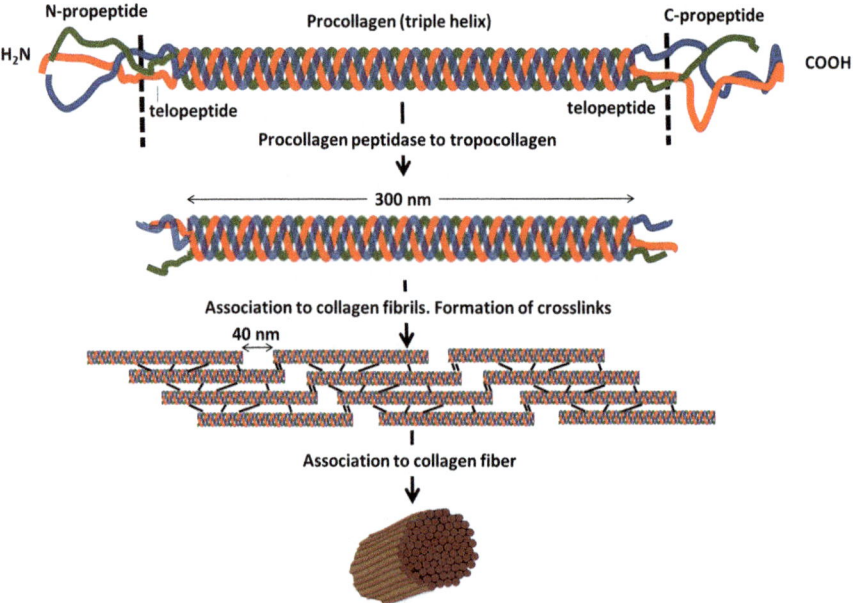

Figure 5.4   Schematic representation of supramolecular assembly of fibril-forming collagen and association to fibers for collagens. The tropocollagen molecular structures are formed from procollagen after cleavage by C- and N-procollagenases. Cross-links are formed mainly between the telopeptide segments between aldehydes formed on lysine and hydroxylysine groups and adjacent lysine groups to form a staggered construct of collagen molecules.

are involved in the formation of sheets or protein membranes that surround tissues and organisms, such as basement membranes (collagen type IV), and collagen type VIII in Descemet's membrane found in the cornea.[38]

5.4.2   Collagens in Biomedical Applications

5.4.2.1   *Natural Collageneous Materials*

The application of natural collageneous materials in the form of decellularized tissues and organs from animals offers large advantages over non-biological materials. For instance, degradable meshes of biological origin are often recommended for the treatment of abdominal wall hernias and pelvic organ prolapse[39] in high infection risk scenarios indicated by comorbidity factors[40] such as smoking and obesity, and is based on the immune-compatible nature and anti-inflammatory character of the material, the physicochemical properties that it provides for use as a scaffold for tailored drug release and the non-adhesive character preventing bacterial biofilm formation.[41] Porcine tissue has also been used in the fabrication of bio-prosthetic heart valves, where the biological nature of the material is claimed to reduce thrombosis, offering improved hemodynamic properties.[42]

In order to use animal-based tissues, they first need to be decellularized because cellular antigens are recognized as foreign by the host and therefore will induce an inflammatory response or an immune-mediated rejection of the tissue.[43] Several methods exist for decellularizing tissues and often comprise a combination of physical, enzymatic and chemical treatments, depending on the tissue of interest.[44] All of the steps in the decellularization protocol will have an impact on the properties of the ECM. Some of them disrupt the collagen network, which will impair the mechanical characteristics, whereas enzymatic treatments might compromise the non-degradability of the material.

5.4.2.2   *Collagen Sponges and Gels*

Collagen sponges and hydrogels [in a pure form or in a blend with synthetic polymers such as poly(vinyl alcohol) and poly(acrylic acid) to enhance the mechanical properties] are often used as drug reservoirs owing to their open structure and opportunities to bind molecules covalently or non-covalently. Typically, collagen molecules to be prepared in a sponge or gel should first be extracted from animal tissue. For instance, collagen type I can be easily extracted from rat tail tendons by removing the skin and immersing the tendons in 0.1 M acetic acid for 2–3 days with stirring. Insoluble material is removed by centrifugation and the soluble fraction remains in solution and can be frozen and lyophilized to obtain collagen sponges. This procedure applied to remove the water will lead to intermolecular cross-linking between the collagen aggregates, forming a dense network of fibers with a pore size that is dependent on the cooling rate.[45] Because stiffness and degradability depend strongly on the formation of covalent intermolecular cross-links between the individual protein subunits, these properties can be further controlled by cross-linking the sponge, *e.g.* by glutaraldehyde, 1-ethyl-3-(3-dimethylaminopropyl)carbodiimide (EDC) or *N*-hydroxysuccinimide (NHS).[46] Collagen sponges allow the absorption of tissue exudates and have been very useful as wound dressings and as drug reservoirs for the delivery of antibiotics. Another form of collagen often used for drug delivery is a hydrogel, prepared from high-concentration collagen solutions at elevated temperatures. Gels are applied, for instance, in ocular shields that protect the cornea from damage after surgery or in the form injectable gels.[35] Gels also can be cross-linked to obtain higher mechanical stiffness and increased resistance against degradation *in vivo*. Collagen hydrogels are very open and release of drugs is characterized by a very rapid depletion, taking place within hours. To control the release rate, several methods are available, including encapsulation of drugs in liposomes, covalently bound to the collagen matrix and released after degradation of the matrix.[47] Another methodology to allow extended release periods is based on increased affinity of drugs with specific macromolecules grafted within the collagen matrix. For instance, heparin has been used to immobilize and control the release of various growth factors (such as fibroblast growth factor involved in angiogenesis and wound healing), which interact non-covalently but highly specifically with heparin.[48]

5.4.2.3   Recombinant Collageneous Materials

Animal-based collagens that are often used today, such as of bovine and porcine origin, have disadvantages, however, such as the risks of immune reactions, transmittance of animal-based viruses and of prions causing Creutzfeld–Jacob disease and the impossibility of 100% control of the presence of impurities and of reproducible composition. Therefore, there is a desire to replace animal-based collagen with a non-animal-based (synthetic) collagen, and the first recombinant collageneous biomaterials are have become commercially available.[49] The term "collageneous" is used on purpose because recombinant material usually lacks the stable triple helix structure. This is a result of non-mammalian expression systems missing the gene encoding for the synthesis of prolyl- and lysylhydroxylase, both responsible for the post-translational hydroxylation of the proline and lysine residues and essential for intramolecular hydrogen bonding. Recently, hydroxyproline-rich collagen molecules have been produced in a new expression system that combines genes for collagen type I and for prolylhydroxylase, which paves the way for the production of recombinant collagens.[50]

5.5   Virus Particle Self-Assembly

The final types of inter-protein assembly to be discussed are virus particles. A virus is an obligate intracellular parasite that contains an RNA or DNA genome protected by a protein coat.[51] The complete and infectious virus particle is called a *virion* and its main goal is to penetrate the host cell in order to express its genome. The term parasite is included in its definition because a virus is inert outside the host cell and it is fully dependent of its energetic mechanism in order to replicate.[52] The protein coat (capsid) of the virus particle not only has the function of protecting the genome from the surrounding environment but is also responsible for attachment and recognition of the virion to the targeted cell receptors.[2]

Although viruses are not as robust as inorganic particles, this characteristic can also be beneficial since it allows for assembly and disassembly, as also observed for ferritin, and hence they have great potential as building blocks and functional bionanoparticles in (biomedical) materials science.[3] Virus particles have one of the most defined structures formed by proteins in the nanometer regime and the size range extends from 30 nm for poliovirus to 300 nm for tobacco mosaic virus, and even larger viruses are available with various morphologies. The capsid consists of many replicas of the same few proteins, sometimes only one type, which assemble in a very accurate three-dimensional structure. The assembly process of virus particles produces perfect structures every time once the biological system is self-regulated. This implies that any defect in the construction will inhibit the formation of a functional particle, and therefore either the structure disassembles in order to repair the error or the whole biological system crashes, and as a result only perfect particles are produced.[2,3]

5.5.1 Virus Morphology

The virus particle structure determines how it is formed, how it enters in the host cell and how it replicates. Viruses are classified based on their morphology, chemical composition and mode of replication. Regarding the morphology, they can be classified into helical and polyhedral, which correspond to rod and spherical shapes, respectively. There are also more complex virus assemblies such as that of the bacteriophages.[2]

Helical virus particles are those in which the coat protein is assembled with helical symmetry. The genome of the virus also follows the proteins and is wrapped as a helical filament inside the capsid.[2] The result of the protein assembly with this symmetry is rod-shaped viruses, as discovered by Watson and Crick in 1956.[51] Representative structures of these types of virus particles are tobacco mosaic virus (TMV) and potato virus X (PVX) (Figure 5.5). TMV is one of simplest viruses known and is the most studied and used. It has a rod-shaped structure 18 nm in diameter and 300 nm long. TMV is built by self-assembly of 2130 copies of the coat protein (MW 17 kDa) into a helical structure with $16\frac{1}{3}$ subunits per layer.[53] The TMV particle has a rigid structure whereas the PVX particle has a more flexible rod structure 515 nm in length and 13 nm in diameter with 1270 assembled coat protein molecules (MW 25 kDa).[54] The bacteriophage M13 also belongs to the rod-shaped virus structures and is 930 nm long and 6.5 nm in diameter.[55,56] Rod-shaped structures are a perfect example of genetic economy because they allow the use of many copies of just one subunit to assemble the virus capsid, which needs only a small sequence in the genome.

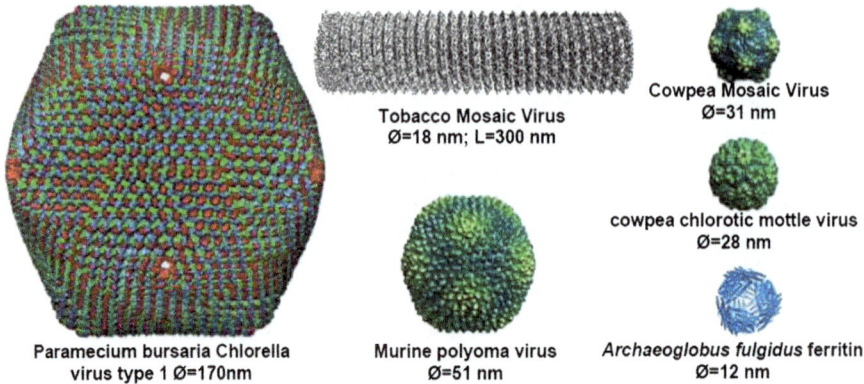

Cowpea Mosaic Virus
Ø=31 nm

Tobacco Mosaic Virus
Ø=18 nm; L=300 nm

cowpea chlorotic mottle virus
Ø=28 nm

Paramecium bursaria Chlorella
virus type 1 Ø=170nm

Murine polyoma virus
Ø=51 nm

Archaeoglobus fulgidus ferritin
Ø=12 nm

Figure 5.5   Structures and dimensions of protein assemblies such as virus particles that vary greatly in size and shape. Adapted from ref. 3.

Spherical viruses such as cowpea mosaic virus (CPMV) and cowpea chlorotic mottle virus (CCMV) (Figure 5.5) possess polyhedral morphology. An icosahedron is the best shape to build up a completely closed shell. Moreover, it is also used to combine larger viruses because the size of the proteins does not change drastically; the strategy is to increase the number of different coat proteins in the capsid.[56,57] The CPMV capsid structure is made by self-assembly of 180 molecules of only one coat protein (MW 20.3 kDa) forming a spherical particle 28.6 nm in diameter whereas CCMV is a particle 30 nm in diameter assembled with 60 subunits of two different coat proteins.[58] Finally, there are the most complex viruses structures, such as the T4 phages, formed by a head, a tail (composed of two hollow tubes) and fiber tails at the end.[2]

The symmetry presented by the virus structures results in each subunit having the same interactions and complementary surface contacts with its neighbors. Repeated interaction patterns lead to a symmetrical structure. In icosahedral viruses with $T > 1$ the positions of the subunits are no longer identical but quasi-equivalent as the interactions are similar, still head-to-head or tail-to-tail, but are not equal. Moreover, all the interactions between the subunits are non-covalent, which means reversible, error-proof and self-regulated. Disassembly is only possible because the constituent parts of the capsid are not covalently bound.

5.5.2   Self-Assembly and Disassembly of Viruses

The formation and the deterioration of the capsid is part of the metastable characteristics of viruses in general and part of the infectious cycle itself. This equilibrium in the assembly process is necessary for the survival of a virus in non-replicating environments for longer periods. TMV is probably one of the most often used and best characterized viral structures and is therefore explored here to a greater extent and used as an example to demonstrate the assembly behavior and applications of virus particles.

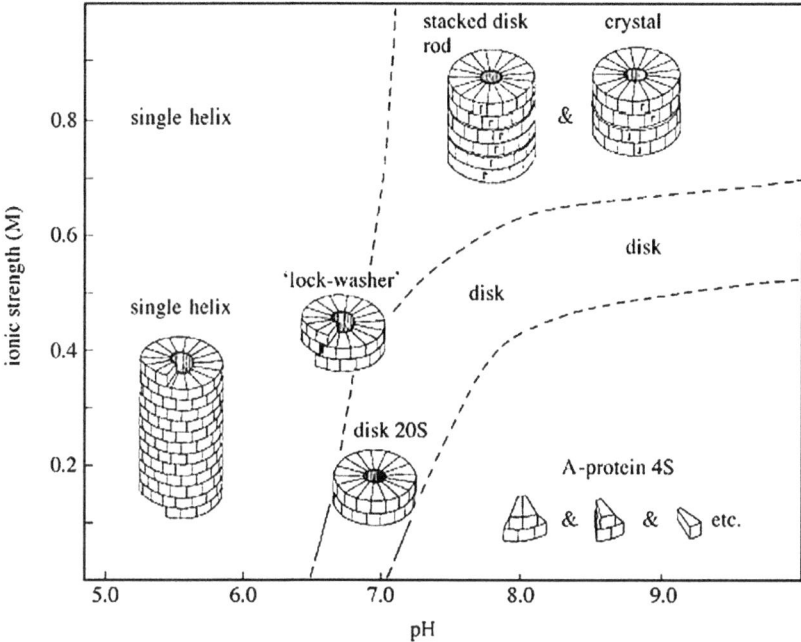

Figure 5.6 Dynamic structures of TMV. The formation of the different structures is reversible and depends on the medium with respect to pH and ionic strength. Reproduced from ref. 53 with permission from the Royal Society.

5.5.2.1 Tobacco Mosaic Virus

In addition to the rod-shaped structure, TMV also presents intermediate structures that can be obtained with changes of pH and ionic strength, as shown in Figure 5.6.[53] The main interactions observed for the TMV particle include a polar region at high radius and an extended salt-bridge region at low radius (close to the center of the rod) stabilized by Ca^{2+} and Mg^{2+} ions. All coat proteins are aligned laterally with their neighbors sharing the same interactions, in a laterally complementary fashion except for those at the end of the particle.

The helix structure is stable at low pH, probably because protonation of carboxyl–carboxylate pairs forming hydrogen bonds, which prevents electrostatic repulsion and further disaggregation. At high pH values, those groups are ionized and disassembly occurs (Figure 5.6). Nonetheless, when under physiological conditions, coat proteins combine into a two-layer disk called a 20S aggregate. The disk is a mandatory intermediate of the TMV assembly process. The 20S particle has two main purposes: first, promoting nucleation in the solution to overcome the entropically unfavorable growth, and second, a recognition element for the nucleic acid, which triggers the assembly of the virus in the helix form. Before the nucleic acid packing, the 20S aggregate has its two layers in contact at the high radius, whereas in the center there is a gap between them, which is waiting for the specific RNA sequence to bind, then it closes and all the structure dislocates to form the helix but at the same time preserving the lateral hydrophobic contacts of the subunits. Moreover, Bloomer *et al.* confirmed that the hydrophobic interactions within subunits occur in the lateral interface, which also suggests that the TMV assembly is an entropically driven system.[2,52,59]

Plant viruses are not infectious against animal cells, and they have been studied since the 1950s, their host, replication and structure being very well known. Moreover, plant viruses do not present an envelope in their final structure and during the construction, the coat proteins self-assemble into the capsid. For these reasons viral particles (VPs) and virus-like particles (VLPs) make very good building blocks for the development of new materials. Recently, a review of the potential of VPs and VLPs in biomedicine applications was published and this is also discussed in chapters 8 and 9.[60] It is possible to use this mechanism in engineering new materials and structures; changes in pH, temperature and ionic strength can dislocate the equilibrium to a desired structure or simply the coat protein alone.

5.5.3   Modifications and Applications of VPs

In order to achieve the purpose of new supramolecules, the VPs often need further modification of their structure. Different approaches can be used depending on the coat protein properties chosen and the desired main application. The functional groups in the surface of the coat proteins are often the target of chemical modification through bioconjugation. Amine, carboxylic acid and thiol groups are extensively used to couple other molecules covalently, which can contribute to obtaining different desired properties of the surface such as charge, size and reactivity. Chemical modifications can also be used to multiply the number of reactive groups on the surface. TMV is easily modified *via* its glutamate residues situated in the center of the rod and also the tyrosine residues located on the outer surface of the particle.[61] Bioconjugation of CPMV was performed by using biotin and streptavidin and applying this coupling to form layer-by-layer structures.[62] Furthermore, new routes of conjugation have also been developed, such as copper(ı)-catalyzed azide–alkyne cycloaddition (CuAAC), the best known of the "click" chemistry reactions with TMV and CPMV, which were discussed in more detail in Chapter 1.[63–65] For convenience with respect to quantification and also potential fluorescence properties, it is also possible to use dyes as the cross-linking reagent in these reactions, which were performed using turnip yellow mosaic virus (TYMV) as the VP.[66]

Most of the modifications to the VP that pursue the aim of creating novel supramolecular biomaterials with a wide range of applications are achieved with different kinds of polymers. Bundles of TMV covered with polyaniline and poly(4-vinylpyridine) stabilized by the same virus particle were obtained, the aim of which was to add electrically conductive properties in order to use these materials in nanoelectronic devices.[67,68] TMV displays good mechanical properties when combined with poly(vinyl alcohol) in electrospinning, the polymer layer covered the VP forming a composite.[69]

In a different area, for drug-delivery systems, polypeptides are added to induce the self-assembly of the coat proteins into a hollow sphere, which can act as a nanocontainer with a cargo of hydrophobic or hydrophilic (only changing the oligonucleotide) drug or imaging agent.[70] These nanocontainers are also attractive for use as nanoreactors, hence the VPs are also promising materials in this area; recent studies showed that it is possible to assemble the coat proteins in a swollen state filled with enzymes.[71] The coat protein was stable and did not break down in the presence of the cargo; moreover, the catalytic activity of the enzyme was not compromised. Additionally, CCMV particles were functionalized with photosensitive cationic dendrons that are electrostatically bound to the negative particle surface. The assembly and disassembly of this material can be tuned with an external optical stimulus and the transport and release of materials of interest can be controlled and selective.[72]

5.6   Discussion and Conclusion

The perfect reproducibility and the precise association of proteins, either intramolecularly or intermolecularly, are of tremendous value for the development of new biomaterials. Here only three main examples have been given, but there are many more that offer a large variety of different approaches for drug delivery, bioconjugation, wound dressings, and so on and, as will become clear later, many other applications in sensing, bioelectronic devices and other biomedical areas can be addressed with these structures. However, in addition to targeting more applications, it is of paramount importance to understand the behavior and mechanisms of protein assemblies either in solution or at interfaces. Although many systems have been characterized very rigorously, there are many more possible variations of conditions, such as pH, ionic strength, temperature and availability of other protein species, so a large amount of work still lies ahead.

For translating systems as discussed here into "real" applications, important factors include upscaling and ensuring that no immune responses will be initiated by using such biomaterials. As already mentioned, using recombinant systems already ensures better compatibility in combination with fewer batch-to-batch variations. However, not all protein systems allow these approaches, hence a lot of work on effective translation is still required by biotechnologists to prepare truly fully applicable systems.

References

1. G. G. Borisy, J. B. Olmsted and R. A. Klugman, *Proc. Natl. Acad. Sci. U. S. A.*, 1972, **69**, 2890–2894.
2. D. J. Kushner, *Bacteriol. Rev.*, 1969, **33**, 302–345.
3. P. van Rijn and A. Böker, *J. Mater. Chem.*, 2011, **21**, 16735.

4. M. Calamai, F. Chiti and C. M. Dobson, *Biophys. J.*, 2005, **89**, 4201–4210.
5. W. Wohlleben, T. Subkowski, C. Bollschweiler, B. von Vacano, Y. Liu, W. Schrepp and U. Baus, *Eur. Biophys. J.*, 2010, **39**, 457–468.
6. M. Braun, F. Endriss, H. Killmann and V. Braun, *J. Bacteriol.*, 2003, **185**, 5508–5518.
7. Y. Xu, J. Ye, H. Liu, E. Cheng, Y. Yang, W. Wang, M. Zhao, D. Zhou, D. Liu and R. Fang, *Chem. Commun.*, 2008, 49.
8. M.-C. Daniel, I. B. Tsvetkova, Z. T. Quinkert, A. Murali, M. De, V. M. Rotello, C. C. Kao and B. Dragnea, *ACS Nano*, 2010, **4**, 3853–3860.
9. C. B. Chang, C. M. Knobler, W. M. Gelbart and T. G. Mason, *ACS Nano*, 2008, **2**, 281–286.
10. C. T. Carlson, S. S. Jena, M. Flenniken, T.-F. Chou, R. A. Siegel and C. R. Wagner, *J. Am. Chem. Soc.*, 2006, **128**, 7630–7638.
11. T. F. Chou, C. So, B. R. White, J. C. T. Carlson and M. Sarikaya, *ACS Nano*, 2008, **2**, 2519–2525.
12. E. R. Ballister, A. H. Lai, R. N. Zuckermann, Y. Cheng and J. D. Mougous, *Proc. Natl. Acad. Sci. U. S. A.*, 2008, **105**, 3733–3738.
13. F. F. Miranda, K. Iwasaki, S. Akashi, K. Sumitomo, M. Kobayashi, I. Yamashita, J. R. H. Tame and J. G. Heddle, *Small*, 2009, **5**, 2077–2084.
14. T. A. Whitehead, A. L. Meadows and D. S. Clark, *Small*, 2008, **4**, 956–960.
15. Q. H. Q. Liu, I. A. Kriksunov, Z. Wang, R. Graeff and H. C. Lee, *J. Phys. Chem. B*, 2008, **112**, 14682–14686.
16. I. Y. Wong and M. J. Footer, *J. Am. Chem. Soc.*, 2008, **130**, 7908–7915.
17. E. T. O'Brien III, M. R. Falvo, D. Millard, B. Eastwood and R. M. Taylor II, *Proc. Natl. Acad. Sci. U. S. A.*, 2008, **105**, 19438–19443.
18. H. Yan, H. Frielinghaus, A. Nykanen, J. Ruokolainen and A. Saianid, *Soft Matter*, 2008, **4**, 1313–1325.
19. P. J. Knowles, T. W. Oppenheim, A. K. Buell and D. Y. Chirgadze, *Nat. Nanotechnol.*, 2010, **5**, 204–207.
20. X. Lin, J. Xie, G. Niu, F. Zhang, H. Gao, M. Yang, Q. Quan, M. A. Aronova, G. Zhang, S. Lee, R. Leapman and X. Chen, *Nano Lett.*, 2011, **11**, 814–819.
21. P. C. Arosio and P. R. Ingrassia, *Biochim. Biophys. Acta*, 2009, **1790**, 589–599.
22. S. C. Andrews, A. K. Robinson and F. Rodríguez-Quiñones, *FEMS Microbiol. Rev.*, 2003, **27**, 215–237.
23. R. R. Crichton and J.-P. Declercq, *Biochim. Biophys. Acta*, 2010, **1800**, 706–718.
24. M. J. Grossman, S. M. Hinton, V. Minak-Bernero, C. Slaughter and E. I. Stiefel, *Proc. Natl. Acad. Sci. U. S. A.*, 1992, **89**, 2419–2423.
25. P. M. Harrison, P. C. Hempstead, P. J. Artymiukand, and S. C. Andrews, Structure–function relationships in the ferritins, in *Metal Ions in Biological Systems, Vol. 35, Iron Transport and Storage in Microorganisms, Plants, and Animals*, ed. A. Sigel and H. Sigel, Marcel Dekker, New York, 1998, pp. 435–478.
26. M. Kim, Y. Rho, K. S. Jin, B. Ahn, S. Jung, H. Kim and M. Ree, *Biomacromolecules*, 2011, **12**, 1629–1640.

27. J. Swift, C. A. Butts, J. Cheung-Lau, V. Yerubandi and I. J. Dmochowski, *Langmuir*, 2009, **25**(9), 5219–5225.

28. Z. Yang, X. Wang, H. Diao, J. Zhang, H. Li, H. Sun and Z. Guo, *Chem. Commun.*, 2007, 3453–3455.

29. L. Zhang, L. Laug, W. Münchgesang, E. Pippel, U. Gösele, M. Brandsch and M. Knez, *Nano Lett.*, 2010, **10**, 219–223.

30. A. R. Arnold and J. K. Barton, *J. Am. Chem. Soc.*, 2013, **135**, 15726–15729.

31. B. Webb, J. Frame, Z. Zhao, M. L. Lee and G. D. Watt, *Arch. Biochem. Biophys.*, 1994, **309**, 178–183.

32. J. M. Anderson, A. Rodriguez and D. T. Chang, *Semin. Immunol.*, 2008, **20**, 86–100.

33. L. Hall-Stoodley, J. W. Costerton and P. Stoodley, *Nat. Rev. Microbiol.*, 2004, **2**, 95–108.

34. T. W. Gilbert, A. M. Stewart-Akers and S. F. Badylak, *Biomaterials*, 2007, **28**, 147–150.

35. C. H. Lee, A. Singla and Y. Lee, *Int. J. Pharm.*, 2001, **221**, 1–22.

36. P. Szpak, *J. Archaeol. Sci.*, 2011, **38**, 3358–3372.

37. K. Gelse, E. Pöschl and T. Aigner, *Adv. Drug Delivery Rev.*, 2003, **55**, 1531–1546.

38. M. van der Rest and R. Garrone, *FASEB J.*, 1991, **5**, 2814–2823.

39. R. D. Rice, F. S. Ayubi, Z. J. Shaub, D. M. Parker, P. J. Armstrong and J. W. Tsai, *Aesthetic Plast. Surg.*, 2010, **34**, 290–296.

40. The Ventral Hernia Working Group, K. Breuing, C. E. Butler, S. Ferzoco, M. Franz, C. S. Hultman, J. F. Kilbridge, M. Rosen, R. P. Silverman and D. Vargo, *Surgery*, 2010, **148**, 544–558.

41. S. Daghighi, J. Sjollema, H. C. van der Mei, H. J. Busscher and E. T. J. Rochford, *Biomaterials*, 2013, **34**, 8013–8017.

42. P. S. Mykén and O. Bech-Hansen, *J. Thorac. Cardiovasc. Surg.*, 2009, **137**, 76–81.

43. D. Wainwright, *Burns*, 1995, **21**, 243–248.

44. T. W. Gilbert, T. L. Sellaro and S. F. Badylak, *Biomaterials*, 2006, **27**, 3675–3683.

45. H. Schoof, J. Apel, I. Heschel and G. Rau, *J. Biomed. Mater. Res.*, 2001, **58**, 352–357.

46. Y. S. Choi, S. P. Hong, Y. M. Lee, K. W. Song, M. H. Park and Y. S. Nam, *Biomaterials*, 1999, **20**, 409–417.

47. I. P. Kaur, A. Garg, A. K. Singla and D. Aggarwal, *Int. J. Pharm.*, 2004, **269**, 1–14.

48. N. X. Wang and H. A. von Recum, *Macromol. Biosci.*, 2011, **11**, 321–332.

49. W. Liu, K. Merrett, M. Griffith, P. Fagerholm, S. Dravida, B. Heyne, J. C. Scaiano, M. A. Watsky, N. Shinozaki, N. Lagali, R. Munger and F. Li, *Biomaterials*, 2008, **29**, 1147–1158.

50. X. Xu, Q. Gan, R. C. Clough, K. M. Pappu, J. A. Howard, J. A. Baez and K. Wang, *BMC Biotechnol.*, 2011, **11**, 69.

51. S. J. Flint, L. W. Enquist, A. M. Skalka and V. R. Racaniello, *Principles of Virology*, American Society for Microbiology, Washington, DC, 3rd edn, 2009.

52. W. H. Roos, R. Bruinsma and G. J. L. Wuite, *Nat. Phys.*, 2010, **6**, 733–743.
53. A. Klug, *Philos. Trans. R. Soc., B*, 1999, **354**, 531–535.
54. N. F. Steinmetz, M. E. Mertens, R. E. Taurog, J. E. Johnson, U. Commandeur, R. Fischer and M. Manchester, *Nano Lett.*, 2010, **10**, 305–312.
55. T. Li, L. Wu, N. Suthiwangcharoen, M. A. Bruckman, D. Cash, J. S. Hudson, S. Ghoshroy and Q. Wang, *Chem. Commun.*, 2009, 2869–2871.
56. C. M. Soto and B. R. Ratna, *Curr. Opin. Biotechnol.*, 2010, **21**, 426–438.
57. A. de la Escosura, R. J. M. Nolte and J. J. L. M. Cornelissen, *J. Mater. Chem.*, 2009, **19**, 2274.
58. L. Lavelle, M. Gingery, M. Phillips, W. M. Gelbart, C. M. Knobler, R. D. Cadena-Nava, J. R. Vega-Acosta, L. A. Pinedo-Torres and J. Ruiz-Garcia, *J. Phys. Chem. B*, 2009, **113**, 3813–3819.
59. A. C. Bloomer, J. N. Champness, G. Bricogne, R. Staden and A. Klug, *Nature*, 1978, **276**, 362–368.
60. C. Lico, A. Schoubben, S. Baschieri, P. Blasi and L. Santi, *Curr. Med. Chem.*, 2013, **20**, 3471–3487.
61. T. L. Schlick, Z. Ding, E. W. Kovacs and M. B. Francis, *J. Am. Chem. Soc.*, 2005, **127**, 3718–3723.
62. N. F. Steinmetz, K. C. Findlay, T. R. Noel, R. Parker, G. P. Lomonossoff and D. J. Evans, *ChemBioChem*, 2008, **9**, 1662–1670.
63. M. A. Bruckman, G. Kaur, L. A. Lee, F. Xie, J. Sepulveda, R. Breitenkamp, X. Zhang, M. Joralemon, T. P. Russell, T. Emrick and Q. Wang, *ChemBioChem*, 2008, **9**, 519–523.
64. S. Sen Gupta, J. Kuzelka, P. Singh, W. G. Lewis, M. Manchester and M. G. Finn, *Bioconjugate Chem.*, 2005, **16**, 1572–1579.
65. G. Destito, R. Yeh, C. S. Rae, M. G. Finn and M. Manchester, *Chem. Biol.*, 2007, **14**, 1152–1162.
66. M. Manchester and P. Singh, *Adv. Drug Delivery Rev.*, 2006, **58**, 1505–1522.
67. G.-P. Jin, X. Peng, Y.-F. Ding, W.-Q. Liu and J.-M. Ye, *J. Solid State Electrochem.*, 2008, **13**, 967–973.
68. C. D. Chin, V. Linder and S. K. Sia, *Lab Chip*, 2007, **7**, 41–57.
69. L. Wu, J. Zang, L. A. Lee, Z. Niu, G. C. Horvatha, V. Braxtona, A. C. Wibowo, M. A. Bruckman, S. Ghoshroy, H.-C. zur Loye, X. Li and Q. Wang, *J. Mater. Chem.*, 2011, **21**, 8550.
70. D. P. Patterson, P. E. Prevelige and T. Douglas, *ACS Nano*, 2012, **26**, 5000–5009.
71. D. P. Patterson, B. Schwarz, K. El-Boubbou, J. van der Oost, P. E. Prevelige and T. Douglas, *Soft Matter*, 2012, **8**, 10158.
72. M. A. Kostiainen, P. Hiekkataipale, J. Á. de la Torre, R. J. M. Nolte and J. J. L. M. Cornelissen, *J. Mater. Chem.*, 2011, **21**, 2112–2117.

CHAPTER 6

Self-Assembling Properties of Bionanoparticles at Liquid/Liquid and Liquid/Air Interfaces

MARINA J. RICHTER[a], PATRICK VAN RIJN[*b], AND
ALEXANDER BÖKER[*a,c]

[a]DWI - Leibniz-Institut für Interaktive Materialien e.V. und IPC, Lehrstuhl
für Makromolekulare Materialien und Oberflächen, RWTH Aachen University,
Forckenbeckstrasse 50, 52056 Aachen, Germany; [b]W. J. Kolff Institute for
Biomedical Engineering and Materials Science, FB-41, University Medical
Center Groningen, University of Groningen, A. Deusinglaan 1, 9713 AV
Groningen, The Netherlands; [c]Fraunhofer-Institut für Angewandte
Polymerforschung, Lehrstuhl für Polymermaterialien und
Polymertechnologie, Universität Potsdam, Geiselbergstraße 69,
14476 Potsdam-Golm, Germany
*E-mail: alexander.boeker@iap.fraunhofer.de, p.van.rijn@umcg.nl

6.1 Introduction

The use of bionanoparticles at interfaces has a long history, ranging from
beer brewing to the material science of natural biological structures. In beer,
the proteins reside at the gas/liquid interface, which provides a nice foam
that is appreciated by many. Also in the case of whipped cream, milk foams
and whipped egg whites, the airy mixture is stabilized due to the high affinity

RSC Smart Materials No. 16
Bio-Synthetic Hybrid Materials and Bionanoparticles: A Biological Chemical Approach
Towards Material Science
Edited by Alexander Böker and Patrick van Rijn
© The Royal Society of Chemistry 2015
Published by the Royal Society of Chemistry, www.rsc.org

of protein structures at polar/apolar interfaces.[1] In natural systems, proteins are also found at interfaces and often at solid/liquid interfaces, as is seen in the build-up of bone and teeth structures, but proteins also play important roles at lipid bilayer interfaces.[2] Because of these large variations in the function of proteins at interfaces, more emphasis is placed on understanding how proteins behave at interfaces with respect to adsorption kinetics/thermodynamics, structural changes, packing and retention of primary function, *e.g.* catalysis. These understandings provide the foundation for biocompatible materials, which are not only interesting for industry to improve food quality, functionalize textiles[3] and to design micro-reactors, but also serve as a platform for biohybrid materials relevant for (bio-)medical applications, *e.g.* drug delivery and materials that can be used in tissue engineering.[2,4–8]

Although, it is still a challenge to design functional materials that are biocompatible and biodegradable with tailor-made properties at best under mild and environmental conditions, using the bionanoparticle interface offers a tremendous wealth of possibilities. In this chapter, we describe materials that are formed based on the interfacial assembly of proteins (bionanoparticles) at liquid/liquid and liquid/gas interfaces. In order to understand the processes associated with material formation, various techniques are provided along with the models for interpretation that describe the behavior of proteins at interfaces not only with respect to packing and inter-particle arrangement but also the intra-particle behavior, namely unfolding and denaturation. While nature uses naturally occurring proteins, in material science biosynthetic hybrid particles are also used in order to influence the protein stability and the overall adsorption behavior.

6.2   Inspired by Nature: From Mollusk Shells to Tailor-Made Functional Materials

Nature provides a great diversity of functionality, intelligent solutions and building principles that take place under mild conditions. Natural systems and structures have developed and optimized over billions of years.

Biominerals are an excellent example of naturally designed materials that have been further developed over a huge time scale. The underlying process is known as biomineralization. To date more than 70 different biologically built minerals are known. These biominerals are highly hierarchically structured and differ from their counterpart minerals. In contrast to inorganic minerals, biominerals consists of an inorganic and an organic part. The inorganic part in 50% of all biominerals consists of calcium minerals such as calcium carbonate and calcium phosphate, whereas proteins, bacteria and algae build the organic part. The organic part serves as a scaffold for the inorganic material by inducing crystallization.[5,9] These composite materials are well known in the form of corals, sponges, mollusk shells[10] and spines of sea urchins in the maritime milieu and bone, teeth and tendons in the human body. They have a defined morphology, size and orientation, which is not the case for the naturally built minerals that are conglomerations of diverse forms and shapes. Minerals in general differ in shape, crystallinity and isotopic and trace element composition.[11] However, the main difference lies in functionality: biominerals are associated with a special function. They serve, for example, for protection against enemies (shells), as tools (teeth), as gravity-sensing devices (otoliths) or as supporting matrices (skeleton). Both bone and teeth consist mainly of calcium phosphate in the mineral phase of hydroxyapatite, but differ in their organic phase. In bone, the mineralization takes place inside the cavities between fibrils (triple helical structure), which are formed by the structural protein collagen, whereas it is considered that the extracellular matrix protein amelogenin self-assembles into nanosphere structures and directs the mineralization of enamel in tooth development.

The hierarchically structured biominerals built under environmentally friendly conditions with moderate pH values and temperatures offer an excellent example of the design of functional composite materials. Therefore, the idea of mimicking these materials to develop new functional materials is very attractive, since it offers the possibility of controlling the properties depending on the type of conditions, *e.g.* type of protein, available mineral phase, pH, temperature. In addition to controlling the properties with respect to the biomimetic material, defined size, shape and morphology are also of great importance and hence new preparative methods and new approaches are being investigated in order to broaden the scope of possible materials.

In the last 30 years, there has been huge progress in biomimetic synthesized materials. The research studies of the Mann, Bäuerlein, Epple, Kniep, Taubert, Culver and Cölfen groups[11,12] have had an immense impact on the exploration of biomineralization.

In an example in which biomineralization mimicking nature was used but employing liquid/liquid instead of solid/liquid interfaces, Böker and co-workers developed a biomimetic synthesis route for hydroxyapatite hollow capsules (Figure 6.1).[2,13-17] The basis of this synthesis route is an oil/water (o/w) emulsion. The oil droplets serve as templates for the mineral capsules and are stabilized by a small globular protein, hydrophobin. Hydrophobins can be used for the stabilization of hydrophobic/hydrophilic interfaces because of their self-assembly properties and amphiphilicity resulting from hydrophilic and hydrophobic patches in the protein structure. In nature, hydrophobins are secreted by filamentous fungi and are involved in fungi growth and

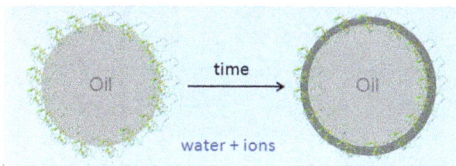

Figure 6.1 Scheme of capsule growth in template-directed synthesis *via* protein-stabilized o/w emulsions. Protein molecules self-assemble at the o/w interface, leading to stabilization of the oil droplets in water. The protein also serves as a scaffold for crystallization: a mineral shell grows at the o/w interface over time.

development. Here, they are integrated in interfacial interactions where they self-assemble and build an amphiphatic membrane. For instance, the surface of fungi becomes hydrophobic because of a hydrophobin layer at the air/fungus interface; moreover, hydrophobins ensure that fungi can escape from their initial aqueous environment and play an important role in the initial steps of fungal pathogenesis. In the synthetic approach of Böker *et al.*, hydrophobin acts in the same way as the protein does in nature: hydrophobin diffuses to the interface, which is in this case o/w, where it self-assembles and lowers the interfacial tension (IFT) stabilizing the oil droplets in water. Self-assembled at the hydrophobic/hydrophilic interface, hydrophobin builds an amphiphatic protein layer because of its hydrophobic and hydrophilic patches, which are exposed to their counterparts: hydrophobic residues are exposed to the oil phase and hydrophilic residues to the water phase. This amphiphatic protein layer, or rather membrane, fulfils two functions at once: on the one hand it stabilizes the o/w emulsion and on the other it serves as a scaffold for subsequent mineralization. In the mineralization step, Ca^{2+} and HPO_4^{2-} can crystallize to calcium phosphate in the form of hydroxyapatite because of the protein matrix covering the oil droplet. After a certain mineralization time, a mineral shell is built at the o/w interface. Finally, mineral capsules can be extracted after evaporation of the inner oil phase (Figure 6.2). This biomimetic synthesis is a representative example of the development of simple syntheses for composite materials in which processes from nature are transferred to the laboratory and integrated in synthesis routes. Furthermore, there are two natural objects combined in one synthesis, although they do not interact in nature in the same way as they do in this type of synthesis: hydrophobin is normally not integrated in biomineralization processes in nature. Further details can be found elsewhere.[2,16,17]

There is a demand for biomimetic synthetic composite materials owing to the mild synthetic conditions and the specific material properties and thus not being limited to one special field of applications.[18] These materials can serve in many industrial sectors as separation membranes for polymers, colloids or cells. Furthermore, in medical and pharmaceutical approaches,

Figure 6.2   Mineral hollow capsule synthesized *via* the biomimetic synthesis route developed by Böker *et al.*

biominerals are used for macroporous medical implants and drug delivery and release systems for viruses, DNA or active substances. Moreover, the interest in the catalytic chemistry is also enhanced in terms of applications in catalyst supports.[4–6,8,19]

6.3 Protein Adsorption – Understanding the Fundamentals

In order to understand the conditions and circumstances for synthesizing composite materials, we have to take a step backwards. It is of fundamental importance to analyze and understand the underlying processes first, before it is possible to transfer these natural systems into the laboratory. A number of groups have been investigating the organic or inorganic part itself, but also their interactions. Here, we focus on the organic part and consider the behavior of bionanoparticles, especially proteins at liquid/liquid and liquid/air interfaces, in more detail.

As we have already learned, proteins have a huge impact on many systems in nature. They usually can be connected with interfacial interactions; for example, they show a high adsorption affinity towards solid/liquid, liquid/liquid and air/liquid interfaces. The food industry has long since identified this high potential and has used proteins as stabilization agents for emulsion or foams to improve food quality. Furthermore, proteins meet the biocompatibility and biodegradability requirements that are essential in this industry sector. For example, β-casein has a high impact on wine, beer and milk processing. Milk is the most common example, where β-casein serves as an emulsifier among whey proteins such as bovine serum albumin (BSA), α-lactalbumin and β-lactoglobulin. In contrast to these whey proteins, β-casein builds micelles forming a colloidal particle, and is thus the dominant species at the o/w interface.[20] Investigations on the function of β-casein in emulsion stabilization lead to optimal control of the properties of milk homogenization and cheese.

Moreover, the field of protein applications is not limited to food; proteins also find uses in medicine, cosmetics, healthcare and the photographic industry. Here, they are involved in chromatography, coating processes, biosensors and drug delivery approaches.[2,4,6–8] Hydrophobins with their high potential owing to their excellent properties and their recombinant production with comparably large amounts, are a promising candidate for numerous applications:

- biomaterials and medical applications
- emulsions
- personal care
- pharmaceutical applications
- food industry
- biosensors and electrodes

- separation methods
- antifouling applications
- gushing factor detection.

Hydrophobin self-assembles at hydrophobic/hydrophilic interfaces and is able to reverse the surface polarity. If hydrophobin is adsorbed on glass, the surface polarity of the glass changes from hydrophilic to hydrophobic. Teflon substrates are another example where the surface polarity changes from hydrophobic to hydrophilic. Hence, there is great potential in antifouling applications, preventing further protein adhesion or growth of bacteria on surfaces. Furthermore, the cosmetic industry uses the adsorption properties to stabilize emulsions in creams or ointments. A common problem with hair care products is their poor adhesion on hair, so that they are easily removed during shampooing. Hydrophobin seems to be a promising additive to prolong the residence time, so that hair care products can survive several washing steps.[21]

In the last 30 years, a number of groups have been working on protein adsorption dynamics to gain a deeper understanding of the processes involved in protein behavior at interfaces. Some of the most frequently studied proteins are human serum albumin (HSA), bovine serum albumin (BSA), β-casein, β-lactoglobulin and lysozyme. Miller is one of the leading researchers in this field, working on dynamic interfacial layers, focusing on the thermodynamics, kinetics and rheology of interfacial layers such as thin liquid layers, foams and emulsions that are stabilized by proteins or surfactants. The investigation of interfacial layers is not limited to the adsorption of single components; moreover, the investigations also involve the adsorption and interaction of two components, proteins and surfactants, both adsorbed at one interface. Dickinson and co-workers have also investigated mixtures of proteins and surfactants or single components at liquid/liquid interfaces. In contrast to Miller, Dickinson focused on interfaces related to food applications and food colloids. For example, there have been numerous investigations of milk proteins building interfacial layers.

Section 6.3.1 presents a short overview of common experimental techniques for protein adsorption analysis, and Section 6.3.2 presents an overview of protein adsorption kinetics.

6.3.1 Experimental Techniques – Analyzing Protein Adsorption

To make use of proteins for new applications, it is fundamental to investigate protein behavior such as adsorption processes or denaturation and reorientation at interfaces. There are several experimental techniques for analyzing proteins in solid/liquid, liquid/liquid and air/liquid systems.

The protein structure can be analyzed by Fourier transform infrared (FTIR) spectroscopy and circular dichroism (CD) spectroscopy in the case of secondary structure investigations and X-ray reflectivity (XRR) in case of tertiary structure investigations. Non-linear optical techniques such as second harmonic generation (SHG) and sum frequency generation (SFG) provide information about the surface-molecular orientation. Using rapid mixing methods or overflowing cylinder techniques, it is possible to determine rate constants. Furthermore, there are several methods for determining the amount of protein adsorbed at interfaces: surface tension measurements (tensiometry, Wilhelmy plate), surface radioactivity measurements, ellipsometry, neutron reflectivity (NR), optical waveguide light-mode spectroscopy (OWLS), surface plasmon resonance (SPR), total internal reflection fluorescence (TIRF), quartz crystal microbalance (QCM) and quartz crystal microbalance with dissipation (QCM-D).[22] Moreover, atomic force microscopy (AFM) offers a useful tool for imaging surface arrangements and provides structural information on the assembled structure.

Surface tension measurement by pendant drop tensiometry is a frequently used technique in applications dealing with emulsions, foams or similar liquid/liquid or liquid/air systems. In principle, a pendant drop at the tip of a capillary is immersed in a cuvette filled with another liquid or air. An integrated video camera captures video images of the drop shape at stated times or over a certain time scale. The drop shape is then fitted to the Young–Laplace equation [eqn (6.1)] and the surface tension of the system can be determined. The Young–Laplace equation, which describes the increase in pressure in a liquid drop resulting from the surface curvature due to the surface pressure, relates the interfacial tension to the drop shape:

$$\Delta p = \gamma \left(\frac{1}{r_1} + \frac{1}{r_2} \right) \tag{6.1}$$

where Δp is the pressure difference across the fluid interface, γ is the surface tension and r_1 and r_2 are the principle radii of curvature.

To gain a deeper understanding of protein denaturation and crystallization, some groups have integrated the obtained analytical results into molecular dynamic (MD) simulations. In the case of the mineral capsule example in Section 6.2, Schulz *et al.*[23] performed additional MD simulations, involving the adsorption of class I and II hydrophobins at hexane/water interfaces. They found that there are differences with respect to behavior and effect on crystallization at liquid/liquid interfaces between two proteins belonging to the same family, but differing in their protein class. They used two hydrophobins: class I hydrophobin EAS and class II hydrophobin HFBII, both of which have one mutual structural unit: they contain eight cysteine residues forming four disulfide bridges. Otherwise, HFBII contains β-sheets and one α-helix, whereas EAS does not contain any α-helix. To expand the difference between EAS and HFBII, it should be mentioned that class I hydrophobins such as EAS tend to build highly insoluble aggregates in the form of rodlets in solution, whereas class II hydrophobins such as HFBII build readily soluble aggregates (no rodlet formation). Schulz *et al.* showed that EAS adsorbed at the interface tends to lose its secondary structure. In contrast, HFBII changes its secondary structure only slightly. If ions come into play, the difference between the structural behaviors of the two proteins is even more pronounced. Ca^{2+} and HPO_4^{2-} ions interact with the protein surface of EAS, whereas they do not interact significantly in the presence of HFBII and remain mostly free in solution. The MD simulations were also run under mineralization conditions and were in good agreement with the experimental result of biomineralization, which is that only EAS is prone to induce mineralization at a liquid/liquid interfaces. It is suspected that EAS fulfils two functions: on the one hand the charged residues on the protein surface increase local concentration and on the other hand EAS pre-organizes ions for the crystallization. Figure 6.3 gives an overview of the changes in the secondary structure of both hydrophobins at the hexane/water interface at different simulation times.[23]

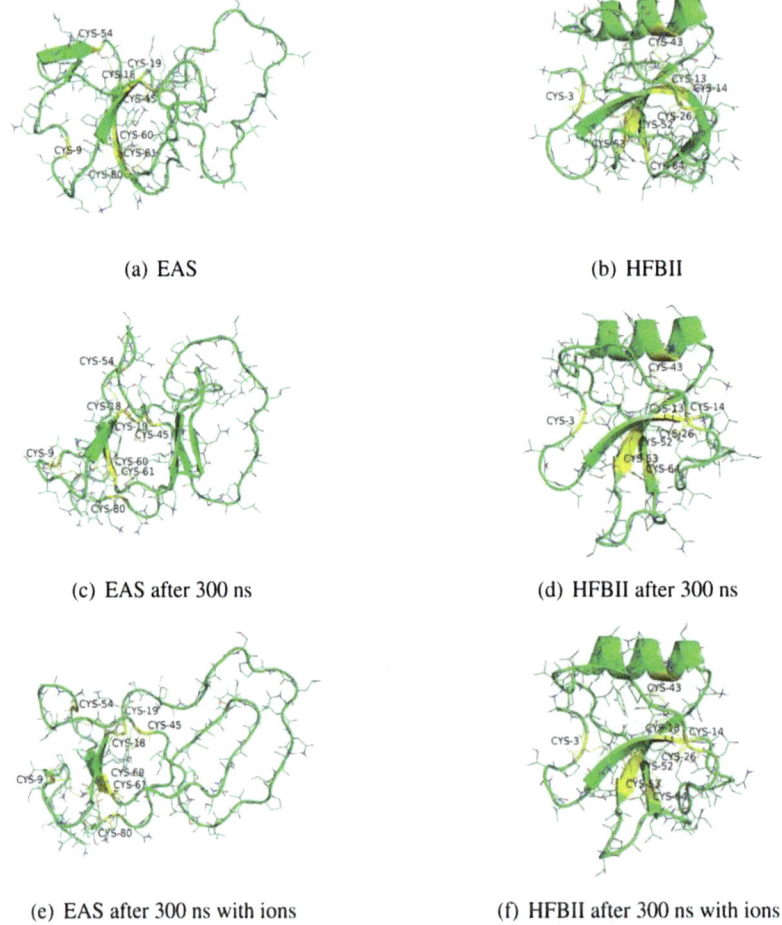

(a) EAS

(b) HFBII

(c) EAS after 300 ns

(d) HFBII after 300 ns

(e) EAS after 300 ns with ions

(f) HFBII after 300 ns with ions

Figure 6.3   Snapshot of protein structure of proteins EAS and HFBII at the beginning of the simulations, after 300 ns without ions and with ions. Reproduced from ref. 23 with permission from The Royal Society of Chemistry.

Schulz *et al.* focused mainly on the protein adsorption and less on the mineralization process. In contrast, Zahn and co-workers developed a specialized simulation method for investigating nanocrystal aggregation from solution.[24] They took a close look at the nucleation mechanism of fluorapatite–collagen composites, in particular aggregate formation and growth control by collagen fibers. MD simulations will not be discussed in further detail; however, it is important to mention that simulations provide important insights into possible mechanisms important for the construction of bio-hybrid materials as discussed here. For more details on MD simulations, the reader is referred to the literature.[25]

6.3.2 Dynamics of Protein Adsorption – Dealing with Kinetics

Nature has taught us that proteins tend to self-assemble at hydrophobic/hydrophilic interfaces; the question is, which sort of processes lie at the foundation of these self-assembly properties? Globular proteins (hydrophobin, lysozyme, BSA) have mainly an apolar protein core, containing most of the apolar amino acid residues, and keep the majority of their polar amino acid residues at the surface. This amphiphilic property is essential for adsorption at hydrophobic/hydrophilic interfaces. During the adsorption at hydrophobic/hydrophilic interfaces, proteins interact with their surroundings. At an o/w interface, for example, proteins have to expose their apolar residues, being predominantly in the protein interior, to interact with the hydrophobic interface. This interaction of the protein's hydrophobic patch requires the protein to unfold or rather denature. During protein unfolding, proteins with an α-helix or random structure in their native conformation tend to unfold at hydrophobic interfaces to β-structures. In comparison, proteins with β-structure contents in the native state tend to lose this structural motif. Furthermore, van der Waals attraction, electrostatic interaction and hydrogen bonding are also involved in the protein adsorption.[26]

In general, several steps are involved in protein adsorption at fluid interfaces:

- bulk diffusion
- barrier overcoming
- adsorption at the interface
- protein unfolding
- diffusion of additional proteins
- protein–protein rearrangements of the protein adsorption layer
- phase transition
- aging of the protein adsorption layer.

Ultimately, all of these individual steps result in a stabilized interface with the morphology and the extent of protein denaturation depending on the protein structure and stability.[26]

6.3.2.1   Time-Dependent Interfacial Tension

The adsorption of proteins at interfaces, either liquid/liquid or liquid/gas, can be monitored by measuring the interfacial tension using pendant-drop surface tension measurements.[27] Beverung *et al.* divided the adsorption of proteins, or rather the dynamic interfacial tension, into three regimes: (I) induction period, (II) monolayer saturation and (III) interfacial gelation.[28] Figure 6.4 illustrates the time-dependent interfacial tension divided into these three regimes with corresponding drawings of the protein assembly at the o/w interface. During the induction period, protein molecules diffuse to the interface and the first conformational changes of adsorbed proteins are initiated to expose apolar residues of the protein's interior. The amount of adsorbed protein is still relatively small, having only a minimal influence on the interfacial tension; consequently, the interfacial tension remains that of the pure solvent. This first regime only occurs with low-concentration protein solutions. This induction time t_{ind} is necessary to achieve minimal coverage of the liquid/liquid or liquid/gas interface. After the induction time, additional protein molecules diffuse to the interface, where they self-assemble and relax from their rigid conformation to less compact structures, increasing the number of contacts between apolar amino acid residues and the hydrophobic surface. Irreversible adsorptions are initiated. The interfacial protein concentration increases over time and the protein monolayer at the o/w interface saturates. The time-dependent interfacial tension shows a sharp decline in this regime. Finally, conformational changes of the initially adsorbed proteins continue, leading to aggregation and branches, resulting in an amorphous gel-like network structure at the

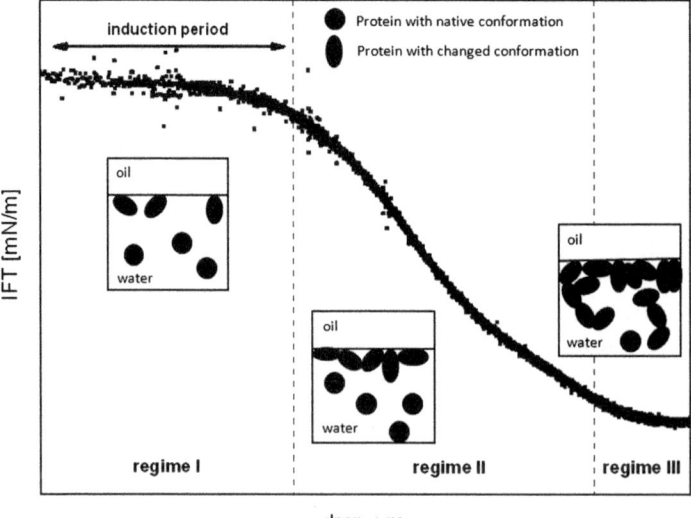

Figure 6.4   Time-dependent IFT divided into three regimes with corresponding drawings of the protein assembly present at the o/w interface.

interface. The interfacial tension levels off to a steady minimum indicating the equilibrium state. It should be stressed that these three regimes only occur for highly diluted protein solutions.

However, protein adsorption is not independent of external conditions; temperature, pH, ionic strength, protein concentration and additives such as salts or organic compounds affect protein adsorption. Temperature influences significantly the formation of protein–substrate bonds. Thermo-dynamically unstable proteins adsorb more strongly at an interface than ther-modynamically stable proteins. The unfolding temperature decreases during protein adsorption, leading to the assumption of reduced thermal stability of the protein. High temperatures, however, increase the interfacial protein concentration. Suitable choice of the pH of the system is essential because the protein charge depends strongly on pH. Proteins are neutral if the pH and the isoelectric point of the protein are equal, whereas the protein is positively charged at pH lower than the isoelectric point and negatively charged at pH above the isoelectric point. The protein charge becomes important when the protein interacts with a charged interface. If the substrate and the protein are charged oppositely, the diffusion to the interface and also the adsorp-tion are much faster than for equally charged protein–substrate systems, provided that electrostatic attraction favors diffusion to the interface. The total amount of protein adsorbed at the interface occurs at the isoelectric point. The concentration of dissolved ions (ionic strength) also affects pro-tein adsorption. Low ionic strength increases the protein–substrate electro-static interaction in the case of oppositely charged substrates. The larger the amount of protein, the higher the protein adsorption rate. Additives affect protein adsorption depending on the protein–additive interaction and the interfacial hydrophobicity/hydrophilicity. Salts have a huge impact on the protein activity. There are two types of salt effects: non-specific effects, result-ing from their ionic properties, which are independent of the salt type, and specific effects, occurring at higher salt concentration, which depend on the salt type. Generally, salts form an electric double layer around the charged protein and thus around the interface. They favor the interaction of proteins with hydrophobic interfaces, whereas protein–hydrophilic interfacial inter-actions (electrostatic interactions) are reduced. In contrast, organic solvents affect the protein stability and have an opposite influence on protein adsorp-tion: protein–hydrophobic interface interactions are reduced, whereas pro-tein–hydrophilic interface interactions are enhanced.[22]

The time-dependent interfacial tension provides information on protein unfolding kinetics and can be determined by, for example, pendant drop measurements. Figure 6.5 shows a characteristically time-dependent inter-facial tension plot. In the literature there are many approaches to theoret-ical models offering a mathematical description for protein adsorption at hydrophobic/hydrophilic interfaces. These approaches differ in the extent to which protein unfolding affects the interfacial properties, which is not yet well understood. On the one hand, it is suggested that the increase in surface pressure is due to protein unfolding at the interface; on the other hand, it is

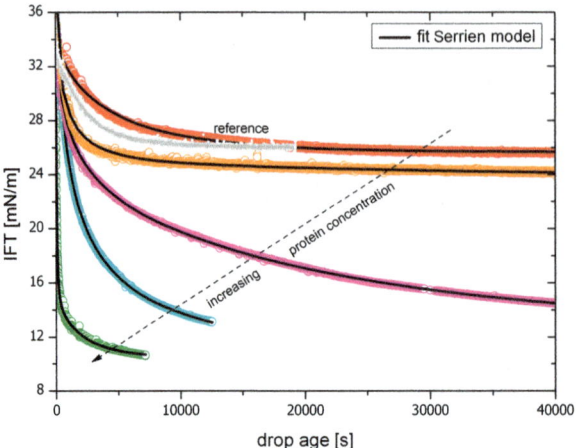

Figure 6.5 Interfacial tension (IFT) *versus* drop age of a silicone oil/water interface stabilized with protein using different protein concentrations. Data points represent the IFT at drop age; black lines represent the fit by the Serrien model.

suggested that only unfolded proteins adsorb at the interface and that the initial adsorption leads to structural stability.[29]

The simple isotherm models as laid down by Langmuir and Henry do not work for protein adsorption kinetics because the conformation of adsorbed molecules at the interface has not been taken into consideration. Singer's equation for surface layers was the first calculation of the total number of various conformations for a flexible-chain polymer which was located within the surface [eqn (6.2)].[30] However, this two-dimensional quasi-crystal model takes only the contribution of configurational entropy to the free energy into consideration, and does not include any intermolecular interactions such as polymer–surface or polymer–solvent interactions.[31]

$$\Pi = \frac{RT}{\omega_0}\left[-\ln\left(1-\theta_1\right) + \frac{z}{2}\left(1-\frac{\omega_0}{\omega_1}\right)\ln\left(1-\frac{2}{z}\theta_1\right)\right] \quad (6.2)$$

where Π is the surface pressure, R is the gas constant, T is the absolute temperature, ω_0 is the molar area of a quasi-crystal cell (water molecule or macromolecule segment), ω_1 is the molar area of the macromolecule unfolded at the surface, $\theta_1 = \omega_1\Gamma_1$ is the surface coverage, Γ_1 is the adsorption and z is the coordination number of the lattice.

In addition to the theories presented in the previous paragraph mainly concerning polymers, other scaling theories or thermodynamic models specially for proteins have been developed. The theories of hard-core particles, which do not interact, changed into attempts suggesting "soft" macromolecules, which are compressible into smaller molecular areas with increasing adsorption. Douillard and co-workers presented some more complex scaling theories, although they do not consider the presence of solvent molecules at the surface.[31–33] In contrast, solvent molecules are taken into consideration in thermodynamic models. These two-dimensional solution treatments connect the surface composition and the surface tension with the thermodynamic potential at the surface, μ_i^s, of both solvent and solute(s) in one mathematical description. Butler developed a fundamental equation on which many following thermodynamic models are based.[34] The chemical potential μ_i for solvent molecules ($i = 0$), surfactant monomers ($i = 1$) and n-mers ($i = n$) in a surface layer is defined as follows:

$$\mu_i = \mu_i^0 + RT\ln(f_i x_i^s) - \gamma\omega_i \qquad (6.3)$$

where μ_i^0 is the standard chemical potential, R is the gas constant, T is the absolute temperature, f_i is the activity coefficient, x_i^s is the molar fraction at the surface, γ is the surface tension and ω_i is the partial molar area. For instance, Joos,[35] Ter-Minassian-Saraga,[36] Lucassen-Reynders[37] and Fainerman[38] used the Butler equation as a starting point for the development of improved, more suitable models. Joos suggested that the degree of surface denaturation decreases with increasing surface pressure and that the adsorption layer thickness increases with increasing protein concentration.[31,35,39] Ter-Minassian-Saraga developed a two-dimensional solution model considering the inter-relation between protein denaturation processes within the surface and the activity of the solvent (water).[36] Lucassen-Reynders[37] studied the effect of the sizes of mixture molecules on the protein surface layer entropy to the first order.

Ward and Tordai derived a diffusion-controlled adsorption model for nonionic surfactants [eqn (6.4)].[40] In general, the adsorption of a surfactant at an interface is divided into two processes: the surfactant diffuses from the bulk solution to the sub-surface before it adsorbs from the sub-surface onto the interface. This equation can also be applied to proteins.

$$\Gamma(t) = 2\sqrt{\frac{D}{\pi}}\left[c_0\sqrt{t} - \int_0^{\sqrt{t}} c\left(0, t - \tau\right)\mathrm{d}\sqrt{t}\right] \qquad (6.4)$$

where $\Gamma(t)$ is the interfacial concentration, D is molecular diffusion coefficient, c_0 is the surfactant concentration in bulk solution, $c(0, t - \tau)$ is the surfactant concentration in the sub-surface, t is time and τ is a variable with units of time.

Although Graham and Philipps[41] carried out the first systematic investigations, the relations between adsorption of proteins that can adsorb with two different modifications with different partial molar areas were not discussed.[37] This aspect was first described by Joos[35] and Serrien[42] and then developed by Fainerman, Lucassen-Reynders and Miller[43].[27] They applied the existing theories considering only two different protein modifications at the surface to a model taking a number of conformations of protein molecules at the surface into account. Furthermore, they transferred known models to the adsorption of surfactants and proteins at interfaces.[26,27,43]

$$\gamma(t) = \gamma_{eq} + \left[\alpha \exp\left(-\frac{\sqrt{4t}}{\pi t} \right) + \beta \right] \exp(-kt) \tag{6.5}$$

where $\gamma(t)$ is the interfacial tension over time, γ_{eq} is the interfacial tension in equilibrium, t is time, $\tau = \Gamma_{eq}^2/4DC^2$ is the relaxation time of the protein with protein concentration C, α describes the importance of the diffusion to the reorientation at the interface, β is the basic velocity independent of the diffusion and k is the rate constant of the reorientation.

The following example demonstrates the potential of interfacial studies and should also give an idea of the practical use of these theories. In the mineral capsule example from Section 6.2, Schulz *et al.* carried out additional tensiometry measurements equivalent to the protein-stabilized o/w emulsion of the biomimetic synthesis route.[16] They investigated the impact of different oils and protein concentrations on the interfacial protein adsorption and the interfacial tension. In the analysis part, they fitted the IFT data by the mathematical expressions of Beverung and Serrien [eqn (6.5)], resulting in following assumptions. They concluded that a higher protein concentration leads to a sharper decline of the IFT over time and that the IFT in equilibrium is lower than for lower protein concentrations. This is a logical observation: the more protein is in solution, the more can be adsorbed at the interface, lowering the IFT. The protein diffusion also seems to be faster for higher protein concentrations. Moreover, they supposed that there is greater aggregate formation at higher protein concentrations, because the importance of diffusion over reorientation increases. The nearly constant rate constant of reorientation values of different protein concentrations indicates that the protein concentration is independent of protein denaturation. Figure 6.5 shows the change in IFT during drop aging for the investigated protein-stabilized o/w system.

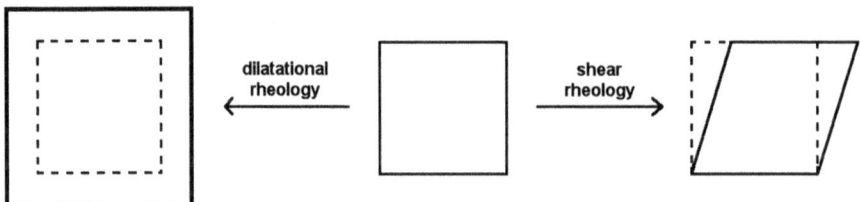

Figure 6.6   Schematic of the differences in shape and area of the sample between dilatational and shear rheology experiments.

6.3.2.2 Interfacial Rheology

Interfacial rheology measurements provide a deeper understanding of the processes at the interface and the final protein layer by studying the flow behavior of different substances or systems. This technique provides information about surfactant or protein concentration, interactions between molecules and structures at the interface. Hence, interfacial rheology measurements are important in emulsion and foam investigations. The interfacial characteristics of stabilized emulsions or foams depend not only on the two phases, but also on the stabilizing agent. For example, in protein-stabilized emulsions, proteins form an immobile viscoelastic interfacial protein layer. The adsorbed protein exposes its hydrophobic parts towards the hydrophobic phase and its hydrophilic parts towards the hydrophilic phase. The structure of adsorbed proteins and the strength of interactions between each protein influences the protein layer characteristics. Additionally, regarding the isoelectric point, the pH also influences the properties of the protein layer.[44] If low molecular weight surfactants (LMWSs) are used as a stabilizing agent instead of proteins, the interfacial layer is mobile with low viscoelastic properties. The surfactant solubility controls the film characteristics; therefore, it is essential to know if the surfactant is more water or oil soluble. The alkyl chain length is another factor that controls the characteristics of the interfacial layer: the longer the surfactant alkyl chain, the more rigid the interfacial layer and the higher the dilatational elasticity.

The principle of interfacial rheology measurements is the deformation of an interface. A special force deforms the interfacial layer built by proteins or surfactants and the response of this interfacial layer is measured. Generally, there are two types of interfacial rheology, depending on the sort of deformation: shear and dilatational rheology (Figure 6.6). The main difference between the two types is the approach if there are changes in shape or area. During dilatational rheology measurements, the shape of the interface is kept constant, while the surface area changes. For shear rheology measurements, it behaves oppositely: the shape changes at constant area. Furthermore, whereas shear rheology represents long-term stability, dilatational rheology is linked to short-term stability. Moreover, the two methods lead to different information about the interfacial layer. Dilatational rheology measurements not only investigate the dynamic properties and the magnitude of interactions at the interface, but also layer composition, competitive adsorption between different adsorbed molecules such as proteins or LMWSs and surfactant or rather protein concentration. In contrast, shear rheology investigates the structures formed at the interface. However, the results of both shear and dilatational rheology measurements have to be combined and analyzed as a whole to obtain evidence about interfacial mechanical properties and the structure of interfacial layers.

According to the previous paragraph, dilatational rheology measurements provide much more information about the interface than shear rheology measurements. In principle, the drop shape or capillary pressure inside a droplet or a bubble at constant shape and changing surface area is measured. A drop/bubble profile tensiometer (DPT) determines the rheology

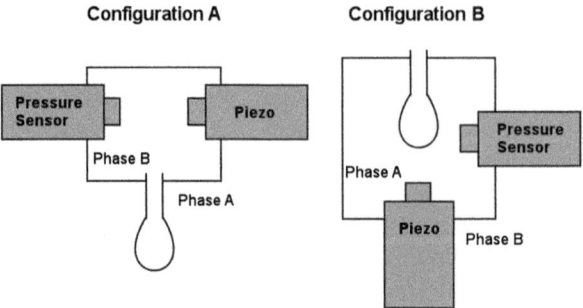

Figure 6.7 Schematic of capillary pressure tensiometry (CPT) with two different configurations, A and B. Reproduced from ref. 45 with permission from Elsevier.

parameters for gas–liquid and liquid–liquid systems and provides information on mechanical properties. A pendant drop or bubble—depending on the system—is inflated at the tip of a capillary and immersed in a liquid or surrounded by air. The drop/bubble oscillates usually with a sinusoidal oscillation caused by a piezo pump. A video camera detects the changing drop/bubble shape and the interfacial tension is calculated from the shape geometry, density difference of the two phases of the system and the Young–Laplace equation [eqn (6.1)]. The capillary pressure tensiometer (CPT) (Figure 6.7) is applicable to systems containing two liquids with small density differences. The instrument comprises two chambers that are connected by a capillary tube. A pressure sensor and a piezoelectric element are both included in one closed chamber, while another chamber is open to atmospheric pressure. As shown in Figure 6.7, there are two possibilities, configuration A or B, for placing the liquids in the chambers: the closed chamber can be filled with either the liquid forming the drop (phase 2, configuration A) or with the liquid surrounding the drop (phase 1, configuration B). The piezoelectric rod controls the drop volume and the pressure sensor detects the present pressure at every time step. A video camera captures the drop/bubble profile and can thus determine the drop/bubble diameter. On the other hand, the diameter can be calculated from the volume of liquid used.[45]

There are two methods for the determination of shear rheology: a *direct* and an *indirect method*. In the *direct method*, commonly, a torsion pendulum surface viscometer, which detects the damping of the torsion pendulum because of the viscous drag of the interface, is used. Those viscometers are available for gas/liquid and liquid/liquid interfaces depending on the geometry of the plate. A sharp-edge plate geometry is suitable for gas/liquid interfaces, whereas a biconal disk plate geometry can also be used for liquid/liquid systems. Figure 6.8 shows three different touching bodies. The touching body is placed at the interface where it rotates or oscillates, while the sample is placed in stationary dish.[46]

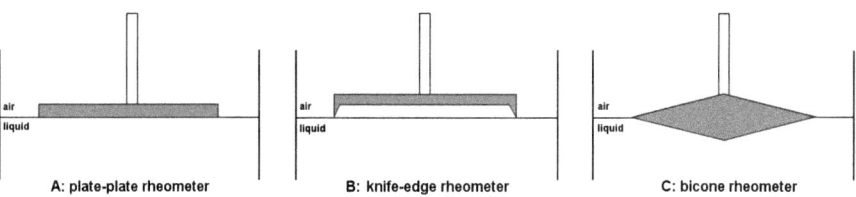

Figure 6.8 Schematic of a plate-plate rheometer (A), a knife-edge rheometer (B) and a bicone rheometer (C). Reproduced from ref. 46 with kind permission from Springer Science and Business Media and from ref. 48 with permission from De Gruyter.

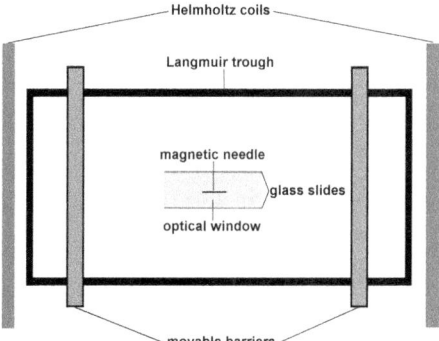

Figure 6.9 Schematic of a magnetic rod viscometer. Reproduced from ref. 46 with kind permission from Springer Science and Business Media and from ref. 48 with permission from De Gruyter.

The magnetic rod viscometer shown in Figure 6.9 is another, more sensitive, interfacial shear rheometer. In contrast to the previously discussed rheometers, the magnetic rod viscometer can be used in combination with a Langmuir trough, providing additional values of the thermodynamic state of the interfacial layer. An external magnetic field applies a force at a magnetic needle to shear the interfacial layer. The needle position is detected by a microscope or a photodiode array, defining the strain rate.[47,48]

In the *indirect method*, inert visible particles are used to measure surface velocity profiles. The deep channel surface viscometer and the channel surface viscometer are two examples of available instruments (Figure 6.10). The deep channel surface viscometer has the advantage that measurements are not limited to gas/liquid interfaces as is the case for the channel surface viscometer; measurements can also be carried out for liquid/liquid systems. More detailed descriptions of the presented measurement techniques can be found elsewhere.[44,46–50]

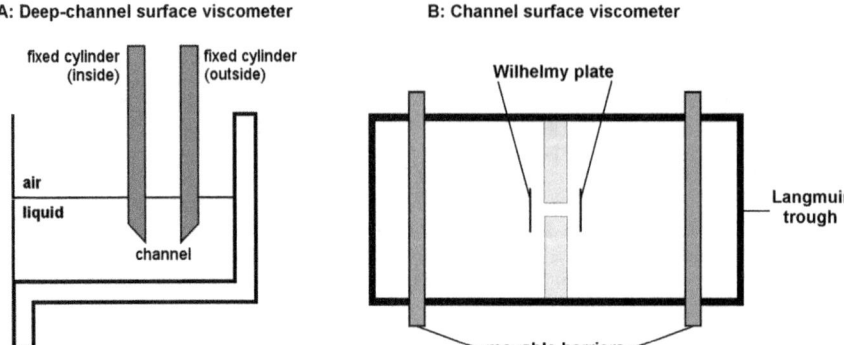

A: Deep-channel surface viscometer   B: Channel surface viscometer

Figure 6.10   Schematics of a deep-channel surface viscometer (A) and a channel surface viscometer (B). Reproduced from ref. 48 with permission from De Gruyter.

6.4 Functional Systems Using Interfacial Self-Assembly

As shown above, bionanoparticles are able to stabilize polar/apolar interfaces and these properties have been used in foods and cosmetic formulations for many years.[51,52] The approach of confining proteins at interfaces is still an active area of research, since it allows for a directed assembly approach and various proteins can be used with a large variety of properties and different sizes.[53,54] Protein properties such as enzymatic activity can be used to induce catalysis at the interface; this will only be possible, of course, when the protein's structural features relevant for catalysis are preserved. In this section, a few recently reported approaches are discussed that combine the directed interfacial assembly together with other aspects such as catalysis[55] and induced denaturation[56–58] to develop new materials and systems.

One approach which has been known for a long time but is still developing is the use of lysozyme to stabilize water/oil interfaces in combination with the catalytic activity of lysozyme to convert the oil phase into an inorganic material.[59] Hydrolases such as lysozyme, silicatein and trypsin are able to convert organosilanes such as tetraethyl orthosilicate (TEOS) into silicon dioxide.[60] Since the conversion occurs at the interface *via* a polycondensation reaction, an inorganic silicon dioxide shell is formed around the stabilized oil droplet and hence capsules emerge. The approach taken here is easy and relatively cheap since lysozyme can very easily be isolated from egg whites and is commercially available on a large scale. Although the approach is easy, over the years this synthetic route towards biomineralized capsules has undergone continuous developments.[59] As mentioned before, ions in the system, pH, temperature, buffer and organic phases present in the system affect the assembly at the interface. This in turn is also shown in the capsule structures as a result of these variations. Conventionally, in the early lysozyme–TEOS systems, glycine buffer was used for the synthesis. However, on changing the buffer to phosphate-buffered saline (PBS) the shell morphology changed from fairly smooth to rough and the size and size distribution changed. These changes are due to the change in ionic strength, which was directly seen from analysis of the chemical composition of the inorganic silicon dioxide–lysozyme shell. Lysozyme is incorporated into the shell and upon analysis with energy-dispersive X-ray spectroscopy the ions present in PBS were also found in the shell of the capsule. This discovery led to the notion that ions can be incorporated easily *via* addition of different salts to the aqueous phase, thereby changing not only the chemical composition but also the overall morphology. This was expanded further by adding charged surfactants, which resulted in highly folded surface structures offering a much higher surface area on the outside of the capsule. In addition to introducing various components to the aqueous phase, it is also possible to add other silicon dioxide precursors to the oil phase. It was found that on adding up to 8% of (3-aminopropyl)trimethoxysilane (APTMS) to the TEOS phase, the surface of the capsule becomes reactive towards chemical modifications owing to the presence of primary amine groups on the surface.[61] The amino groups can be functionalized *via* simple peptide coupling, as also mentioned in Chapter 1.

As mentioned previously, proteins are very efficient in foams; especially when heated or mechanically disturbed, denaturation of the protein occurs, forming gel-like layers.[62] This can also be done in combination with lysozyme, which is probably the most commonly used since it is very abundant in egg whites. While in protein foams, initially the protein stabilizes the interface, and after denaturation they tend to stick together and provide mechanical strength.[1] This approach was also used to form biodegradable protein capsules.[63] Again, an emulsion template system was used where lysozyme in PBS was combined with a small amount of oil. Emulsification was performed *via* ultrasound treatment, which breaks up the two phases to produce small oil droplets which are then stabilized by the lysozyme, but additionally the lysozyme is extensively denatured by the ultrasound, forming a gel-like layer around the oil droplet, creating a capsule (Figure 6.11). The capsule is

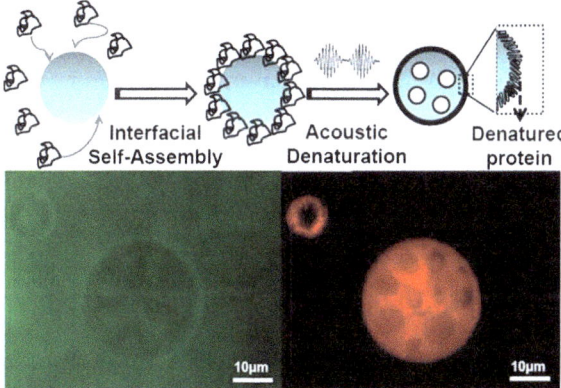

Figure 6.11 Schematic representation of lysozyme-based capsules *via* interfacial self-assembly followed by ultrasound-induced denaturation. Reprinted from ref. 61.

composed only of agglomerated peptide strands originating from the lyso-zyme, hence introducing a peptidase such as trypsin will induce digestion of the peptide strands. This digestion results in decomposition of the cap-sules under mild conditions, which are also found inside the body. Such approaches provide non-toxic and easy to form systems for possible applica-tions in foods, cosmetics and pharmaceuticals.

When considering proteins at interfaces, especially liquid/liquid inter-faces, capsule formation is the most obvious possible material or system to utilize. This approach has been adopted extensively with many types of bionanoparticles.[64] In the above-mentioned examples, the proteins were also used or transformed in order to add stability. In the case of lysozyme, both catalytic activity and gel formation *via* denaturation were involved. It is also possible to use native bionanoparticles that do not become altered or have no additional specific functionality. Virus particles have been used for this purpose and since they have many different shapes and sizes they offer the possibility of obtaining various structured surfaces. Initially tobacco mosaic virus (TMV) and turnip yellow mosaic virus (TYMV) were used, which differ considerably in overall morphology since TMV is rod shaped whereas TYMV is globular.[65,66] The approach can also be used in combination with inter-particle cross-linking, which in the case of cowpea mosaic virus (CPMV) was done with glutaraldehyde.[67] This provides a bion-anoparticle-based membrane around the interface and displays a higher stability than non-cross-linked structures. The confinement of different bionanoparticles such as virus particles can be used in very sophisticated scavenging systems. When oil droplets are stabilized by bionanoparticles and subsequently the oil phase is solidified, *e.g.* polymerized when the oil phase is a styrene droplet or another polymerizable hydrophobic liquid, after removing the protein particle from the solidified oil phase an imprint is left behind. This imprint can then be used to take up particles from solution.[68–70]

Even though the bionanoparticles remain globular and retain their over-all shape, it can still be expected that some denaturation and unfolding occurs. This can be circumvented by adding a protective polymer shell. Such so-called protein–polymer hybrid bionanoparticles can be formed *via* syn-thetic modifications, as was discussed in Chapter 1. The polymer shell also greatly influences the interfacial activity. This was shown by comparing the cage protein ferritin with ferritin–polyisopropylacrylamide conjugates and polyisopropylacrylamide alone. The covalently linked combination of pro-tein and polymer was more efficient in adhering to polar/apolar interfaces and provided a much denser packing.[71] The advantage of adding polymers is that additional functionalities can be incorporated, such as temperature responsiveness, and again undergo inter-particle cross-linking.[72,73] These hybrid structures are not only suited for capsule formation but can also be used as macroscopic materials with some additional manipulations, *e.g.* extrusion, which produced capsules embedded in protein–polymer fiber networks.[74]

6.5 Outlook and Future Perspectives

Interfacial directed self-assembly of bionanoparticles offers tremendous possibilities for developing new materials that are biocompatible and biodegradable and have additional specific properties such as catalysis. Although the uses of directed interfacial self-assembly preceded the understanding of the precise phenomena that occur at the interface, additional interpretation along with the development of new models in combination with MD simulations allow us to understand the processes better. It is expected that once a deeper knowledge about the underlying processes of protein behavior, interface interactions, buffer composition and apolar phase composition is obtained, more sophisticated and better structured systems can be developed based on a much higher rational design. These developments are being stimulated by the changing perspective with respect to proteins. Formerly, proteins were regarded as delicate structures that are easily destroyed and should perform a specific function, *e.g.* catalysis, storage, transmembrane channels, *etc.*, and are only used by biologists and biochemists and in biocatalysis. Nowadays, however, material scientists look at proteins as particles that are produced with high accuracy and substantial chemical functionality, which can be targeted chemically and have a variety of sizes and potential functions. Hence protein structures are entering many disciplines where they serve new purposes.[75–78]

References

1. P. A. Wierenga and H. Gruppen, *Curr. Opin. Colloid Interface Sci.*, 2010, **15**(5), 365.
2. A. Schulz, *et al.*, *J. Mater. Chem.*, 2011, **21**(47), 18903.
3. I. Shahidul, M. Shahid and F. Mohammad, *Ind. Eng. Chem. Res.*, 2013, **52**(15), 5245.
4. E. Ruiz-Hitzky, *et al.*, *Adv. Mater.*, 2010, **22**(3), 323.
5. N. Nassif and J. Livage, *Chem. Soc. Rev.*, 2011, **40**(2), 849.
6. Z. Xia and M. Wei, *J. Biomater. Tissue Eng.*, 2013, **3**(4), 369.
7. N. J. M. Sanghamitra and T. Ueno, *Chem. Commun.*, 2013, **49**(39), 4114.
8. A. George and S. Ravindran, *Nano Today*, 2010, **5**(4), 254.
9. S. V. Dorozhkin and M. Epple, *Angew. Chem., Int. Ed.*, 2002, **41**(17), 3130.
10. M. Suzuki and H. Nagasawa, *Can. J. Zool.*, 2013, **91**(6), 349.
11. A.-W. Xu, Y. Ma and H. Cölfen, *J. Mater. Chem.*, 2007, **17**(5), 415.
12. M. Kellermeier, H. Cölfen and J. M. García-Ruiz, *Eur. J. Inorg. Chem.*, 2012, **32**, 5123.
13. G. Jutz and A. Böker, *J. Mater. Chem.*, 2010, **20**(21), 4299.
14. G. Jutz and A. Böker, *Polymer*, 2011, **52**(2), 211.
15. A. Schulz, S. Hiltl, P. van Rijn and A. Böker, in *Biomaterials Surface Science*, ed. A. Taubert, J. F. Mano and J. C. Rodríguez-Cabello, Wiley-VCH, Weinheim, 2013, pp. 3–22.
16. A. Schulz, *et al.*, *J. Mater. Chem.*, 2011, **21**(26), 9731.

17. A. Schulz, *et al.*, *J. Mater. Chem. B*, 2013, **1**(8), 1190.
18. S. Kim and C. B. Park, *Adv. Funct. Mater.*, 2013, **23**(1), 10.
19. W. E. G. Müller, *et al.*, *IUBMB Life*, 2013, **65**(5), 382.
20. V. Raikos, *Food Hydrocolloids*, 2010, **24**(4), 259.
21. H. J. Hektor and K. Scholtmeijer, *Curr. Opin. Biotechnol.*, 2005, **16**(4), 434.
22. F. Y. Yohko, *J. Phys.: Condens. Matter*, 2012, **24**(50), 503101.
23. A. Schulz, *et al.*, *Soft Matter*, 2012, 8(44), 11343.
24. H. Tlatlik, *et al.*, *Angew. Chem., Int. Ed.*, 2006, **45**(12), 1905.
25. D. Frenkel and B. Smit, *Understanding Molecular Simulation: From Algorithms to Applications, Computational science series*, Academic Press Elsevier Science, 2001, **44**, p. 143.
26. G. Yampolskaya and D. Platikanov, *Adv. Colloid Interface Sci.*, 2006, **128–130**, 159.
27. V. B. Fainerman, E. H. Lucassen-Reynders and R. Miller, *Adv. Colloid Interface Sci.*, 2003, **106**(1–3), 237.
28. C. J. Beverung, C. J. Radke and H. W. Blanch, *Biophys. Chem.*, 1999, **81**(1), 59.
29. P. A. Wierenga, *et al.*, *J. Colloid Interface Sci.*, 2006, **299**(2), 850.
30. S. J. Singer, *J. Chem. Phys.*, 1948, **16**, 872.
31. V. B. Fainerman, E. H. Lucassen-Reynders and R. Miller, *Adv. Colloid Interface Sci.*, 2003, **106**, 237.
32. R. Douillard and J. Lefebvre, *J. Colloid Interface Sci.*, 1990, **139**(2), 488.
33. R. Douillard, *et al.*, *J. Colloid Interface Sci.*, 1994, **163**(2), 277.
34. J. A. V. Butler, *Proc. R. Soc. London, Ser. A*, 1932, **138**, 348.
35. P. Joos, *Biochim. Biophys. Acta*, 1975, **375**, 1.
36. L. Ter-Minassian-Saraga, *J. Colloid Interface Sci.*, 1981, **80**(2), 393.
37. E. H. Lucassen-Reynders, *Colloids Surf., A*, 1994, **91**, 79–88; E. H. Lucassen-Reynders and J. Benjamins, in: *Food Emulsions and Foams: Interfaces, Interactions and Stability*, ed. E. Dickinson and J. R. Patino, Royal Society of Chemistry, Cambridge, 1999, p. 195.
38. V. B. Fainerman, R. Miller and R. Wüstneck, *J. Colloid Interface Sci.*, 1996, **183**(1), 26–34.
39. R. Miller, *et al.*, *Adv. Colloid Interface Sci.*, 2000, **86**(1–2), 39.
40. A. F. H. Ward and L. Tordai, *J. Chem. Phys.*, 1946, **14**, 543.
41. D. E. Graham and M. C. Phillips, *J. Colloid Interface Sci.*, 1979, **70**(3), 403–414.
42. G. Serrien, *et al.*, *Colloids Surf.*, 1992, **68**(4), 219.
43. V. B. Fainerman, E. H. Lucassen-Reynders and R. Miller, *Colloids Surf., A*, 1998, **143**(2–3), 141.
44. M. A. Bos and T. van Vliet, *Adv. Colloid Interface Sci.*, 2001, **91**(3), 437.
45. F. Ravera, *et al.*, *Colloids Surf., A*, 2010, **365**(1–3), 2.
46. R. Miller, *et al.*, *Colloid Polym. Sci.*, 2010, **288**(9), 937.
47. S. Reynaert, *et al.*, *J. Rheol.*, 2008, **52**(1), 261.
48. J. Pelipenko, *et al.*, *Acta Pharm.*, 2012, **62**(2), 123.
49. F. Ravera, *et al.*, *Adv. Colloid Interface Sci.*, 2005, **117**(1–3), 75.
50. L. Liggieri, *et al.*, *J. Colloid Interface Sci.*, 2002, **255**(2), 225.

51. M. A. Augustin and Y. Hemar, *Chem. Soc. Rev.*, 2009, **38**, 902.
52. A. Drakos and V. Kiosseoglou, *J. Agric. Food Chem.*, 2006, **54**, 10164.
53. S. Fujii, *et al.*, *J. Colloid Interface Sci.*, 2009, **338**, 222.
54. E. Silletti, *et al.*, *J. Colloid Interface Sci.*, 2007, **313**, 485.
55. X. Wang, *et al.*, *J. Colloid Interface Sci.*, 2009, **332**, 96.
56. B. Lei, *et al.*, *Food Chem.*, 2011, **124**, 808.
57. P. B. Stathopulos, *et al.*, *Protein Sci.*, 2004, **13**, 3017.
58. X. Wang, *et al.*, *Biomaterials*, 2008, **29**, 1054.
59. T. M. Garakani, *et al.*, *Chem. Commun.*, 2012, **48**, 10210.
60. R. L. Brutchey and D. E. Morse, *Chem. Rev.*, 2008, **108**, 4915.
61. H. Wang, *et al.*, *J. Mater. Chem. B*, 2013, **1**, 6427.
62. A. J. Green, *et al.*, *Curr. Opin. Colloid Interface Sci.*, 2013, **18**(4), 292.
63. P. van Rijn, H. Wang and A. Böker, *Soft Matter*, 2011, **7**, 5274.
64. A. Böker, *et al.*, *Soft Matter*, 2007, **3**, 1231.
65. J. He, *et al.*, *Langmuir*, 2009, **25**, 4979.
66. G. Kaur, *et al.*, *Langmuir*, 2009, **25**, 5168.
67. J. T. Russell, *et al.*, *Angew. Chem., Int. Ed.*, 2005, **44**, 2420.
68. E. Verheyen, *et al.*, *Biomaterials*, 2011, **32**, 3008.
69. A. Cumbo, *et al.*, *Nat. Commun.*, 2013, **4**, 1503.
70. P. van Rijn, *Polymers*, 2013, **5**(2), 576.
71. P. van Rijn, *et al.*, *Langmuir*, 2013, **29**, 276.
72. N. C. Mougin, *et al.*, *Adv. Funct. Mater.*, 2011, **21**, 2470.
73. P. van Rijn, *et al.*, *Chem. Commun.*, 2011, **47**, 8376.
74. P. van Rijn, N. C. Mougin and A. Böker, *Polymer*, 2012, **53**, 6045.
75. P. van Rijn and A. Böker, *J. Mater. Chem.*, 2011, **21**, 16735.
76. P. van Rijn, *et al.*, *Chem. Soc. Rev.*, 2013, **42**, 6578.
77. D. J. Evans, *J. Mater. Chem.*, 2008, **18**, 3746.
78. X. Huang, *et al.*, *Nat. Commun.*, 2013, **4**, 2239.

CHAPTER 7

Protein Interactions at Liquid/Solid Interfaces: Protein Interactions with Colloidal Alumina Particles Functionalized with Amino, Carboxyl, Sulfonate and Phosphate Groups

LAURA TRECCANI[†*a], FABIAN MEDER[†b], AND KUROSCH REZWAN[a]

[a]Advanced Ceramics, IW3 Building, University of Bremen, Am Biologischen Garten 2, 28359 Bremen, Germany; [b]Centre for BioNano Interactions (CBNI), School of Chemistry and Chemical Biology, University College Dublin, Belfield, Dublin 4, Ireland
*E-mail: treccani@uni-bremen.de

7.1 Introduction

Protein adsorption on solid material surfaces is a widespread, common phenomenon, playing a fundamental role in many natural and biological processes. It has therefore attracted broad interest throughout many disciplines, including

---

[†]These authors contributed equally to this work.

RSC Smart Materials No. 16
Bio-Synthetic Hybrid Materials and Bionanoparticles: A Biological Chemical Approach Towards Material Science
Edited by Alexander Böker and Patrick van Rijn
© The Royal Society of Chemistry 2015
Published by the Royal Society of Chemistry, www.rsc.org

medicine, biology, pharmaceutical and food science, and has been the subject of numerous experimental, theoretical and computational investigations.

Protein adsorption on engineered nano- and colloidal particles is known to strongly determine particle biological fate and functionality, such as particle agglomeration, biodistribution, trafficking, cellular uptake and their overall cytotoxicity level.[1–3]

On artificial tissues and engineered scaffolds, protein adsorption has an important effect and is determinant for their integration, biocompatibility and functionality.[4] The adsorption of serum proteins can lead to fouling processes, blood clot formation (thrombosis) and immunoreactions; it influences the adhesion of blood cells and macrophages or platelets, thereby guiding inflammation and vascularization processes.[5,6]

Protein adsorption influences bacterial adhesion and subsequent biofilm formation. Attached proteins form a "preconditioned" surface that alters the intrinsic physical and chemical properties and facilitates, *e.g.*, bacterial attachment and growth.[7,8] Biofilms are considered a serious problem for public health as they may cause epidemics or medical device-related infections. Additionally, biofilms cause problems in water treatment processes and heating systems, as they may induce loss of process efficiency, corrosion and loss of heat transfer.[9,10]

In the analytical sciences, non-specific protein adsorption and protein assembly on sensors, separation systems, bioreactors and microfluidic devices are a significant problem affecting the specificity, sensitivity, separation efficiency and durability of the devices.[5,11,12]

The formation of biominerals, such as sponges, bones and mollusc shells, is tightly regulated by surface adsorption and subsequent protein incorporation into the crystal structure.[13–19] Understanding how macromolecules interact with inorganic crystals is crucial for creating bioinspired or biomimetic synthetic materials with advanced properties.

Hence it is clear that the interaction of biomolecules such as proteins with artificial materials, either nano-, micro- or macrosized, is an overarching phenomenon and is vital for the development of efficient materials and systems for a variety of applications.

Despite considerable progress, protein adsorption processes on surfaces remain substantially unclear. Protein–surface interactions are very complex and depend on several parameters and forces that are generated between the material surfaces and the proteins. There are still widely differing and even contradictory opinions on protein adsorption kinetics, orientation and structural rearrangements and cooperative and competitive adsorption. The disagreements are partly related to highly differing and arbitrary experimental parameters and incomplete material surface analysis.[5]

It has been recognized that, irrespective of the material, the interaction of proteins with surfaces is largely a function of the material surface chemistry, in addition to that of the proteins themselves.[3,20–25] At the protein/material interface, different interactions, including electrostatic, ionic, van der Waals, solvation and donor–acceptor interactions, overlap or combine to achieve a more thermodynamically favoured state.[26]

Material surfaces can be modified by altering the charge and electric potential of the surface and thereby the surface–protein interaction can be tuned. A feasible strategy to elucidate the effect of surface functionalities and to achieve a better understanding and control of protein–surface inter-dependency is the introduction of relative simple functional groups such as carboxyl (−COOH), sulfonate (−SO$_3$H), phosphonate (−PO$_3$H$_2$) and amine (−NH$_2$). Such negatively and positively charged functional groups can be used to impart a specific net charge to the surface and promote interactions with count-er-charged groups of the proteins.[27–30] They are commonly used for the modi-fication of the capacity of the surface of chromatographic matrixes to enhance selectivity and efficiency,[31] and immobilizing proteins or anchoring ligands on nanomedical drug delivery vehicles, in biochips[32] and in microfluidic devices.[12]

However, the interactions of proteins and surfaces cannot only be explained in terms of surface net charge. Other factors such as the material surface chemistry on a more molecular level, determined by type, concentration and distribution of the surface functional groups, are known to determine pro-tein adsorption behaviour, and also protein conformation and activity.[12] Nev-ertheless, their specific effects are still generally unknown.[4,33–35]

Hence it emerges that the selection and close control of appropriate sur-face chemistries and choice of functional groups can be used to influence the protein adsorption behaviour and to tune the properties of the adsorbed layer. In this chapter, we summarize our recent work investigating the inter-action of proteins with surface functional groups, with special focus on the interaction of proteins with alumina (Al$_2$O$_3$) colloidal particles bearing sur-face acidic (−COOH, −SO$_3$H, −PO$_3$H$_2$) and basic (−NH$_2$) groups.

Al$_2$O$_3$ colloidal particles can be easily modified and are considered an ideal substrate for studying and describing the interaction of proteins as a func-tion of differing surface chemistries. Al$_2$O$_3$ features high chemical stability, bioinertness and low susceptibility to hydrolysis and is easy to functionalize by wet chemical methods.[36,37] Moreover, Al$_2$O$_3$ is widely used as a substrate or starting material for the fabrication of sensors, protein separation and purifi-cation devices and biocatalysts for biomedical, biotechnological and environ-mental applications and is therefore of particular technical relevance.[21,38–44]

In this chapter, first some fundamental considerations on phenomena and factors that influence protein adsorption on material surfaces are pre-sented. Next, we introduce and discuss some key results showing how the adsorption of proteins with distinct features such as bovine serum albumin (BSA), lysozyme (LSZ) and trypsin (TRY) can be guided at material inter-faces through the introduction of fine control of −COOH, −SO$_3$H, −PO$_3$H$_2$ and −NH$_2$ surface functional groups. In addition, the adsorption of three viruses, hepatitis A virus (HAV) and the phages PhiX174 and MS2, which can be considered as large protein assemblies, on surface-functionalized Al$_2$O$_3$ particles is considered. The interaction of viruses with functionalized surfaces is not only of fundamental relevance but also provides examples of practical interest, *e.g.* for medical and remediation purposes or hybrid, bioinspired materials.

We present some exemplary and experimental and theoretical approaches that can contribute to achieving a better understanding of the "protein adsorption problem" and may facilitate the design and development of highly efficient, advanced materials and devices for biomedical, biotechnological, analytical and environmental purposes.

It must be emphasized that the examples and findings presented here relate only to the above-mentioned materials, functional groups, proteins and viruses. Direct transfer of these results to other types of materials, functional groups, proteins and viruses tested under different experimental conditions should be made with great caution to avoid misinterpretations.

7.2 Protein Adsorption at Liquid/Solid Interface

7.2.1 Phenomena at the Protein/Material Interface

Among the various biopolymers, proteins are the most surface active. Whenever a surface is exposed to a protein-containing liquid, proteins spontaneously accumulate at the liquid/solid interface and different events can occur. These are governed by a complex interplay of electrostatic and hydrophobic, van der Waals and hydrogen bonding interaction forces and are affected by several factors, such as surface chemistry, surface charge, solvation, donor–acceptor interactions, hydrophobicity/hydrophilicity of surfaces and proteins, a protein's hard or soft structure and the properties of the surrounding media.[4,45] An overview of the parameters that influence protein adsorption and events on the surface are shown schematically in Figure 7.1(A) and (B), respectively.

The tendency of proteins to accumulate at solid/liquid interfaces is determined not only by the material surface properties, but also by the nature of the solvent, the presence of other solutes, pH, ionic strength and temperature. Protein adsorption is accompanied by changes in the Gibbs energy of the system. If the Gibbs energy of adsorption (ΔG_{ads}) decreases, protein adsorption occurs and large negative values of ΔG_{ads} typically indicate high adsorption affinity.

In aqueous media, metal oxides and proteins feature a surface charge, which can lead to electrostatic interactions. They are strong and effective over large distances and considered one of the most important driving forces in many biological interactions and adsorption phenomena.[46] As soon as an oxidic surface is exposed to water, chemical surface reactions take place between the oxygen ions and the water molecules, resulting in a surface charge. Surface charges on oxide surfaces in solutions are predominantly induced by the ion dissociation on the surface and the protonation and dissociation of surface groups. The protonation/deprotonation of surface groups on metal oxides such as Al_2O_3 is shown schematically in Figure 7.2(A). On the surface of metal oxides (MO), such as Al_2O_3, surface hydroxyl groups ($-OH$) can be deprotonated (negatively charged) or protonated (positively charged),

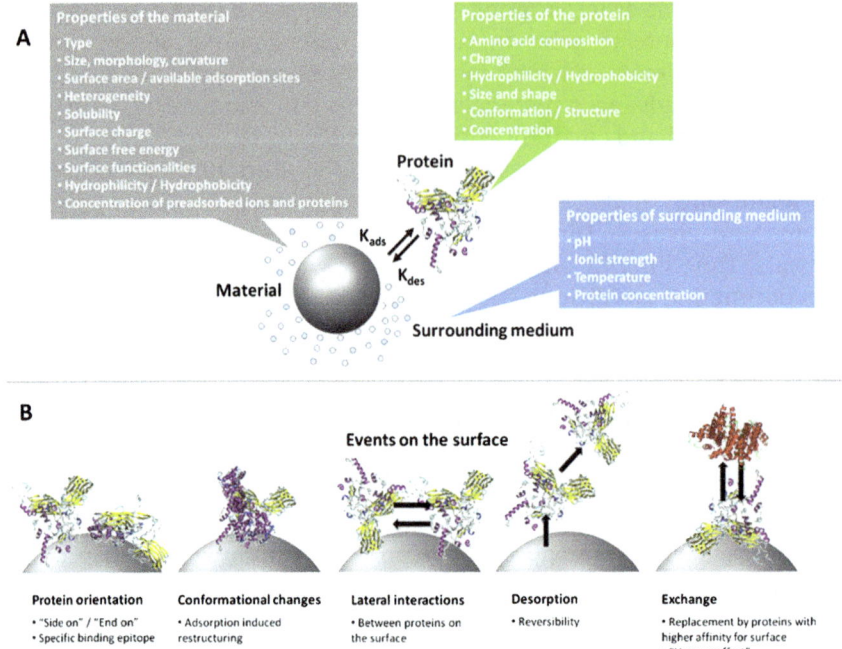

Figure 7.1   (A) Properties of particles, proteins and the surrounding media that influence protein adsorption. (B) Events on the material surface that may occur during and after protein surface adsorption.

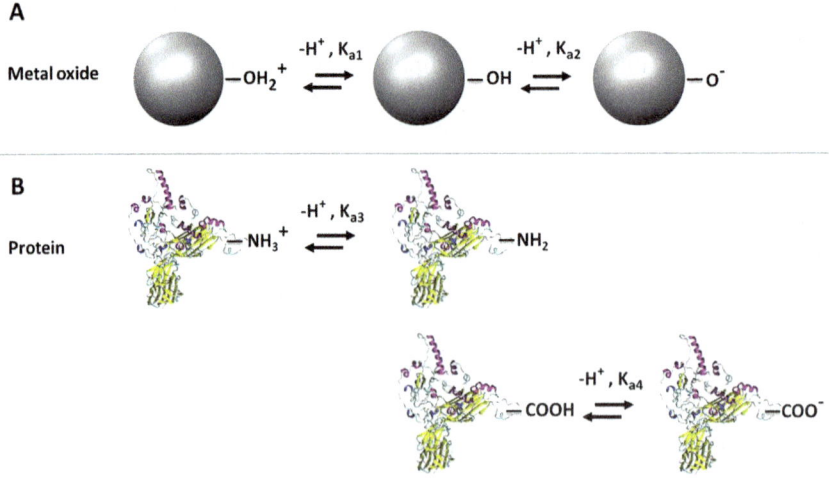

Figure 7.2   Protonation and deprotonation of surface groups on metal oxide surfaces (A) and proteins (B).

depending on the pH of the medium. The total surface charge depends on various parameters such as the number of the protonatable/deprotonatable surface sites per unit surface area and the acidity/basicity of the MO surface. The interplay is fairly complex considering that unfunctionalized MO materials may already have several different types of hydroxyl groups with varying dissociation behaviours, inhomogeneous charge distributions due to surface defects and/or crystallinity.[47]

Similarly, $-NH_2$ and $-COOH$ functional groups present on the surface of proteins can protonate or deprotonate, respectively (Figure 7.2(B)). This leads to positively charged $-NH_3^+$ or negatively charged $-COO^-$ groups.[48] The total surface or protein charge at given pH and ionic strength can be calculated when considering (and knowing) the sum of its dissociated surface groups.[49,50]

In the following, the most important types of interaction are briefly outlined. A deep and detailed description of the fundamental processes is outwith the focus of this chapter and details can be found in specific publications dedicated to this topic.[27,51–53]

7.2.2 Electrostatic Interactions

As previously described, in solutions both the protein molecules and most particle surfaces are electrostatically charged, if not specifically functionalized with non-dissociable molecules. Charged surfaces are surrounded by counter ions that neutralize surface charges and form an electrical double layer.[48] The electrostatic potential is maximal at the particle surface and decays linearly in the Stern layer and exponentially in the diffuse layer with distance from the particle surface.[46] The accumulation of charged species at charged interfaces can be described by the Poisson–Boltzmann equation, which allows the calculation of the potential distribution in the diffuse layer as a function of the distance from the particle surface. This model is relatively simple for planar surfaces or spheres with evenly distributed net charges but becomes more and more complex for inhomogeneous surfaces. Nevertheless, it has found large consensus and use for understanding the electrostatic interactions between, *e.g.*, colloids and biological systems.[46,54]

7.2.3 Hydration and Hydrophobic Interactions

Hydration and hydrophobic forces depend on the interaction of the particle surface with water molecules. They are also of fundamental importance for many biological interactions, *e.g.* for the formation of the intrinsic protein structure and biomolecular interactions such as antibody–antigen binding. The water layers on hydrophilic surfaces lead to repulsive forces, which are known as "hydration forces." The replacement of water is easier on hydrophobic surfaces and when, *e.g.*, hydrophobic particles adhere to each other as their free energy decreases through a reduction of the surface area exposed to the water. There is general agreement that hydrophobic surfaces have a higher adsorbent capacity than an equal surface area of a hydrophilic surface.[4] Nevertheless, hydration and hydrophobic interactions are not completely understood[48,51] and specific experimental characterization of surface hydrophobic and hydrophilic properties is therefore required.[4,20,55,56]

7.2.4 Steric Forces

In addition to van der Waals and electric double-layer forces, hydration and hydrophobic forces and repulsive steric forces are known to influence inter-particle interactions[46,51] Repulsive steric forces result from exposed molecules (often long-chain polymers) that protrude from the surface of colloidal particles.[57] Hydrophilic polymers such as poly(ethylene glycol) (PEG) can sterically hinder the approach of two particle surfaces due to entropic or osmotic repulsive forces and are widely used to reduce or completely suppress protein adsorption. However, depending on the type of polymer on, *e.g.*, particle surfaces, the steric forces may also bridge the particles and thus lead to attraction.[46] The exact role of such polymers and effectivity towards protein repulsion are still unclear and under discussion.[21]

7.2.5 Events on the Surface

As shown in Figure 7.1(B), once a protein comes in contact with a particle surface, different events can occur, including protein orientation, conformational changes, protein–protein interactions, exchange and desorption. Proteins do not behave as rigid particles but adapt a specific orientation corresponding to their free energy minimum, arising from Coulomb and van der Waals interactions, hydrogen bonds, the entropy of the solvents and counter ions.[5] The orientation of a protein on a surface determines which part of the protein is exposed to the bulk solutions and, therefore, protein activity.[5,53,58] For instance, asymmetric proteins (well-known examples are Y-shaped antibodies or rod-like membrane proteins,[59] but also high-abundance blood proteins such as albumins are not perfectly symmetrical) may orient with their short end ("end-on") or long side ("side-on") with respect to the particle surface, or certain proteins may bind only *via* specific amino acids (binding epitopes).[5,11,60] The protein orientation also determines the area that the proteins occupy on the particle surface and, thus, the protein surface density and the final thickness of a protein layer.[5,61,62] Depending on orientation, shape and amount, preadsorbed biomolecules such as proteins, peptides, amino acids, lipids and sugars can sterically hinder further adsorption.

Once a protein molecule is attached to the surface, it relaxes towards its equilibrium structure and may also undergo conformational changes. Upon adsorption, the protein conformation might differ from its native structure in the solution and can influence the protein function or enzymatic activity.[63] These can also be affected by the lateral interactions between adsorbed proteins that can either stabilize or destabilize the adsorbed protein layer.

Protein adsorption and layer properties are also affected by the protein concentration in the bulk solution. At low concentrations, the surface coverage increases slowly, whereas at high concentrations the surface is rapidly covered and structural changes are possibly hindered. The protein layer structure can be either dense or loosely packed and monolayers or multilayers can form. Monolayers typically form when protein–protein attractions are only weak or repulsive and multilayers are found under specific conditions that promote protein aggregation or repress inter-protein repulsion.[5] If several proteins are present, the most abundant proteins will adsorb on the surface in the initial stage; however, over time they will be replaced or exchanged by higher affinity proteins.[30,64] The protein layer structure and assembly of the proteins on the surface can be predicted from the protein dimensions.[61,65–67] An estimation of the side-on and end-on cross-sectional areas can be calculated from the three-dimensional structure files from X-ray diffraction in protein databases under the assumption that the protein size in solution and on the surface does not vary significantly. Using the cross-sectional areas, the theoretical protein content in a monolayer on the surface can be calculated assuming either regular adsorption (packing density of 78.5%) or random adsorption (packing density of 54.7%).[55,61,65] More detailed descriptions and definitions of end-on and side-on and also regular and random monolayers are given in Section 7.4.

Protein adsorption is a reversible phenomenon and dependent on changes in the media conditions or on adsorption and replacement of a second protein with a higher affinity for the surface. The time-dependent exchange of proteins on particle surfaces is known as the Vroman effect.[68]

7.3   Surface Functionalization

It is recognized that the surface chemistry of a material and its specific alteration by chemical functionalization influence protein–surface interactions.[53,60,66,69,70] Hydrophilic surface molecules, such as PEG, can reduce but never completely suppress protein adsorption,[43,71] which on the other hand can be increased by hydrophobic groups.[72] Positively or negatively charged anionic, cationic, proton donor and acceptor groups can interact with the oppositely charged counter groups of the proteins.[27] Negatively charged functional surface groups, such as $-COO^-$ and $-SO_3^-$, can be used to confer a net negative charge on the surface, but the two groups lead to different protein–surface interactions.[28–30] This clearly suggests that simplistic assumptions about the effect of surface net charge or overall hydrophilicity/hydrophobicity do not provide a complete picture of how such functional groups can influence or can be applied to guide protein adsorption. It has been recognized that other factors, such as the concentration and distribution of the surface functional groups, determine protein interaction behaviour.[73] Nevertheless, their specific effect is still largely unknown.[30,33,34]

Although several highly sophisticated surface modification approaches have been reported throughout the literature,[21] a simple functionalization strategy to understand protein–surface interactions consists of the introduction of basic (−NH₂) and acidic (−COOH, −SO₃H, −PO₃H₂) functional groups. Such surface functionalities are often employed either for stabilizing colloidal particles in biological media or for conjugation of biomolecules.[21,32,74–78]

Amino functional groups are widely applied to functionalize material surfaces for medical purposes, *e.g.* biosensors, cell and organ imaging, drug delivery, implants and implant coatings.[21,79] For instance, amino-functionalized particles can be used to target specific cell types by conjugating a cell line-specific ligand or antibody in a relatively easy way and for drug and gene delivery, plasmid DNA transport and synthesis of silica–polypeptide composite particles.[80] Positive surface charges appear to improve the particle efficacy; however, a higher cytotoxicity of such constructs has been reported.[79] Carboxyl groups (−COOH) are the most demanded functionality in biomaterial and chromatography applications[76] and in diagnostic applications owing to its well-known coupling chemistry, which requires the use of a carbodiimide as an intermediate.[77,79] Sulfonate groups and sulfonic acids (*e.g.* taurine) are known to be involved in many biochemical reactions and can bind to proteins.[81] In recent decades, organophosphorus compounds have attracted increasing attention as surface modifiers for drug delivery systems for the treatment of bone diseases,[82] gas-sensing probes and dye-sensitized solar cells.[83] Bisphosphonates[84] and pyrophosphate,[85] which have a high affinity towards basic proteins, can significantly enhance the adsorption of the basic protein LSZ compared with acidic proteins such as BSA.[86] These groups can be created by wet-chemical approaches employing organosilanes and phosphonate precursors to any metal oxide material and support.[20–22,56,87,88]

It must be pointed out that in addition to the above-mentioned charged functional groups, hydrophobic functionalities can be used as functional groups for stabilizing particles in hydrophobic media. However, hydrophobic functionalities will be only marginally considered in this chapter.

In Figure 7.3, some functionalized metal oxide particles, exemplary widely applied and commercially available precursors and relative characteristics are presented.

As shown, silanization can be carried out with organosilane precursors bearing different functionalities.[20–22,55,56,87,89] By controlling the reaction parameters and the initial precursor concentration, amino, carboxy and sulfonate groups can be introduced in a desired and controlled manner without altering morphological parameters such as size and particle shape.[20] Representative transmission electron microscopy (TEM) and high-resolution TEM (HR-TEM) images of native (unfunctionalized) polycrystalline α-Al₂O₃ particles before and after functionalization with the precursor 3-(trihydroxysilyl)-1-propanesulfonic acid (Figure 7.3) are shown in Figure 7.4(A) and (B) and Figure 4(C) and (D), respectively.

By selecting the initial precursor concentration, is it possible to obtain particles with either maximum or partial functional group coverage. An

Surface functionalised MO particles	Precursor name	Precursor pK$_a$	Abbreviation
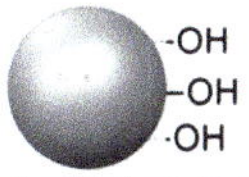 -OH -OH -OH	Native surface (no functionalisation)		MO-OH
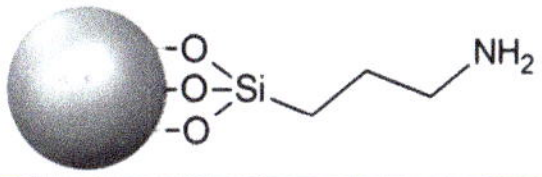 -O -O-Si / NH$_2$ -O	APTES 3-Aminopropyltriethoxysilane	10.6 (propylamine)	-NH$_2$
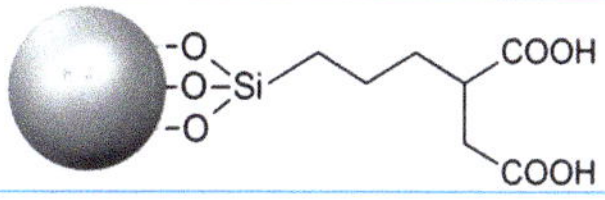 -O -O-Si / COOH -O COOH	TESPSA 3-(Triethoxysilyl)propylsuccinic anhydride	4.13 (methylsuccinic acid)	-COOH
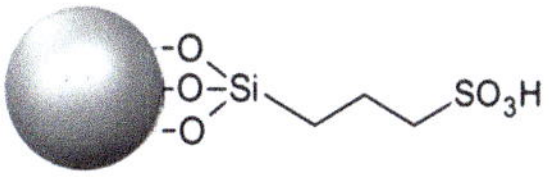 -O -O-Si / SO$_3$H -O	HSPSA 3-(Trihydroxysilyl)-1-propanesulfonic acid	1.53 (propylsulfonic acid)	-SO$_3$
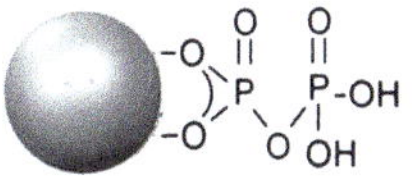 -O P P-OH -O O OH	PPA Pyrophosphoric acid	2.16 (phosphoric acid)	-PO$_3$H$_2$
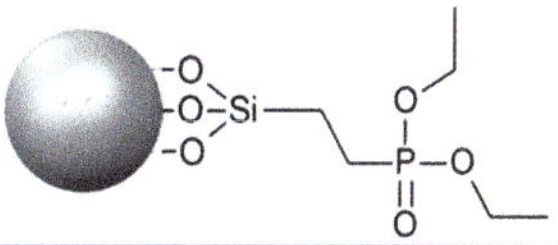 -O -O-Si — P—O -O O	(Diethylphosphatoethyl-triethoxysilane)		-PO$_3$(C$_2$H$_5$)$_2$
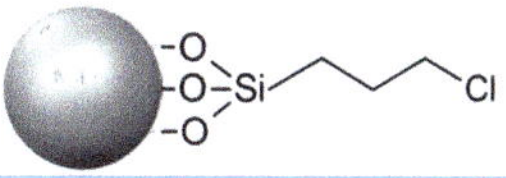 -O -O-Si / Cl -O	CPTES 3-(Chloropropyl)triethoxysilane		-Cl

Figure 7.3 Examples of surface-functionalized metal oxide (MO) particles modified with different precursors, giving the molecular structure, precursor name and, if available, relative dissociation constant (pK$_a$). Adapted from ref. 20 and 87.

estimation of the degree of surface coverage can be obtained by zeta potential measurements as a function of pH and IEP shift. For maximum surface coverage (Figure 7.5(A)), the IEP of Al$_2$O$_3$ at pH 9.3 shifts towards a higher or lower values close to the acid dissociation constant (pK$_a$) of the functional group: at pH 10 for −NH$_2$ (pK$_a$ = 10.6), at pH 3.2 for −COOH (pK$_a$ = 4.13), at pH 1.5 for −SO$_3$ (pK$_a$ = 1.53) and at pH 2.4 for −PO$_3$H$_2$ (pK$_a$ = 2.15). For selected precursors, the dissociation constants are given in Figure 7.3. By measuring the IEP shift and combining it with zeta potential, surface elemental composition, specific surface area and water or heptane adsorption capacity, a firm and well-defined picture of the surface properties after functionalization can be obtained. In other studies, it was shown that by simply varying the initial precursor concentration, particles with either maximum or partial coverage of functional groups could be easily synthesized.[22,55] By increasing the −SO$_3$H surface coverage, the IEP shifts towards more acidic values close to the precursor dissociation constant (Figure 7.5(B)). This allows fine adjustment of surface charges within a broad range and the synthesis of tailored particles

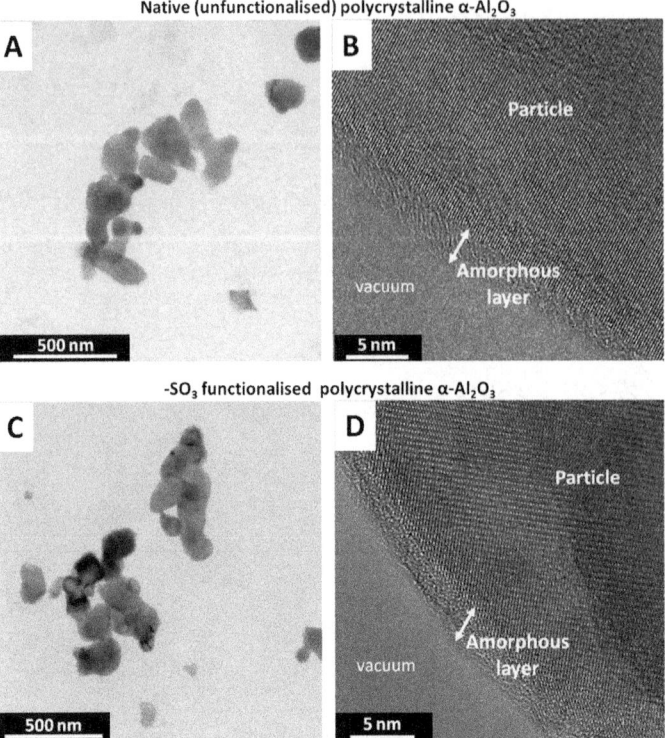

Figure 7.4   TEM images of α-Al$_2$O$_3$ particles (A) before and (C) after functionalization with 3-(trihydroxysilyl)-1-propanesulfonic acid. By controlling the
reaction parameters, the particle shape, size and curvature remain
unchanged after functionalization. HR-TEM images of α-Al$_2$O$_3$ particle
surfaces showing the crystal structure of the core particle and native
surface amorphous layer (B) before and (D) after functionalization.
(A and C) Reprinted from ref. 22 (supporting information), Copyright
2013, with permission from Elsevier. (B and C) adapted from ref. 20,
Copyright 2012, with permission from Elsevier.

with multifunctional surface chemistries made up of coexisting −OH and
−SO$_3$H[22] or −OH and −COOH[55] functional groups.

Alternatively to silane chemistry, hydroxyl-terminated particle surfaces
can be readily and stably functionalized with organophosphates and phosphonate precursors.[20,21,36,37] Owing to its versatility, phosphorus chemistry
allows the attachment of different organic residues and functional groups
to a phosphate-based tail consisting of phosphonic or phosphoric acid
groups.[36,90] For instance, a terminated phosphate group surface can be
obtained by simply using pyrophosphoric acid. One of the two phosphate
groups of the pyrophosphoric acid binds to the surface, whereas the other
one introduce a terminal phosphate group onto the material surface.[91] In
comparison with silanes, the condensation and cross-linking between the
phosphonate precursor molecules are significantly lower, and thereby more
homogeneous monolayers can be obtained.[36,92]

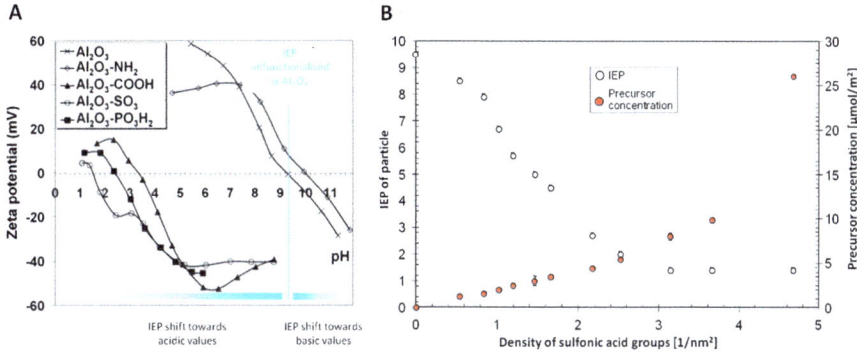

Figure 7.5 Zeta potential of alumina particles functionalized with $-NH_2$, $-COOH$, $-SO_3$ and $-PO_3H_2$ groups as a function of pH. (A) IEP shift of α-Al_2O_3 particles with maximum surface functional group coverage. A maximum surface coverage leads to a maximized shift of the IEP of native α-Al_2O_3 (IEP = 9.3), towards more acidic or basic values, close to the acid dissociation constant (pK_a) of the respective attached functional group. Image adapted from ref. 20. (B) Example of particles functionalized with varying $-SO_3H$ surface coverage. By increasing the precursor concentration, the number (density) of $-SO_3H$ increases. A higher $-SO_3H$ surface coverage leads to a gradual IEP shift towards more acidic values close to the acid dissociation constant (pK_a = 1.53) of the attached $-SO_3H$ group. The $-SO_3H$ surface coverage and amount of $-SO_3H$ can be determined by elemental analysis and acid–base titration and correlated with the particle specific surface area. Adapted from ref. 22, Copyright 2013, with permission from Elsevier.

7.4 Modelling and Visualization of Protein Surface Properties

To understand and predict protein adsorption and orientation on charged, functionalized surfaces, models considering the protein spatial surface potential distributions can be applied. Proteins are macromolecular polymers composed of acidic, basic, polar or non-polar amino acids, which determine the protein's structure, function, hydrophobicity and charge.[93] Proteins in solution may adopt various structures, depending mainly on the surrounding pH, ionic strength and temperature. The types of amino acids present or accessible on the protein's surface determine the protein functions and properties, including their surface charge, IEP and hydrophilicity/hydrophobicity, and thereby their interaction with material surfaces. A reasonable approach to determine the surface charge distribution and net charge of a protein relies on its correlation with the number of solvent-exposed charged amino acid end-groups. Charge distribution models of proteins can be calculated from the protein structure experimentally determined by X-ray crystallography or NMR spectroscopy by considering the pK_a of the amino acids, the pH and the ionic strength and using the Poisson–Boltzmann equation (Figure 7.6).[49,93,94] The model assumes a "static" conformation of the proteins

Secondary structure   **CPK model**

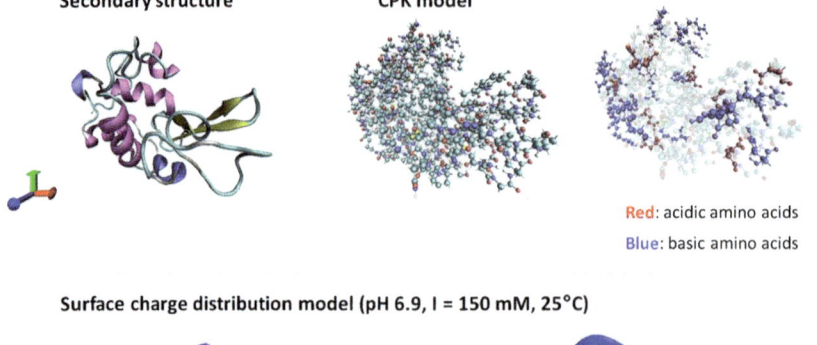

Red: acidic amino acids
Blue: basic amino acids

Surface charge distribution model (pH 6.9, I = 150 mM, 25°C)

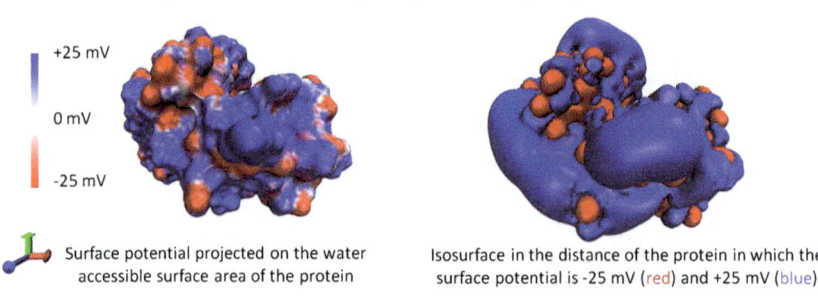

+25 mV

0 mV

-25 mV

Surface potential projected on the water   Isosurface in the distance of the protein in which the
accessible surface area of the protein   surface potential is -25 mV (red) and +25 mV (blue)

Figure 7.6   Secondary structure, CPK model and surface charge distribution calculated for LSZ using the crystal structure PDB ID 1GXV in the Protein Data
Bank (PDB) file (http://www.rcsb.org). The surface potentials were calculated using the PDB2PQR program and adaptive Poisson–Boltzmann
solver (APBS) and visual molecular dynamic (VMD) software.[49,94,96]

and the number of accessible ionizable groups and charge distributions may
differ slightly if protein conformational changes occur in solution or in the
adsorbed state.[95]

Using the surface charge distribution protein models, a simplified
approach to estimating and predicting the orientation of the proteins on
the surface was described by Meder *et al.*[20] It is hypothesized that a protein region with the same sign as the surface potential (either positive or
negative) will rather not orientate towards the surface, whereas protein
regions that have the opposite sign of the particle surface might increasingly interact with it, especially during the coming together of the protein
and the surface and/or the first protein–surface contacts before any other
reactions, *e.g.* conformational changes or protein–protein interactions,
occur.

Therefore, proteins can be simplified to symmetrical three-dimensional
objects, *e.g.* cuboids with six faces, as shown in Figure 7.7(A). More complicated geometric objects may be used to increase the accuracy of finding preferred interaction sites. The predominance of positive or negative potentials
on each of the six faces of the cuboid can be calculated by considering its
total charge distribution at a given pH and ionic strength. Therefore, images
of each of the six sides can be analysed regarding the charge distribution and

A

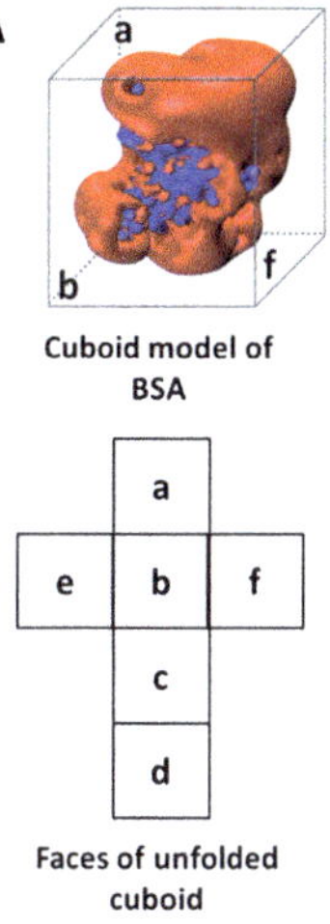

Cuboid model of BSA

Faces of unfolded cuboid

B

Cuboid face	Pixel type	BSA Pixel amount	%	LSZ Pixel amount	%	TRY Pixel amount	%
a	blue	85143	14.2	401542	95.6	292303	81.1
	red	513627	85.8	18466	4.4	68122	18.9
	total	598770		420008		360425	
b	blue	40125	8.2	855427	99.9	253630	68.0
	red	452003	91.8	1256	0.1	119552	32.0
	total	492128		856683		373182	
c	blue	34055	5.7	417687	100.0	266316	74.3
	red	564621	94.3	0	0.0	92165	25.7
	total	598676		417687		358481	
d	blue	42539	8.6	460367	94.1	312888	84.0
	red	450750	91.4	29124	5.9	59705	16.0
	total	493289		489491		372593	
e	blue	88066	15.2	556353	99.6	295427	75.8
	red	489497	84.8	2029	0.4	94493	24.2
	total	577563		558382		389920	
f	blue	29522	5.2	441223	96.1	308377	79.5
	red	535093	94.8	17730	3.9	79527	20.5
	total	564615		458953		387904	

Figure 7.7   (A) Exemplary cuboid model of BSA. (B) Estimation of the surface potential distribution of single protein sites by enumeration of the red and blue pixels projected on corresponding cuboid faces a–f of BSA, LSZ and TRY by using the spatial surface potential representations. BSA, LSZ and TRY secondary structures and properties are shown in Figure 7.8.

total charges by, *e.g.*, counting the number of blue and red pixels as shown in Figure 7.7(B) using image analysis software.

The advantage of the simplified model is that it considers charge distributions rather than net charges and it is easy to apply to all proteins whose structures are known. It is furthermore extendable to large molecules such as high molecular weight proteins and viruses.[56] A detailed computational analysis of specific interaction sites is often more time consuming and limited to smaller proteins and peptides.

Once the charge distribution on each cuboid side has been determined, it is then possible to predict protein "favoured" or "unfavoured" orientations towards a charged surface considering the assumptions made by the model. "Favoured" and "unfavoured" adsorption of BSA, LSZ and TRY on functionalized, charged surfaces are shown in Figure 7.8. For visual guidance, the amino acid residue Lys313 of BSA, catalytic centres of LSZ (residues Glu35 and Asp52) and catalytic centre of TRY (residues His57, Ser195 and Asp102) are highlighted. For BSA dimers, a side-on adsorption on positively charged surfaces is suggested to be the electrostatically "favoured" adsorption mode. For the enzymes LSZ and TRY, the catalytic centre is oriented towards the surface in the case of "unfavoured" adsorption on positively charged surfaces, but is on top in the case of "favoured" adsorption on negatively charged surfaces, which may allow one to estimate on which surface the enzyme might have higher activity. The suggestions resulting from the model were in agreement with experimental analysis of BSA, LSZ and TRY adsorption using protein assays and zeta potential measurements.

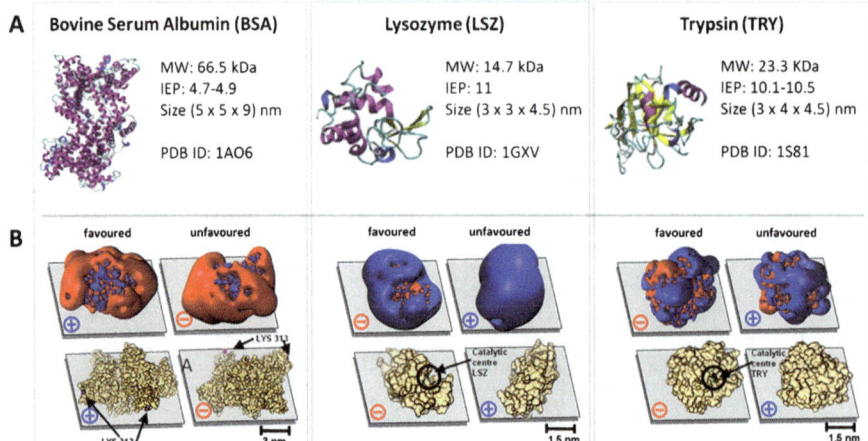

Figure 7.8 (A) Secondary structure molecular weight (MW), IEP and dimensions of the proteins BSA, LSZ and TRY. (B) Possible protein adsorption sites on positively and negatively charged surfaces. 'Favoured" adsorption on the oppositely charged surface or "unfavoured" adsorption on a similarly charged surface is hypothesized by analysing the spatial surface potential distribution of the proteins. For BSA, the arrows point to amino acid Lys313 for visual guidance. For TRY and LSZ, the catalytic centres are highlighted with arrows. Crystal structure and PDB files were taken from the PDB file (http://www.rcsb.org). (B) Reprinted from ref. 20, Copyright 2012, with permission from Elsevier.

7.5 Effect of Amino, Carboxyl, Sulfonate and Phosphate Functional Groups on Protein Adsorption

Colloidal oxide particles dispersed in biological media immediately become coated with proteins, leading to the formation of protein–particle conjugates.[97,98] The type of proteins that adsorb is known to be strongly influenced by the surface properties of the particles; nonetheless, the fundamental correlations between surface characteristics and the, so far mainly uncontrollable, protein adsorption are still not clear.

In a systematic study, we investigated the effect of different surface functional groups on the adsorption of three proteins, BSA, LSZ and TRY, on 180 nm alumina particles functionalized with $-NH_2$, $-COOH$, $-SO_3H$ and $-PO_3H_2$ groups.[20] Surfaces were functionalized using some of the precursors given in Figure 7.3.

Under well-defined experimental conditions (pH 6.9 ± 0.3, ionic strength I = 2.75 mM, 25 °C), the adsorption of BSA, LSZ and TRY was found to be mainly governed by electrostatic forces and to depend on the type of functional group, and also by the intrinsic protein characteristics and charge themselves. The negatively charged BSA (IEP = 4.7–4.9) was found to adsorb preferentially on positively charged surfaces such as native Al_2O_3 (IEP = 9.3) and NH_2-functionalized Al_2O_3 (IEP = 10). Introduction of negatively charged −COOH, −SO_3H and −PO_3H_2 groups decreased the adsorption rates immensely. The positively charged LSZ (IEP = 11) and TRY (IEP = 10.1–10.5) were found to adsorb preferentially on negatively charged particle surfaces bearing −COOH (IEP = 3.2), −SO_3H (IEP = 1.5) and −PO_3H_2 (IEP = 2.15) groups, but not on native Al_2O_3 and NH_2-functionalized Al_2O_3.

In addition to electrostatic effects, the dissociation rates of the functional groups and their acidity were observed to be a potential factor directly influencing protein adsorption behaviour. Although the particle zeta potentials of −SO_3H, −COOH and −PO_3H_2 were of similar magnitudes, LSZ adsorption was lower on strongly acidic −SO_3H particles, whereas TRY adsorption was higher on more weakly acidic −COOH and −PO_3H_2 particles.

Under the reported conditions, the different hydrophilic/hydrophobic properties of the functionalized particles did not affect the amount of protein adsorbed. For instance, the amounts of BSA, LSZ and TRY adsorbed on −COOH- and −PO_3H_2-functionalized particles did not vary significantly, although particles functionalized with −COOH groups were found to be more hydrophobic than those functionalized with −PO_3H_2 groups. The hydrophilic/hydrophobic properties of the particles were confirmed by water vapour and heptane gas adsorption isotherms. In contrast, on native Al_2O_3 and NH_2-functionalized particles featuring different hydrophilic/hydrophobic properties, the same protein adsorption behaviour was observed. The orientation of BSA, LSZ and TRY on the charged surface was predicted as described in Section 7.4.

7.6 Controlling Protein Adsorption by Tailoring Surface Functional Group Density

In addition to the influence of the type of functional group, it has been recognized that the functional group surface density can itself strongly influence protein adsorption behaviour. Electrostatic forces are known to play a determining role, but protein–surface interactions are not always simply a matter of attraction between oppositely charged proteins and particles.[33] Several studies have shown that protein–particle adsorption can deviate greatly from predictions made considering net charges and IEP of the protein and the particle.[54,99] For instance, Milani *et al.*[30] showed that when exposed to plasma, preadsorbed transferrin (IEP = 5.2–5.5[100]) is completely removed from −COOH-functionalized particles, but remains on −SO_3H-functionalized particles, although both functional groups have a net negative charge. The advantages of tuning material surface charge have been addressed elsewhere. Wang *et al.* showed that by balancing the charges of polymer-based nano drug vehicles, safer drug carriers with longer circulation times and low systemic toxicity were obtained.[101] However, comparative studies aimed at investigating the influence of these parameters are still rare.

7.6.1 Protein Adsorption on Colloidal Alumina Particles with Tailored Sulfonate Group Density

The adsorption behaviour of BSA, LSZ and TRY as a function of fine-tailored, graded $-SO_3H$ surface functional group density was investigated in a recent study.[22] Colloidal alumina particles were functionalized with $-SO_3H$ groups in different amounts to vary finely the outer exposed surface charge carriers within a broad range. $-SO_3H$ groups are deprotonated and strongly negatively charged above pH $\approx$ 1.5 due to their acid dissociation constant $pK_a = 1.53$.[102]

The introduction of an $-SO_3H$ monolayer on colloidal alumina can change the protein adsorption pattern and reduce the adsorption of negatively charged proteins such as BSA, while promoting that of positively charged proteins.[20]

By varying the initial precursor concentration, the $-SO_3H$ surface density can be tuned with an accuracy between 0 and 4.7 $-SO_3H$ nm^{-2} (Figure 7.9(A)), thus Al_2O_3 particles with either maximum or partial $-SO_3H$ coverage can be obtained. By introducing $-SO_3H$ groups, the particle IEP decreases by ~2.8 pH units per $-SO_3H$ nm^{-2} until it reaches saturation above ~3 $-SO_3H$ nm^{-2} at about pH 1.4. Zeta potential measurements confirmed that the particle surface chemistries are either determined by $-OH$ groups of the native alumina particles or $-SO_3H$ surface functions or by their coexistence. As shown in Figure 7.9(B), three surface chemistry categories can be identified: (I) native alumina particles (Al–OH); (II) particles with coexisting Al–OH and $-SO_3H$

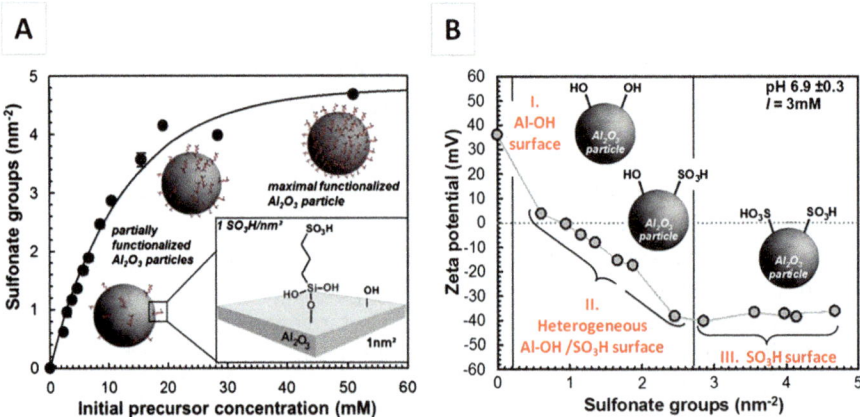

Figure 7.9 (A) Sulfonate group surface density on alumina particles as a function of the initial precursor concentration. (B) Zeta potentials of the particles as a function of the $-SO_3H$ surface density at pH 6.9 ± 0.3 and I = 3 mM. Three surface chemistry categories are defined: (I) native alumina particles (Al–OH); (II) particles with coexisting Al–OH and $-SO_3H$ surface functions and adjustable zeta potential; and (III) $-SO_3H$-dominated particles with a constant zeta potential. Error bars are about the size of the symbols. Reprinted from ref. 22, Copyright 2013, with permission from Elsevier.

surface functions and adjustable zeta potential; and (III) $-SO_3H$-dominated particles with a constant zeta potential.

On such highly tuned functionalized particles, the adsorption of BSA, LSZ and TRY was found to correlate directly with the $-SO_3H$ density (Figure 7.10(A)) and the amounts of BSA, LSZ and TRY adsorbed could be adjusted between almost no coverage and a theoretical monolayer. Different protein adsorption behaviour was observed in each particle category (defined in Figure 9(B)). The unfunctionalized alumina particles (category I) adsorb mainly BSA (~100%), ~37% TRY and marginal LSZ, as expected from electrostatic protein–particle interactions according to previous results.[20,66] By incrementally increasing the $-SO_3H$ surface density, BSA adsorption continuously declined, whereas LSZ and TRY adsorption continuously increased with increasing $-SO_3H$ surface density. In category II, protein adsorption correlates linearly with the $-SO_3H$ surface density. The sign and magnitude of the slope of the linear relationship seem to depend (at least qualitatively) on specific protein properties and are related to the amount and distribution of charged amino acids in the protein structure. For instance, the adsorption of BSA, which bears negatively charged amino acids in excess, declined, whereas the adsorption of LSZ and TRY with an excess of positively charged amino acids increased with increasing $-SO_3H$ surface density but with different slopes.

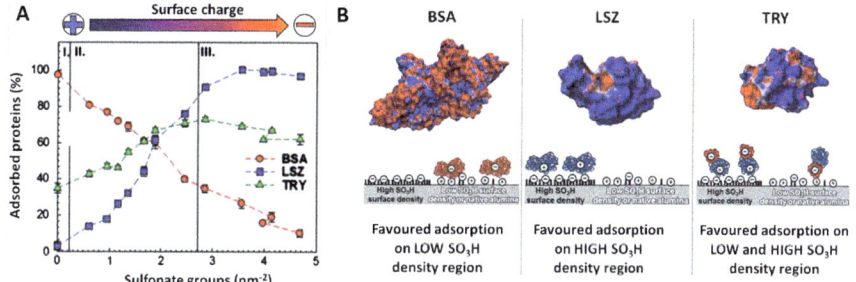

Figure 7.10   (A) Relative amounts of BSA, LSZ and TRY after 20 h of incubation at pH 6.9 ± 0.3 and $I = 3$ mM as a function of the $-SO_3H$ surface density on alumina particles. 100% refers to a theoretical regular end-on monolayer of each protein. I, II and III correspond to the categories defined in Figure 7.9(B). (B) Schematic of the adsorption BSA, LSZ and TRY on regions with different $-SO_3H$ surface densities coexisting on the particle surface. Local accumulations of $-SO_3H$ groups and native or regions with low $-SO_3H$ surface density may potentially provide local charge optima for electrostatically driven protein adsorption even though the net zeta potentials of the particle and protein are similar. Protein surface potential distributions were obtained from the crystal structure using the PDB file (http://www.rcsb.org), PDB2PQR and APBS and visualized by VMD.[49,94,96] Protein surface potentials were calculated at pH 6.9 and $I = 3$ mM. Adapted from ref. 22 Copyright 2013, with permission from Elsevier.

The adsorption mechanisms on these particles can be described by the model presented in Section 7.4 and considering the charge distributions on particles and proteins. The positively charged LSZ and TRY can adsorb on particles with low $-SO_3H$ densities and bearing the same net zeta potential as the proteins. Different mechanisms for this observation can be hypothesized. For instance, local positively and negatively charged regions on the protein surface could be involved in the adsorption;[4,54,99] nonetheless, the net charge suggests an electrostatic repulsion. Owing to the polycrystalline nature, Al_2O_3 core particles may have defects or steps and crystal planes on the surface can be differently exposed (see Figure 7.4).[47] This may lead to a locally different surface reactivity and/or an inhomogeneous distribution of the $-SO_3H$ groups. Exposed $-OH$ groups and the $-SO_3H$ precursor molecules might react with only some of them and assemble in highly specific regions on the particle surface. Thus regions with higher $-SO_3H$ group density and regions with low $-SO_3H$ group density may coexist with the native Al_2O_3 particle surface. If these regions of lower (or higher) $-SO_3H$ surface densities are large enough to allow for protein adsorption, they may provide local charge optima for protein adsorption, as shown schematically in Figure 7.10(B).

7.6.2 Controlling Mixed-Protein Adsorption Layers by Tailoring Carboxyl and Native Hydroxyl Surface Groups

Fine control over the type and amount of surface functional groups not only allows control over the amount of adsorbed proteins, but also permits controlled adsorption of different types of proteins and formation of mixed adsorption layers. The formation of specific protein layers with tailored compositions on colloidal particles is crucial, *e.g.*, when the particles are used as carriers for immunoassays, as biosensors, for cell targeting, as enzyme carriers or for protein purification and separation, imaging and drug delivery.[43,64,103,104] The composition of the protein layer determines the particles' selectivity, toxicity and biocompatibility.

As previously described, charged functional surface groups can direct the adsorption of oppositely charged proteins, hydrophilic/hydrophobic groups guide protein particle adsorption by their interaction with water molecules and larger molecules can sterically reduce protein adsorption.[20,29,72] However, the protein adsorption behaviour often differs from predictions that consider only the net charges or overall hydrophobicity, but the molecular composition of the particle surface plays a crucial role.[22,99] Particles with a multifunctional surface chemistry may contain diverse functional groups with different affinities for certain proteins.[47,105] In addition, even small variations in the concentration of a specific particle surface functional group have been shown to change drastically the adsorption of particular proteins.[20,33,34] It has been observed that certain surface chemistries, such as with protein-imprinted polymeric particles, can have a high specific protein affinity comparable to those of antigen–antibody complexes.[35,64]

The formation of a mixed protein adsorption layer with a defined protein ratio can be induced when carboxyl groups, which deprotonate and introduce negatively charged sites ($-COO^-$), are present together with native aluminium hydroxyl (Al–OH) groups.[55] Al–OH can protonate and introduce positively charged sites ($Al-OH_2^+$) on the particle surface.[76] −COOH groups can be introduced on the particle surface in a tailorable manner using the silane precursor 3-(triethoxysilyl)propylsuccinic anhydride (TESPSA).[20,55,87] By controlling and varying the initial amount of precursor, the amount of −COOH groups per unit area could be increased from 0 (native particles) to 4.4 ± 0.2 −COOH nm^{-2}. The presence of −COOH groups induces an IEP shift from ~9 towards acidic pH regions.[20] The particles' zeta potentials at neutral pH revealed the presence of deprotonated and negatively charged $-COO^-$ groups on the particle surfaces. At the highest precursor concentration, the particle IEP is close to the pK_a of −COOH groups,[20] suggesting that the −COOH groups govern the electrokinetic properties of the particles. In contrast, the IEP of the particles functionalized with a lower precursor concentration is higher (IEP = 6.2), indicating that the particle electrokinetic properties are determined by both native Al_2O_3 and −COOH groups coexisting on the surface. An overview of the particle properties is given in Figure 7.11.

The direct effects of the functional group surface density and ratio on protein adsorption were shown with oppositely charged BSA and LSZ. A comparison of single-component and sequential protein adsorption revealed that BSA and LSZ have specific and preferential adsorption sites. BSA adsorbs primarily *via* −OH groups, whereas LSZ adsorbs only *via* −COOH groups. Thereby, one or two −COOH groups on the particle surface have been shown to be sufficient to bind one LSZ molecule.

Surface functionalised Al_2O_3 particles	-COOH groups per nm^{-2}	IEP	Number of titratable −OH and -COOH groups (nm^{-2})	Abbreviation
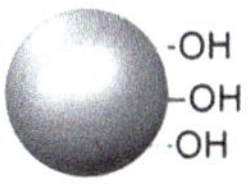 -OH, -OH, -OH	0	9.3	2.97 ± 0.19 / 0	Al_2O_3
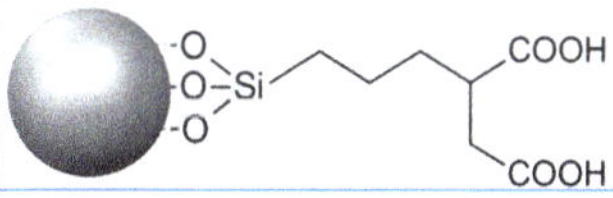 -O, -O–Si, -O, COOH, COOH	1.6	6.2	1.77 ± 0.13 / 0.02 ± 0.007	Al_2O_3-COOH_1.6
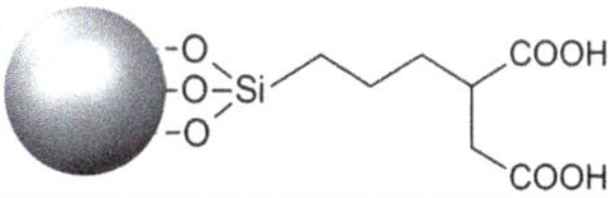 -O, -O–Si, -O, COOH, COOH	4.4	3.3	0.21 ± 0.13 / 2.09 ± 0.23	Al_2O_3-COOH_4.4

Figure 7.11 Schematics of the surface functionalized particles, number of functional −COOH groups per nm^{-2} (surface densities), IEPs and number of titratable Al–OH and −COOH groups on native and functionalized surfaces. −COOH surface densities can be calculated from carbon analysis and the number of titratable groups can be determined by acid–base titration measurements.[55]

A feasible approach to understand protein adsorption on the functionalized particles is to consider the active functional surface groups, which are able to protonate/deprotonate as potential protein adsorption sites. Active functional surface groups are positively charged Al—OH and negatively charged —COOH and they can be experimentally determined by combining elemental analysis, potentiometric acid–base and zeta potential titration measurements. Functional Al—OH and —COOH groups that do not protonate/deprotonate are considered to be non-active. A comparison of the total and active functional groups suggests that certain —COOH groups remain inactive on the particle surface, *e.g.*, by binding to the surface.[67] This is an important detail when considering protein adsorption as a function of surface chemistry.

A sensitive *in situ* determination of LSZ and BSA adsorption can be performed by electroacoustic zeta potential measurements. Owing to the proteins' low density compared with the particles (which are close to the surrounding aqueous medium), only the proteins bound to the particles are measured and unbound proteins in solution are "ignored" by the method. A separation of the particle–protein complexes from unbound proteins is therefore not necessary.[67,106] The impact of sequential LSZ and BSA addition on the particle zeta potential is shown in Figure 7.12. Clear variations of the zeta potential can be observed when the two oppositely charged proteins adsorb. The shifts in the zeta potential correlate well with quantitative analysis of the adsorbed proteins followed by *ex situ* UV/VIS spectroscopy and sodium dodecyl sulfate polyacrylamide gel electrophoresis (SDS-PAGE).

Interestingly, the zeta potential approaches 0 mV after the addition of BSA and LSZ and the effect was found to be independent of surface functionalization. This effect can be described in terms of "compensation and neutralization" of surface charges during protein adsorption and to be due to, *e.g.*, the interaction of a positively charged protein residue with a negatively charged particle surface —COO$^-$ group. This can explain why LSZ interacts predominantly with deprotonated —COO$^-$ groups on the particles and that BSA interacts with the protonated Al—OH$_2^+$ groups.[20,54,107] This correlation was observed for BSA and LSZ and it was independent of the fact that the counterpart protein was already present in solution, as confirmed by sequential protein addition experiments. This strongly suggests that BSA and LSZ do not compete for the same adsorption sites. Al—OH and —COOH groups show a specific and selective adsorption response of BSA and LSZ. The findings are in agreement with the fact that BSA and LSZ have different and complementary properties. Other proteins with slightly different properties or similar net charges might show lower selectivity for such surface groups.[108] Under the selected experimental conditions and with binary protein mixtures, it was furthermore observed that individual protein adsorption was only slightly influenced by the presence of a second protein as a counterpart. *Vice versa*, protein–protein (BSA–LSZ) interactions on the particle surface were apparently marginal, despite their complementary charge, and interactions between the proteins and the particle surface clearly dominated the adsorption process.

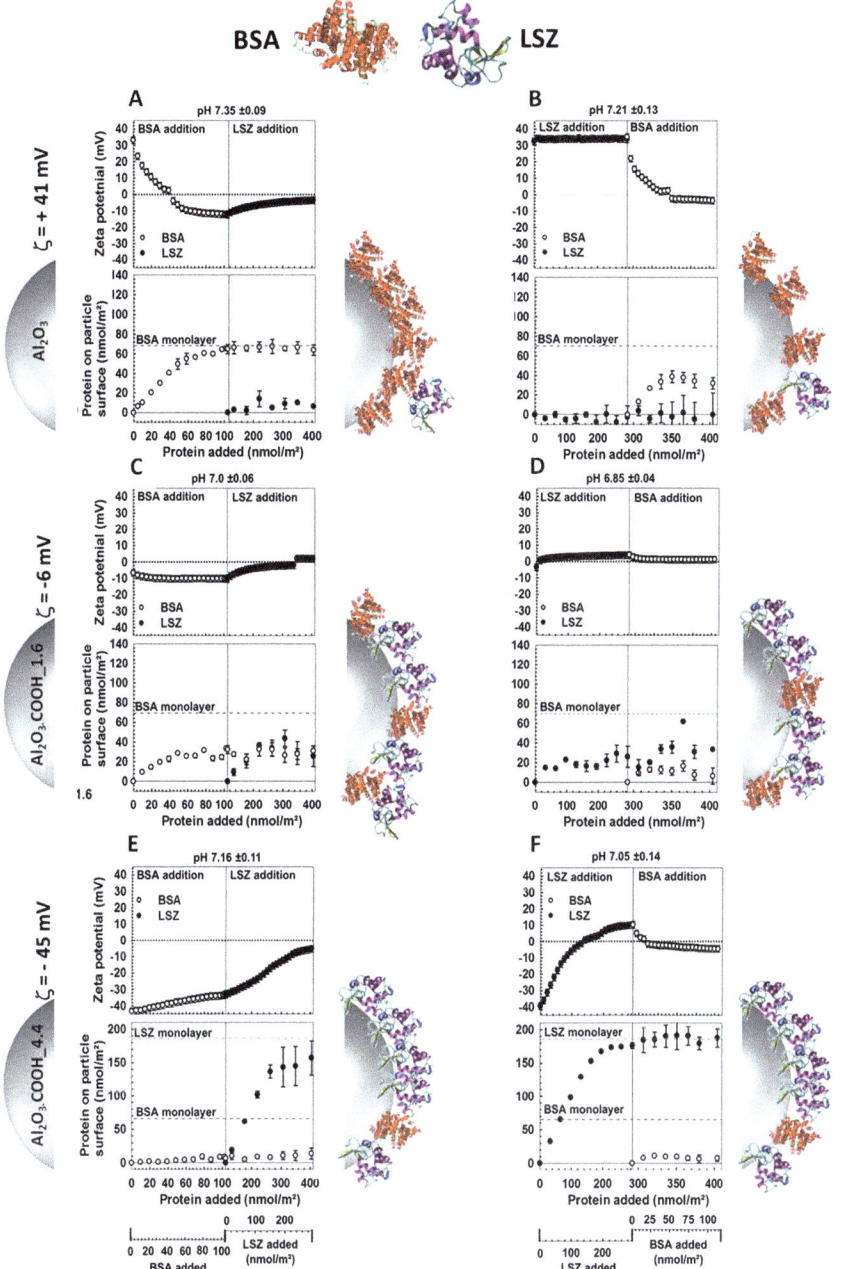

Figure 7.12 Sequential BSA and LSZ adsorption as a function of −COOH surface density and added protein amount, as determined by zeta potential (ζ) measurements (upper panels) and quantified by UV/VIS spectroscopy and SDS-PAGE (lower panels). In (A), (C) and (E), BSA was added first, followed by LSZ, and in (B), (D) and (F), LSZ was added first, followed by BSA. The *x*-axis shows the total protein content of the sample, and the second *x*-axis, at the bottom of the figure, shows the corresponding concentrations of BSA and LSZ. Images adapted with permission from ref. 55, Copyright 2013 American Chemical Society.

7.7 Surface Functional Groups to Control Interactions with Viruses

As shown in the previous sections, the interactions of proteins with colloidal particles can be studied and controlled by specifically tuning the particle surface chemistry, in particular by introducing $-NH_2$, $-COOH$, $-PO_3H_2$ and $-SO_3H$ surface functional groups at well-controlled surface concentrations.

In a further step, the feasibility and possibilities offered by this approach have been studied with viruses. Viruses can be considered as biological nano- to microsized objects (~25–200 nm) that consist to a large extent of highly specific and functional proteins. Viruses are made up of a genome packaged in a symmetrical and stable protein shell (capsid), sometimes enveloped by a membrane.[109] As single proteins, the proteins forming the virus structure show a pH-dependent surface charge in polar media such as water and viruses bear a characteristic IEP.[110] Most viruses often have, independent of their different types, a negative net surface charge at neutral pH and an IEP of <7.[111] This electrostatic charge mainly determines the virus colloidal behaviour, which plays a major role in virus sorption processes.[110] However, here also charge distributions on the virus surface likely contribute to their overall adsorption behaviour.

Such virus–material interactions are particularly important as viruses have unique application potential, *e.g.* as molecular recognition elements in biosensors,[112,113] as drug-delivery objects,[114,115] in bone regeneration applications,[116] as antibacterial surface modifiers[117] and as templates for material synthesis.[118] As viruses represent one the major contaminants of drinking water, causing diseases worldwide, and owing to their vanishingly small sizes, their removal from water still poses particular technological challenge.[110,119] Here, material surfaces that strongly attract viruses are particularly interesting for water purification systems and for designing effective wastewater filter surfaces.[119] It is clear that many of these applications can be significantly improved when the materials used obey controlled and tailor-made virus interactions.

It is suggested that the interaction of viruses and material surfaces are mainly controlled by electrostatic, van der Waals and/or hydrophobic interactions. Several studies have reported that viruses, owing to their net negative surface charge, can be attracted by positively charged surfaces in water.[110,119-122] Electrostatic interactions develop between the charge that exists on the material surface and the ionizable acidic and basic functional groups on the protein capsid of the viruses.[123] Therefore, viruses behave like proteins and with regard to their electrokinetic properties as amphoteric electrolytes.[124]

The possibility of using functionalized particles for designing materials for virus adsorption purposes and for improving the understanding of virus adsorption on inorganic surfaces in the aquatic environment has been studied.[56] To understand at which level viruses can recognize such variations of the material surface chemistry, alumina particles were functionalized with positive $-NH_2$, negative $-COOH$, $-PO_3H_2$ and $-SO_3H$ and hydrophobic chloropropyl ($-Cl$) groups. In contrast to $-NH_2$, $-COOH$, $-PO_3H_2$ and $-SO_3H$ groups,

−Cl groups are uncharged and do not show a pH-dependent charge but can be used to introduce hydrophobic surface functionalities.

The interaction of the functionalized particles was studied with HAV and phages MS2 and PhiX174, which have comparable size and morphology and a net negative surface charge, but feature differing capsid conformations and distributions of charged and hydrophilic/hydrophobic amino acids on their surfaces.

The effect of the different surface groups on the virus adsorption behaviour can be determined by measuring the concentration of the virus titre and by calculating the decrease in the virus titre, in logarithmic terms also called the $\log_{10}$ reduction value (LRV) or $\log_{10}$ clearance.[89] For instance, for the positively charged native and amino-functionalized alumina particles, and also with hydrophobic −Cl functionalized surfaces, an LRV of 5, which corresponds to a 99.999% reduction of the titre, was observed for HAV. In contrast, the functionalized −COOH, −PO$_3$H$_2$ and −SO$_3$H groups did not show any effects and HAV was completely recovered after incubation (Figure 7.13(A)). An LRV of ~4 (reduction of 99.99%) of phage PhiX174 titre was found after incubation with native Al$_2$O$_3$ and Al$_2$O$_3$−Cl (Figure 7.13(B)). For phage MS2, Al$_2$O$_3$, Al$_2$O$_3$−Cl and Al$_2$O$_3$−NH$_2$ led to an LRV of ~3.5 (reduction of 99.95%).

It was observed that viral adsorption was dependent on the surface density of −SO$_3$H groups, similarly to that of proteins as reported in Section 7.3.[22] For example, a similar reduction in HAV titre was observed for unfunctionalized alumina (0 −SO$_3$H nm^{-2}) and on particles with 1.20 −SO$_3$H nm^{-2}, whereas an −SO$_3$H surface density of 2.5 −SO$_3$H nm^{-2} led to an HAV recovery, which is about 2 LVR higher. Higher −SO$_3$H surface densities can inhibit the reduction in HAV titre (Figure 7.14(A)). Similarly, 1.2 −SO$_3$H nm^{-2} reduced the PhiX174 LVR of 2.5, whereas for MS2 the reduction was unchanged. An increment of

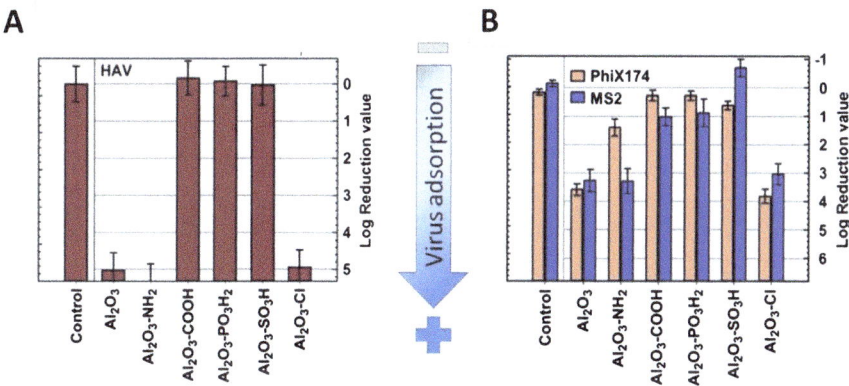

Figure 7.13 Virus adsorption as a function of the log reduction value (LRV) of HAV (A) and phages PhiX174 and MS2 (B) after incubation of the functionalized particles. Control samples were phage PhiX174 or MS2 incubated with the medium without particles. Reprinted from ref. 56, Copyright 2013, with permission from Elsevier.

$-SO_3H$ surface density to 2.5 $-SO_3H$ nm^{-2} reduced the LVR to 2.0–2.5 for both PhiX174 and MS2. A 3.6 $-SO_3H$ nm^{-2} was then sufficient to inhibit completely the PhiX174 log reduction. However, MS2 still shows a 2 log reduced titre at 3.6 $-SO_3H$ surface density, whereas the reduction for MS2 is completely inhibited at 4.7 $-SO_3H$ nm^{-2} (Figure 7.14(B)).

An overview of the LRV for HAV, MS2 and PhiX174 as a function of the type and surface density of particle functional groups and the particles' zeta potential is presented in Figure 7.15. This gives an idea of how the complex virus–surface interactions are guidable by specifically tailoring the surface chemistry. It can be recognized that virus–surface interactions cannot be explained only in terms of surface charges. For example, by introducing positive $-NH_2$ groups onto alumina and increasing the particle zeta potential, the adsorption of HAV and MS2 cannot not significantly altered. This is in disagreement with previous studies, which showed that the introduction of positively charged groups, such as $-NH_2$, and a surface zeta potential increment can favour virus adsorption.[117,125-127] The use of negatively charged functional groups ($-COOH$, $-PO_3H_2$ and $-SO_3H$) with different acidities ($-COOH < -PO_3H_2 < -SO_3H$[20,128]) can suppress viral adsorption and allow the complete recovery of HAV and PhiX174. This is in good agreement with an electrostatic repulsion approach between the negatively charged viruses and negatively functionalized particles. Nevertheless, these particles feature a negative zeta potential, and complete virus removal, as in the case of phage MS2, cannot be excluded. The reasons for the differing, unexpected and, at first glance, contrasting results are still unclear. It can be hypothesized that viral adsorption is related to the different acidities and protonation states of the $-COOH$, $-PO_3H_2$ and $-SO_3H$ functional groups, but also to the intrinsic virus properties, such as conformation and surface charge distribution.

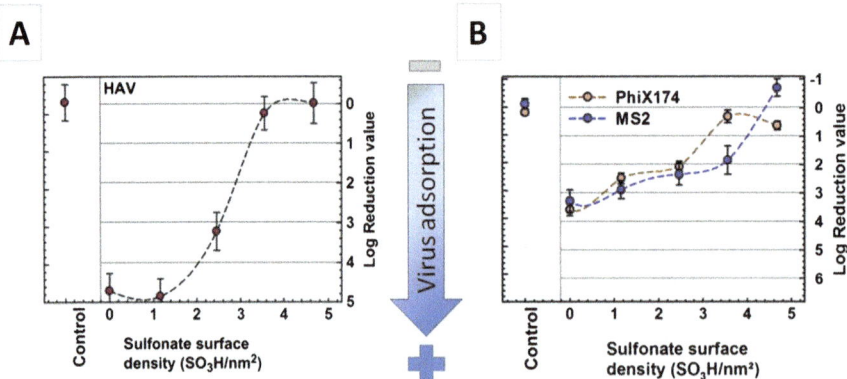

Figure 7.14   Influence of the surface density of $-SO_3H$ groups on virus adsorption and log reduction value (LRV) of HAV (A) and PhiX174 and MS2 (B). Each sample was analysed three times and experiments were conducted with two independent replicates. Symbols and error bars indicate average and standard deviation. Reprinted from ref. 56, Copyright 2013, with permission from Elsevier.

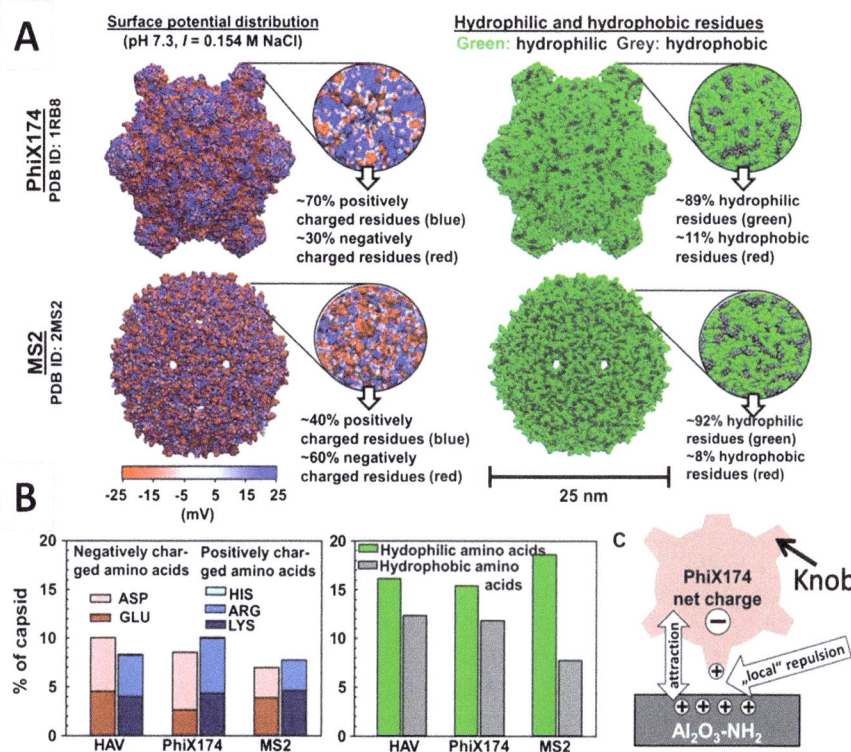

Figure 7.15 (A) Capsid structure, distribution of hydrophilic/hydrophobic amino acids and surface potential distribution of phages MS2 and PhiX174. (B) Relative amounts of charged (at pH 7.3) and hydrophilic/hydrophobic amino acids in the capsids of HAV, phage PhiX174 and phage MS2 (His, light blue, is only partially charged at pH 7.3 and the contribution is <1%). (C) Illustration of the hypothesized mechanism for the rather repulsive interaction between PhiX174 capsid and Al_2O_3–NH_2 due to protruding positively charged knobs of the PhiX174 capsid that might repulse PhiX174 from Al_2O_3–NH_2, although strong attractive forces are expected by the PhiX174 net charge. Surface potential distributions and relative amounts of charged and hydrophilic/hydrophobic amino acids were calculated at pH 7.3, I = 0.154 M at 25 °C on the solvent-accessible surface using the PDB2PQR program and APBS and VMD software. Adapted from ref. 56, Copyright 2013, with permission from Elsevier.

To understand the influence of the intrinsic physicochemical properties of viruses on their interactions with material surfaces, a simplified model that depicts charge and the zeta potential distribution on the virus surface can be applied. This, in combination with an exhaustive surface characterization, may lead to a more detailed picture and understanding of virus–material interactions.

For modelling the virus surface charge and zeta potential distribution, an approach similar to that presented in Section 7.4 can be used. Using the

PDB2PQR program for assigning atomic charges and APBS and VMD software,[49,94,96] the surface potentials can be projected onto the virus's water-accessible surface area. The surface charges can be calculated by considering the amount of positively charged amino acids (Lys, Arg and His) and negatively charged amino acids (Asp and Glu). The surface potential distribution on the surface of MS2 and PhiX174 calculated for pH 7.3 and $I = 0.154$ M at 25 °C and using the Viper database VIPERdb (http://viperdb.scripps.edu, entry 2ms2 files and 1rb8 for PhiX174) are shown in Figure 7.15(A). It must be noted that this approach is only possible if structure files are readily available. In the case of HAV, so far only sequences and no structure files are available. Additionally this approach provides only an estimation of the capsid surface properties as some amino acids are located in the capsid interior.

The percentages of charged and hydrophilic/hydrophobic amino acids in the virus capsids are summarized in Figure 7.15(B). Under the experimental conditions (pH 7.3), the negative charges are introduced by the almost completely deprotonated Asp ($pK_a = 4.1$) and Glu ($pK_a = 4.1$), whereas Arg ($pK_a = 12.5$) and Lys ($pK_a = 10.8$) are almost completely protonated and are positively charged. Under these conditions, His ($pK_a = 6$) is only partially protonated and contributes only marginally to the total surface charge of the virus capsids. HAV has an excess of negatively charged amino acids in correlation with its IEP. PhiX174 and MS2 have an excess of positively charged amino acids, in contrast to their IEPs and to their negative zeta potential. This can be related to the presence of negatively charged nucleic acids in the capsid interior.[129,130] During a virus–particle adsorption, the amino acids of the viral capsid of non-enveloped viruses are suggested to interact directly with the particle surface and influence the specific adsorption of the viruses. Compared with HAV, the MS2 capsid has less negatively charged amino acids but a similar amount of positively charged amino acids (Figure 7.15(B)), which might explain the partial attraction of MS2 onto negative −COOH and −SO$_3$H_3.6, not observed for HAV. Analogously, the higher amount of positively charged amino acids in the capsid of PhiX174 could explain the lower LRV of PhiX174 after incubation with Al$_2$O$_3$−NH$_2$ than with Al$_2$O$_3$. The presence of surface amino groups on the particle may repulse the similarly charged amino acids in the PhiX174 capsid. A similar effect cannot be seen on Al$_2$O$_3$, possibly owing to the lower zeta potential. This hypothesis can be confirmed by taking into account the surface potential distribution and the positively charged protrusions (knobs) on the capsid of PhiX174. When PhiX174 comes close enough to a surface, these knobs may be the first to recognize the particle surface and interact with it, and the positively charged −NH$_3^+$ groups on the particle surface may then repel these knobs and hinder an adsorption of PhiX174, although the net charges of particle and virus suggest attraction (Figure 7.15(C)). In contrast, for the strongly negatively charged particles (Al$_2$O$_3$−SO$_3$H, −COOH, −PO$_3$H$_2$, −SO$_3$H_3.6), the repulsive net charges between PhiX174 and the particles seem to dominate and hinder PhiX174 approaching the particle surface. For MS2, which has a smoother capsid conformation, such behaviour was not observed and the MS2 surface charges (about 40% positive and 60% negative) were more evenly distributed. HAV, PhiX174 and MS2 feature an excess of hydrophilic amino acids in their capsids (Figure 7.15(B)). This is in agreement with water contact angle measurements on these viruses found in the literature. Even though the contact angle values are widely scattered, they are mostly in the hydrophilic range for these viruses. Nevertheless, the capsids of HAV, PhiX174 and MS2 have hydrophobic portions of more than 7% that are distributed on the viral capsids. Interactions with these hydrophobic regions might be the reason for the high LRV of HAV, PhiX174 and MS2 on Al_2O_3–Cl.

The adsorption of viruses on surfaces involves a delicate interplay of several interconnected factors, such as virus type, type and surface density of particle functional groups and the surface zeta potential, but is potentially guidable by tailoring the material surface chemistry. In Figure 7.16, the effects of each surface functionality on the LRV of HAV, PhiX174 and MS2 are compared. Under the experimental conditions (pH 7.3 ± 0.3 and 0.154 M NaCl) the particles' zeta potentials at that pH (pH of PhiX174 and MS2 experiments) is in good correlation with the type and surface density of functional groups and the particles' IEP. However, the IEP of Al_2O_3–SO_3H (IEP = 1.5) is different from that of Al_2O_3–COOH (IEP = 3.2), although both particle types have a similar zeta potential. The zeta potential is, in contrast to the IEP, determined by the amount of charged sites on the particle surface. The similar surface densities

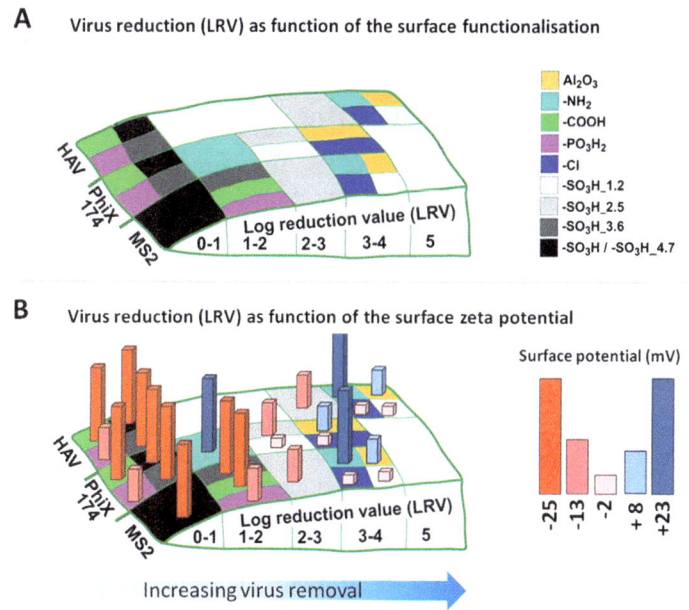

Figure 7.16   Comparison of the LRVs of HAV, phage PhiX174 and phage MS2 as a function of (A) the surface functionalization type and density and (B) the particle zeta potential under the experimental conditions. Adapted from ref. 56, Copyright 2014, with permission from Elsevier.

of functional groups on Al_2O_3–COOH and Al_2O_3–SO_3H (4.4 and 4.7 groups nm^{-2}, respectively) lead to similar amounts of negative charge carriers if all groups are deprotonated, which likely explains the comparable zeta potential despite the different IEPs. The same effect might explain why Al_2O_3–PO_3H_2 features a lower zeta potential although its IEP is smaller than that of Al_2O_3–COOH. It is likely that the phosphate group in Al_2O_3–PO_3H_2 deprotonates only once ($-PO_3H^-$) and the second deprotonation ($-PO_3^{2-}$) is minimal at pH 7.3 ± 0.3. This may lead to fewer negatively charged sites for Al_2O_3–PO_3H_2 than for Al_2O_3–COOH when comparing the surfaces with different −COOH group densities (2.3 and 4.4 groups nm^{-2}, respectively).

It must be noted that the medium itself can influence the particle properties and therefore virus behaviour. For instance, the addition of protein-containing cell culture media such as Dulbecco's modified Eagle's medium (DMEM) supplied with foetal calf serum (FCS) can lead to a shift of the particle zeta potential. This was observed, for instance, for Al_2O_3, Al_2O_3–NH_2, Al_2O_3–Cl and Al_2O_3–SO_3H_1.2 particles, which also clearly adsorbed proteins contained in the FCS. The presence of serum proteins, even at low concentrations, can be sufficient to cover the particle surface completely and vary the particle zeta potentials.

We have briefly give an example of how fine control over the surface chemistry of colloidal particles can be applied to study and control virus adsorption. Nonetheless, virus–particle interactions are highly complex and dictated by both particle and virus surface charge, molecular composition, net zeta potential and overall hydrophobicity.

7.8 Conclusions and Outlook

In this chapter, we aimed to elucidate the fundamental interaction forces that influence protein adsorption on functionalized aluminium oxide surfaces. Protein adsorption on oxide/materials surfaces is widespread and omnipresent in biomedical applications and biosensor technology, thereby having a significant impact on the functionality and long-term usability of technical devices.

We have shown that by using different functional groups, such as −NH_2, −COOH, −PO_3H_2 and −SO_3H, featuring different charges and dissociation constants, the interaction of three model proteins (BSA, LSZ and TRY) with colloidal alumina particles can be controlled at a high degree and in a precise manner. We showed that electrostatic forces dominate protein–surface interactions. Nonetheless, a description of the protein adsorption behaviour considering only the particles' and proteins' net charges is inappropriate. A protein can interact differently with functional groups that have a similar surface charge but with different dissociation constant (pK_a) values. Moreover, by tuning the amount of surface functional groups, the protein surface coverage can be controlled. This effect can furthermore be employed to control the formation of mixed protein layers with a well-defined composition. In fact, we showed that the protein adsorption behaviour is greatly affected by the type, concentration and distribution of differently surface-charged groups. These parameters have to be determined as precisely as possible when characterizing and designing surfaces for predictable and controlled protein interactions. By using a simplified model, based on electrostatic interactions, the protein orientation on a charged surface can be predicted.

We demonstrated that control of surface composition is highly feasible for the specific control of protein– and virus–surface interactions. It must be considered that the more proteins and the more complex the composition of the biological environment is, the more overlaying effects and competing reactions will occur that might influence the described processes drastically. The extension of this approach to other materials classes and proteins may contribute to and yield a more rational and customized design of materials that can efficiently increase or decrease protein adsorption at interfaces and are relevant for biological, biomedical and environmental technologies.

Nomenclature and Abbreviations

Al_2O_3	Aluminium oxide, alumina
Al−OH	Hydroxyl groups on alumina surface
APBS	Adaptive Poisson–Boltzmann solver
APTES	(3-Aminopropyl)triethoxysilane
Arg (ARG)	Arginine
Asp (ASP)	Aspartic acid
BSA	Bovine serum albumin
−COOH	Carboxyl groups
CPTES	3-Chloropropyltriethoxysilane
DMEM	Dulbecco's modified Eagle's medium
FCS	Foetal calf serum
Glu (GLU)	Glutamic acid
HAV	Hepatitis A virus
His (HIS)	Histidine
HR-TEM	High-resolution transmission electron microscopy
HSPSA	3-(Trihydroxysilyl)-1-propanesulfonic acid
LSZ	Lysozyme
I	Ionic strength
IEP	Isoelectric point
kDa	Kilodalton, mass unit ($1 Da = 1 g mol^{-1}$)
Lys (LYS)	Lysine
LVR	Log reduction value
MO	Metal oxide
MS2	Bacteriophage MS2
−NH$_2$	Amino group
−OH	Hydroxyl group
PDB	Protein Data Bank identification number
PhiX174	Bacteriophage PhiX174

pK_a	Acid dissociation constant
−PO$_3$H$_2$	Phosphate group
Ser (SER)	Serine
SEM	Scanning electron microscopy
−SO$_3$H	Sulfonate group
TEM	Transmission electron microscopy
TESPSA	3-(Triethoxysilyl)propylsuccinic anhydride
TRY	Trypsin
VMD	Visual molecular dynamic software

Acknowledgements

The authors would like to all colleagues and students for their important contributions to this work, Alessandro Paroli for graphical support and the European Research Council for financial support within the BiocerEng project.

References

1. M. S. Ehrenberg, A. E. Friedman, J. N. Finkelstein, G. Oberdorster and J. L. McGrath, *Biomaterials*, 2009, **30**, 603–610.
2. M. P. Monopoli, D. Walczyk, A. Campbell, G. Elia, I. Lynch, F. B. Bombelli and K. A. Dawson, *J. Am. Chem. Soc.*, 2011, **133**, 2525–2534.
3. M. Lundqvist, I. Sethson and B. H. Jonsson, *Langmuir*, 2004, **20**, 10639–10647.
4. E. A. Vogler, *Biomaterials*, 2012, **33**, 1201–1237.
5. M. Rabe, D. Verdes and S. Seeger, *Adv. Colloid Interface Sci.*, 2011, **162**, 87–106.
6. J. J. Gray, *Curr. Opin. Struct. Biol.*, 2004, **14**, 110–115.
7. L. D. Renner and D. B. Weibel, *MRS Bull.*, 2011, **36**, 347–355.
8. A. V. Singh, V. Vyas, R. Patil, V. Sharma, P. E. Scopelliti, G. Bongiorno, A. Podesta, C. Lenardi, W. N. Gade and P. Milani, *PLoS One*, 2011, **6**, e25029.
9. R. M. Donlan, *Clin. Infect. Dis.*, 2001, **33**, 1387–1392.
10. B. V. Kjellerup, T. R. Thomsen, J. L. Nielsen, B. H. Olesen, B. Frolund and P. H. Nielsen, *Biofouling*, 2005, **21**, 19–29.
11. K. Nakanishi, T. Sakiyama and K. Imamura, *J. Biosci. Bioeng.*, 2001, **91**, 233–244.
12. S. Hosseini, F. Ibrahim, I. Djordjevic and L. H. Koole, *Analyst*, 2014, **139**, 2933–2943.
13. L. A. Touryan, G. Baneyx and V. Vogel, *Colloids Surf., B*, 2009, **74**, 401–409.
14. L. Wang and M. Nilsen-Hamilton, *Front. Biol.*, 2013, **8**, 234–246.
15. S. Blank, M. Arnoldi, S. Khoshnavaz, L. Treccani, M. Kuntz, K. Mann, G. Grathwohl and M. Fritz, *J. Microsc.*, 2003, **212**, 280–291.
16. F. Heinemann, L. Treccani and M. Fritz, *Biochem. Biophys. Res. Commun.*, 2006, **344**, 45–49.

17. K. Mann, F. Siedler, L. Treccani, F. Heinemann and M. Fritz, *Biophys. J.*, 2007, **93**, 1246–1254.
18. L. Treccani, S. Khoshnavaz, S. Blank, K. von Roden, U. Schulz, I. Weiss, K. Mann, M. Radmacher and M. Fritz, in *Biopolymers, Polyamides and Complex Proteinaceous Materials II*, ed. S. Fahnestock and A. Stein-büchel, Wiley-VCH, Weinheim, 2003, vol. 8, pp. 289–322.
19. L. Treccani, K. Mann, F. Heinemann and M. Fritz, *Biophys. J.*, 2006, **91**, 2601–2608.
20. F. Meder, T. Daberkow, L. Treccani, M. Wilhelm, M. Schowalter, A. Rosenauer, L. Mädler and K. Rezwan, *Acta Biomater.*, 2012, **8**, 1221–1229.
21. L. Treccani, T. Y. Klein, F. Meder, K. Pardun and K. Rezwan, *Acta Biomater.*, 2013, **9**, 7115–7150.
22. F. Meder, C. Brandes, L. Treccani and K. Rezwan, *Acta Biomater.*, 2013, **9**, 5780–5787.
23. S. Sakata, Y. Inoue and K. Ishihara, *Langmuir*, 2014, **30**, 2745–2751.
24. R. Podila, R. Chen, P. C. Ke, J. M. Brown and A. M. Rao, *Appl. Phys. Lett.*, 2012, **101**(26), 263701.
25. D. G. Castner and B. D. Ratner, *Surf. Sci.*, 2002, **500**, 28–60.
26. C. A. Haynes and W. Norde, *Colloids Surf., B*, 1994, **2**, 517–566.
27. M. Mahmoudi, I. Lynch, M. R. Ejtehadi, M. P. Monopoli, F. B. Bombelli and S. Laurent, *Chem. Rev.*, 2011, **111**, 5610–5637.
28. G. Baier, C. Costa, A. Zeller, D. Baumann, C. Sayer, P. H. H. Araujo, V. Mailander, A. Musyanovych and K. Landfester, *Macromol. Biosci.*, 2011, **11**, 628–638.
29. A. Gessner, A. Lieske, B. R. Paulke and R. H. Muller, *J. Biomed. Mater. Res., Part A*, 2003, **65A**, 319–326.
30. S. Milani, F. B. Bombelli, A. S. Pitek, K. A. Dawson and J. Radler, *ACS Nano*, 2012, **6**, 2532–2541.
31. V. Y. Davydov, M. Elizalde Gonzalez, A. V. Kiselev and K. Lenda, *Chromatographia*, 1981, **14**, 13–18.
32. P. Jonkheijm, D. Weinrich, H. Schroder, C. M. Niemeyer and H. Waldmann, *Angew. Chem., Int. Ed.*, 2008, **47**, 9618–9647.
33. C. D. Walkey, J. B. Olsen, H. Guo, A. Emili and W. C. Chan, *J. Am. Chem. Soc.*, 2012, **134**, 2139–2147.
34. A. Gessner, A. Lieske, B. Paulke and R. Muller, *Eur. J. Pharm. Biopharm.*, 2002, **54**, 165–170.
35. Y. Hoshino, H. Koide, T. Urakami, H. Kanazawa, T. Kodama, N. Oku and K. J. Shea, *J. Am. Chem. Soc.*, 2010, **132**, 6644–6645.
36. P. H. Mutin, G. Guerrero and A. Vioux, *J. Mater. Chem.*, 2005, **15**, 3761–3768.
37. M. A. Neouze and U. Schubert, *Monatsh. Chem.*, 2008, **139**, 183–195.
38. N. Kossovsky, A. Gelman, E. E. Sponsler, H. J. Hnatyszyn, S. Rajguru, M. Torres, M. Pham, J. Crowder, J. Zemanovich, A. Chung and R. Shah, *Biomaterials*, 1994, **15**, 1201–1207.
39. J. B. Largueze, K. El Kirat and S. Morandat, *Colloids Surf., B*, 2010, **79**, 33–40.

40. T. D. Lazzara, D. Behn, T. T. Kliesch, A. Janshoff and C. Steinem, *J. Colloid Interface Sci.*, 2012, **366**, 57–63.

41. H. Y. Li, Y. H. Li, J. Jiao and H. M. Hu, *Nat. Nanotechnol.*, 2011, **6**, 645–650.

42. J. Li, J. Q. Wang, V. G. Gavalas, D. A. Atwood and L. G. Bachas, *Nano Lett.*, 2003, **3**, 55–58.

43. E. Mahon, A. Salvati, F. B. Bombelli, I. Lynch and K. A. Dawson, *J. Controlled Release*, 2012, **161**, 164–174.

44. A. Maquieira, E. M. Brun, M. Garces-Garcia and R. Puchades, *Anal. Chem.*, 2012, **84**, 9340–9348.

45. W. Norde, J. Buijs and H. Lyklema, in *Fundamentals of Interface and Colloid Science*, ed. J. Lyklema, Academic Press, London, 2005, vol. 5, pp. 1–59.

46. J. N. Israelachvili, *Intermolecular and Surface Forces*, Academic Press, Waltham, MA, 3rd edn, 2011.

47. G. V. Franks and Y. Gan, *J. Am. Ceram. Soc.*, 2007, **90**, 3373–3388.

48. H.-J. R. Butt, M. Kappl and K. Graf, *Physics and Chemistry of Interfaces*, Wiley-VCH, Weinheim, 2003.

49. N. A. Baker, D. Sept, S. Joseph, M. J. Holst and J. A. McCammon, *Proc. Natl. Acad. Sci. U. S. A.*, 2001, **98**, 10037–10041.

50. I. T. Lucas, S. Durand-Vidal, E. Dubois, J. Chevalet and P. Turq, *J. Phys. Chem. C*, 2007, **111**, 18568–18576.

51. C. J. van Oss, *Interfacial Forces in Aqueous Media*, Taylor & Francis, 2nd edn, 2006.

52. J. N. Israelachvili, *Intermolecular and Surface Forces*, Academic Press, Waltham, MA, 3rd edn, 2011.

53. I. Fenoglio, B. Fubini, E. M. Ghibaudi and F. Turci, *Adv. Drug Delivery Rev.*, 2011, **63**, 1186–1209.

54. R. A. Hartvig, M. van de Weert, J. Ostergaard, L. Jorgensen and H. Jensen, *Langmuir*, 2011, **27**, 2634–2643.

55. F. Meder, S. Kaur, L. Treccani and K. Rezwan, *Langmuir*, 2013, **29**, 12502–12510.

56. F. Meder, J. Wehling, A. Fink, B. Piel, K. Li, K. Frank, A. Rosenauer, L. Treccani, S. Koeppen, A. Dotzauer and K. Rezwan, *Biomaterials*, 2013, **34**, 4203–4213.

57. H. J. Butt and M. Kappl, *Surface and Interfacial Forces*, Wiley-VCH, Weinheim, 2010.

58. A. A. Vertegel, R. W. Siegel and J. S. Dordick, *Langmuir*, 2004, **20**, 6800–6807.

59. B. J. Peter, H. M. Kent, I. G. Mills, Y. Vallis, P. J. G. Butler, P. R. Evans and H. T. McMahon, *Science*, 2004, **303**, 495–499.

60. A. A. Shemetov, I. Nabiev and A. Sukhanova, *ACS Nano*, 2012, **6**, 4585–4602.

61. B. Mueller, M. Zacharias and K. Rezwan, *Adv. Eng. Mater.*, 2010, **12**, B53–B61.

62. K. Rezwan, A. R. Studart, J. Voros and L. J. Gauckler, *J. Phys. Chem. B*, 2005, **109**, 14469–14474.

63. Y. F. Yano, *J. Phys.: Condens. Matter*, 2012, **24**, 503101.
64. M. P. Monopoli, C. Aberg, A. Salvati and K. A. Dawson, *Nat. Nanotechnol.*, 2012, **7**, 779–786.
65. Z. Adamczyk, *Particles at Interfaces: Interactions, Deposition, Structure*, Elsevier, Amsterdam, 2006.
66. K. Rezwan, L. P. Meier and L. J. Gauckler, *Biomaterials*, 2005, **26**, 4351–4357.
67. K. Rezwan, L. P. Meier and L. J. Gauckler, *Langmuir*, 2005, **21**, 3493–3497.
68. L. Vroman and A. L. Adams, *Surf. Sci.*, 1969, **16**, 438–446.
69. A. E. Nel, L. Madler, D. Velegol, T. Xia, E. M. V. Hoek, P. Somasundaran, F. Klaessig, V. Castranova and M. Thompson, *Nat. Mater.*, 2009, **8**, 543–557.
70. P. Aggarwal, J. B. Hall, C. B. McLeland, M. A. Dobrovolskaia and S. E. McNeil, *Adv. Drug Delivery Rev.*, 2009, **61**, 428–437.
71. C. D. Walkey, J. B. Olsen, H. B. Guo, A. Emili and W. C. W. Chan, *J. Am. Chem. Soc.*, 2012, **134**, 2139–2147.
72. A. Gessner, R. Waicz, A. Lieske, B. R. Paulke, K. Mader and R. H. Muller, *Int. J. Pharm.*, 2000, **196**, 245–249.
73. J. L. Townson, Y. S. Lin, J. O. Agola, E. C. Carnes, H. S. Leong, J. D. Lewis, C. L. Haynes and C. J. Brinker, *J. Am. Chem. Soc.*, 2013, **135**, 16030–16033.
74. E. Duguet, M. Treguer-Delapierre and M. H. Delville, in *Nanoscience*, ed. P. Boisseau, P. Houdy and M. Lahmani, Springer, Berlin, 2009, pp. 129–170.
75. N. T. K. Thanh and L. A. W. Green, *Nano Today*, 2010, **5**, 213–230.
76. S. Bertazzo and K. Rezwan, *Langmuir*, 2010, **26**, 3364–3371.
77. G. T. Hermanson, *Bioconjugate Techniques*, Elsevier, Amsterdam, 3rd edn, 2013.
78. E. P. Plueddemann, *Silane Coupling Agents*, Plenum Press, New York, 1991.
79. E. Frohlich, *Int. J. Nanomed.*, 2012, **7**, 5577–5591.
80. E. Soto-Cantu, R. Cueto, J. Koch and P. S. Russo, *Langmuir*, 2012, **28**, 5562–5569.
81. A. Hawe, M. Sutter and W. Jiskoot, *Pharm. Res.*, 2008, **25**, 1487–1499.
82. J. N. Yewle, D. A. Puleo and L. G. Bachas, *Bioconjugate Chem.*, 2011, **22**, 2496–2506.
83. A. Ide, G. L. Drisko, N. Scales, V. Luca, C. H. Schiesser and R. A. Caruso, *Langmuir*, 2011, **27**, 12985–12995.
84. M. Iafisco, B. Palazzo, G. Falini, M. D. Foggia, S. Bonora, S. Nicolis, L. Casella and N. Roveri, *Langmuir*, 2008, **24**, 4924–4930.
85. K. Kandori, S. Oda and S. Tsuyama, *J. Phys. Chem. B*, 2008, **112**, 2542–2547.
86. K. Kandori, S. Tsuyama, H. Tanaka and T. Ishikawa, *Colloids Surf., B*, 2007, **58**, 98–104.
87. M. Schneider, F. Meder, A. Haiss, L. Treccani, K. Rezwan and K. Kummerer, *Chemosphere*, 2014, **99**, 96–101.
88. A. Clearfield, in *Progress in Inorganic Chemistry*, ed. K. D. Karlin, Wiley, New York, 1998, vol. 47, pp. 371–510.

89. F. Meder, PhD thesis, University of Bremen, 2013.
90. G. Alberti, U. Costantino, C. Dionigi, S. MurciaMascaros and R. Vivani, *Supramol. Chem.*, 1995, **6**, 29–40.
91. H. Tanaka, M. Futaoka, R. Hino, K. Kandori and T. Ishikawa, *J. Colloid Interface Sci.*, 2005, **283**, 609–612.
92. M. A. Colettiprevieno and A. Previero, *Anal. Biochem.*, 1989, **180**, 1–10.
93. J. M. Berg, J. L. Tymoczko and L. Stryer, *Biochemistry*, W. H. Freeman, Basingstoke, 7th edn, 2012.
94. T. J. Dolinsky, J. E. Nielsen, J. A. McCammon and N. A. Baker, *Nucleic Acids Res.*, 2004, **32**, W665–W667.
95. N. Brandes, P. B. Welzel, C. Werner and L. W. Kroh, *J. Colloid Interface Sci.*, 2006, **299**, 56–69.
96. W. Humphrey, A. Dalke and K. Schulten, *J. Mol. Graphics Modell.*, 1996, **14**, 33–38.
97. D. Walczyk, F. B. Bombelli, M. P. Monopoli, I. Lynch and K. A. Dawson, *J. Am. Chem. Soc.*, 2010, **132**, 5761–5768.
98. D. A. Puleo and R. Bizios, *Biological Interactions on Materials Surfaces: Understanding and Controlling Protein, Cell and Tissue Responses*, Springer, Berlin, 2009.
99. H. Noh, S. T. Yohe and E. A. Vogler, *Biomaterials*, 2008, **29**, 2033–2048.
100. A. G. Hovanessian and Z. L. Awdeh, *Eur. J. Biochem.*, 1976, **68**, 333–338.
101. Z. Wang, G. L. Ma, J. Zhang, W. F. Lin, F. Q. Ji, M. T. Bernards and S. F. Chen, *Langmuir*, 2014, **30**, 3764–3774.
102. A. E. Martell and R. M. Smith, *Critical Stability Constants*, Plenum Press, New York, 1977.
103. S. A. Ansari and Q. Husain, *Biotechnol. Adv.*, 2012, **30**, 512–523.
104. L. Jia, Y. S. Lu, J. W. Shao, X. J. Liang and Y. Xu, *Trends Biotechnol.*, 2013, **31**, 99–107.
105. D. Rothenstein, B. Claasen, B. Omiecienski, P. Lammel and J. Bill, *J. Am. Chem. Soc.*, 2012, **134**, 12547–12556.
106. Z. Adamczyk, M. Nattich, M. Wasilewska and M. Zaucha, *Adv. Colloid Interface Sci.*, 2011, **168**, 3–28.
107. A. Rosengren, E. Pavlovic, S. Oscarsson, A. Krajewski, A. Ravaglioli and A. Piancastelli, *Biomaterials*, 2002, **23**, 1237–1247.
108. K. M. Chen, Y. S. Xu, S. Rana, O. R. Miranda, P. L. Dubin, V. M. Rotello, L. H. Sun and X. H. Guo, *Biomacromolecules*, 2011, **12**, 2552–2561.
109. T. Douglas and M. Young, *Science*, 2006, **312**, 873–875.
110. M. Wegmann, B. Michen and T. Graule, *J. Eur. Ceram. Soc.*, 2008, **28**, 1603–1612.
111. B. Michen and T. Graule, *J. Appl. Microbiol.*, 2010, **109**, 388–397.
112. V. Nanduri, I. B. Sorokulova, A. M. Samoylov, A. L. Simonian, V. A. Petrenko and V. Vodyanoy, *Biosens. Bioelectron.*, 2007, **22**, 986–992.
113. T. T. N. Binh, A. E. K. Peh, C. Y. L. Chee, K. Fink, V. T. K. Chow, M. M. L. Ng and C. S. Toh, *Bioelectrochemistry*, 2012, **88**, 15–21.
114. C. E. Ashley, E. C. Carnes, G. K. Phillips, P. N. Durfee, M. D. Buley, C. A. Lino, D. P. Padilla, B. Phillips, M. B. Carter, C. L. Willman, C. J. Brinker,

J. D. Caldeira, B. Chackerian, W. Wharton and D. S. Peabody, *ACS Nano*, 2011, **5**, 5729–5745.

115. Y. J. Ma, R. J. M. Nolte and J. J. L. M. Cornelissen, *Adv. Drug Delivery Rev.*, 2012, **64**, 811–825.

116. J. E. Phillips, C. A. Gersbach and A. J. Garcia, *Biomaterials*, 2007, **28**, 211–229.

117. R. Cademartiri, H. Anany, I. Gross, R. Bhayani, M. Griffiths and M. A. Brook, *Biomaterials*, 2010, **31**, 1904–1910.

118. P. J. Yoo, K. T. Nam, J. F. Qi, S. K. Lee, J. Park, A. M. Belcher and P. T. Hammond, *Nat. Mater.*, 2006, **5**, 234–240.

119. L. Gutierrez, X. Li, J. Wang, G. Nangmenyi, J. Economy, T. B. Kuhlenschmidt, M. S. Kuhlenschmidt and T. H. Nguyen, *Water Res.*, 2009, **43**, 5198–5208.

120. M. Wegmann, B. Michen, T. Luxbacher, J. Fritsch and T. Graule, *Water Res.*, 2008, **42**, 1726–1734.

121. R. Attinti, J. Wei, K. Kniel, J. T. Sims and Y. Jin, *Environ. Sci. Technol.*, 2010, **44**, 2426–2432.

122. K. S. Zerda, C. P. Gerba, K. C. Hou and S. M. Goyal, *Appl. Environ. Microb.*, 1985, **49**, 91–95.

123. Y. Jin and M. V. Yates, *Virus Behavior in Saturated and Unsaturated Subsurface Media*, AWWA Research Foundation and American Water Works Association, Denver, CO, 2002.

124. T. C. John, Jr, *J. Am. Water Works Assoc.*, 1969, **61**, 52–56.

125. P. J. Majewski, *Sep. Purif. Technol.*, 2007, **57**, 283–288.

126. Z. L. Chen, F. C. Hsu, D. Battigelli and H. C. Chang, *Anal. Chim. Acta*, 2006, **569**, 76–82.

127. P. Majewski and A. Keegan, *Appl. Surf. Sci.*, 2012, **258**, 2454–2458.

128. *CRC Handbook of Chemistry and Physics*, ed. W. M. Haynes, CRC Press, Boca Raton, FL, 92nd edn, 2011.

129. C. M. Schaldach, W. L. Bourcier, H. F. Shaw, B. E. Viani and W. D. Wilson, *J. Colloid Interface Sci.*, 2006, **294**, 1–10.

130. C. Dika, J. F. L. Duval, H. M. Ly-Chatain, C. Merlin and C. Gantzer, *Appl. Environ. Microb.*, 2011, 77, 4939–4948.

Development of Hybridized Materials Using Tobacco Mosaic Virus as Building Block

LIN LU[a] AND QIAN WANG*[a]

[a]Department of Chemistry and Biochemistry, University of South Carolina, 631 Sumter Street, Columbia, SC 29208, USA
*E-mail: wang263@mailbox.sc.edu

8.1   Introduction

Enzyme complexes, ferritins, heat shock proteins, viruses and virus-like particles are typical natural bionanoparticles (BNPs). A characteristic feature of BNPs is that multiple proteins are organized through non-covalent interactions to form highly organized nanostructures. The unique advantages of BNPs for materials development rely on their molecular precision and intrinsic genetic programmability, where a specific codon modification translates to the corresponding change in amino acid residue with unsurpassed fidelity, consistency and spatial resolution. In addition, these particles exhibit distinctive structural symmetries and yet are stable enough to be extensively modified by bioconjugation chemistry, which allows for the presentation of functional units in a spatially selective manner. Owing to these valuable properties, BNPs have recently been employed as templates or building blocks to fabricate novel functional materials for a variety of applications.[1–4]

RSC Smart Materials No. 16
Bio-Synthetic Hybrid Materials and Bionanoparticles: A Biological Chemical Approach Towards Material Science
Edited by Alexander Böker and Patrick van Rijn
© The Royal Society of Chemistry 2015
Published by the Royal Society of Chemistry, www.rsc.org

Typical examples of BNPs include ferritin, bacteriophages MS2 and M13 and some plant viruses, such as spherical cowpea mosaic virus (CPMV), cowpea chlorotic mottle virus (CCMV), turnip yellow mosaic virus (TYMV) and rod-shaped tobacco mosaic virus (TMV). These particles have good solubility and stability in aqueous solution and are highly uniform scaffolds with a size distribution in the nanometer range. Therefore, in the past decade, these particles have been employed to construct drug/gene delivery vehicles,[5–8] vaccine carriers,[9,10] nanowires[11,12] and composite materials.[13–16] The aim of this chapter is not to summarize comprehensively every single aspect of BNP-based hybrid materials, but to use a model plant virus, TMV, to highlight some of the recent studies in this area. For readers who are interested in obtaining more knowledge about hybrid bionanoparticles, other reviews are recommended.[1–4,17–23]

8.2   Tobacco Mosaic Virus (TMV)

TMV was the first virus to be isolated and is one of the most extensively studied plant viruses so far. After its first isolation by Stanley in 1935,[24] scientists from different fields, including biochemistry, molecular biology and biological physics, have made important contributions to the study of its configuration, structure, chemical composition and assembly, *etc.*[25,26] In 1982, the complete genome sequence of TMV was determined.[27] Later, the structure of TMV was acquired by X-ray fiber diffraction methods at a resolution of 2.9 Å,[28] and the coat protein disk aggregate crystal structure was also available at a resolution of 2.4 Å.[29] The native TMV particle has an exquisitely organized architecture. As a rod-shaped virus, with length 300 nm, outer diameter 18 nm and inner channel width 4 nm, TMV consists of 2130 identical coat protein subunits arranged helically around a right-handed positive-sense RNA strand that is 6395 nucleotides long (Figure 8.1).[25]

The *in vitro* self-assembly of TMV coat proteins, either with or without the genomic RNA as a template, has been studied extensively.[30] The process is primarily driven by hydrophobic interactions, which can be influenced by temperature, ionic strength and pH.[30] Under different conditions, the assembly of coat protein will lead to different structures, *i.e.* helical rods of various lengths occurring at acidic pH, a "two-layer" disk (20S disk) consisting of 34 monomers at neutral pH and protein A, a dynamic equilibrium between monomers, trimers and pentamers, at basic pH.[30]

Although it was traditionally studied for pathology purposes and the development of antiviral agents, TMV has recently emerged as a favorite BNP for materials development. Increasing numbers of studies have utilized TMV as the building block for the fabrication of hybrid materials owing to several inherent merits of TMV. First, the monodisperse TMV particle has a unique shape and size and a highly ordered architecture at near-atomic level. Second, TMV is fairly stable. For example, it can be treated with more than 30% of organic solvents at temperatures up to 90 °C and in a pH range of 3.5–9 without compromising its structural integrity.[31,32] Third, both the interior and exterior surfaces of TMV can be chemically and genetically modified

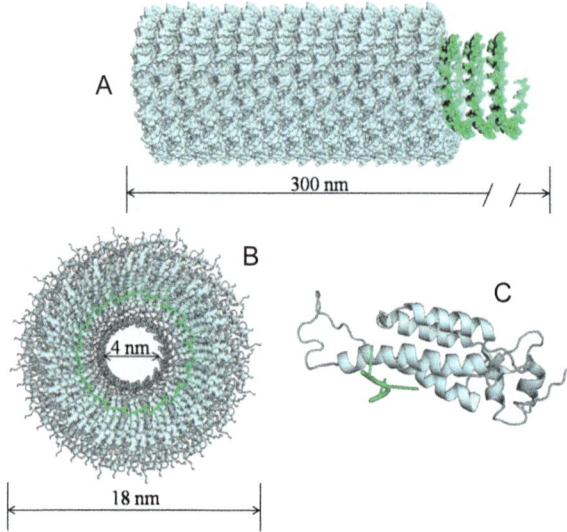

Figure 8.1 (A) Overview and (B) top view of a native TMV particle. (C) Ribbon representation of a TMV coat protein monomer. The gray color is TMV coat protein and the green color is the RNA genome. Models were generated using PyMol (http:\\www.pymol.org) with coordinates obtained from the RCSB Protein Data Bank (http:\\www.pdb.org).

with functional groups to afford new properties. Finally, it is non-pathogenic to humans and animals and it is economically available because large quantities of viruses can be readily purified from the infected tobacco leaves through classical virus purification protocols.

8.3   Surface Modification of TMV by Bioconjugation

8.3.1   Modification on the External Surface at Tyr139

Structural analysis revealed that there are no exposed lysine, cysteine or carboxylate residues on the external surface of native TMV. Francis and co-workers found that the phenol group of Tyr139, which is located on the exterior surface, could be readily derivatized with an electrophilic diazonium salt reagent.[33] Using this reaction, more than 95% of the coat proteins could be modified with ketone motifs, which could be further tailored with different alkoxyamines under mild reaction conditions through the formation of an oxime linkage. For example, biotin and poly(ethylene glycol) (PEG) were attached to the exterior surface of TMV with high efficiency without breaking the structural integrity of the TMV particles.[33]

However, although the diazonium coupling reaction is efficient in targeting the tyrosine residues, it has some apparent limitations, *e.g.* acid-labile groups containing electron-enriched diazonium salts are not good precursors for this reaction. Wang and co-workers tried to combine the coupling reaction of diazonium salts with the Cu(I)-catalyzed azide–alkyne cycload-dition (CuAAC) reaction,[34] typical "click" chemistry.[35,36] As the azide and alkyne groups are absent from natural biomolecules, the CuAAC reaction has been widely utilized in biological systems to modify biomacromolecules, viruses and cells under mild reaction conditions with high efficiency and selectivity.[37] To target the Tyr139 of TMV, Wang and co-workers first treated TMV with a diazonium salt to anchor an alkynyl group quantitatively,[34] and this group was subsequently reacted with different kinds of azide-bearing reagents ranging from small molecules to peptides and polymers under catalysis by $CuSO_4$–sodium ascorbate. This two-step strategy was found to be a highly efficient way to program functionalities on the external surface of TMV without special requirements regarding their electronic properties. For example, CuAAC-modified TMV with RGD peptides at the Tyr139 residue could serve as a multivalent scaffold to promote cell binding with NIH-3T3 fibroblasts.[34] Furthermore, when it is modified with Tn antigen (GalNAc-α-O-Ser/Thr) at Tyr139 or at an inserted cysteine at the N-terminus, TMV capsid can be a promising carrier of weakly immunogenic tumor-associated carbo-hydrate antigens (TACAs) for antitumor vaccine development.[38]

8.3.2   Modification on the Internal Surface at Glu97 and Glu106

The carboxyl residues from Glu97 and Glu106 are exposed on the internal surface of the TMV particle. Francis and co-workers reported the attachment of biotin, chromophores and crown ethers on the internal surface of TMV particles through a carbodiimide coupling reaction on Glu97 and Glu106 without destructing the rod-like structure of TMV.[33] Mass spectrometric analysis of trypsin-digested fragments supported the regional selectivity of the coupling reaction. Although both residues are exposed similarly on the internal surface of the capsid, Glu97 was found to be the primary site of con-jugation. In addition, modification of exterior aspartic acids and the carboxyl terminus was not observed. Using internal modification, wild-type (wt) TMV particles could carry different cargos or probes internally for materials and biomedical applications.[39]

8.3.3   Genetic Modification Followed by Subsequent Bioconjugation Reactions

Based on the gene sequence and detailed structural information of TMV, it is possible to identify the accurate position of functional residues, either solvent accessible or essential to the formation of the capsid. Therefore, researchers could readily predict the potential locations of genetic manip-ulations that are not detrimental to overall virus structures and biological functions. For example, based on the genome information, the insertion of functional polypeptides at the N- or C-terminus or site-directed mutation at certain non-essential locations have been achieved *via* routine molecular cloning approaches.[13,40–43] As TMV protein capsid is composed of multiple copies of identical subunits, once a single genetically engineered coat protein has been obtained, the entire capsid would be modified symmetrically. Therefore, by combining chemical conjugation strategies and genetic methods, TMV can be diversely modified and endowed with more properties and functionalities.

As it is known that there is no cysteine exposed on either the inner or outer surface of coat protein, Francis and co-workers created an S123C mutant and utilized it for the construction of light-harvesting systems through the *in vitro* self-assembly of the modified coat proteins.[44] The mutation of serine to cysteine introduced a thiol group, which reacted with Oregon Green 488 maleimide and *N,N,N',N'*-tetramethylrhodamine maleimide, respectively, as the primary and intermediate donor, or with Alexa Fluor 594 maleimide as the acceptor. By controlling the pH and ionic strength of buffer solution, the modified proteins could self-assemble into long helical fibers that are capable of positioning the chromophores for efficient energy transfer.[44] Majima's group constructed another effective energy transfer and light harvesting system.[45] In order to incorporate porphyrin moieties in TMV coat proteins, double mutations were applied to change the native Cys27 to Ala and Asn127 to Cys. Zinc-coordinated porphyrin (ZnP) donor and free-base porphyrin (FbP) acceptor were site selectively tagged with cysteine. Under appropriate conditions, those coating proteins were assembled and the light energy transfer efficiency was investigated.[45] In addition, Culver and co-workers generated a special TMV cysteine mutant that can propagate in tobacco plants with high efficiency.[46,47] This mutant has been used by many groups in materials applications, which will be discussed in later sections.

Due to the lack of surface reactive amino groups, Demir and Stowell carried out a single site mutation (T158K) at the C-terminus to introduce lysine residues into native TMV.[48] This mutant provides a versatile template for the NHS coupling reaction to afford chemo-specific nanotubular materials. In another study, Wang's group prepared a TMV–EPMK mutant that incorporates a reactive lysine residue and can be employed to attach a variety of functional groups quantitatively.[21] All the surface modifications of TMV protein through bioconjugation reactions mentioned above are summarized in Figure 8.2.

8.4 Fabrication of Hybridized Materials Using TMV as Template and Building Block

8.4.1 Templated Synthesis of Hybrid Materials

TMV has been widely used to template the synthesis of inorganic nanowires and nanofibers. The isoelectric point (p*I*) of native TMV is about 3.4. At pH 2.5, surface arginines and lysines are positively charged and they are able to bind with anionic silicate species from hydrolysis of tetraethoxysilane (TEOS)[49] and produce a silica shell to cover the TMV particles. In addition,

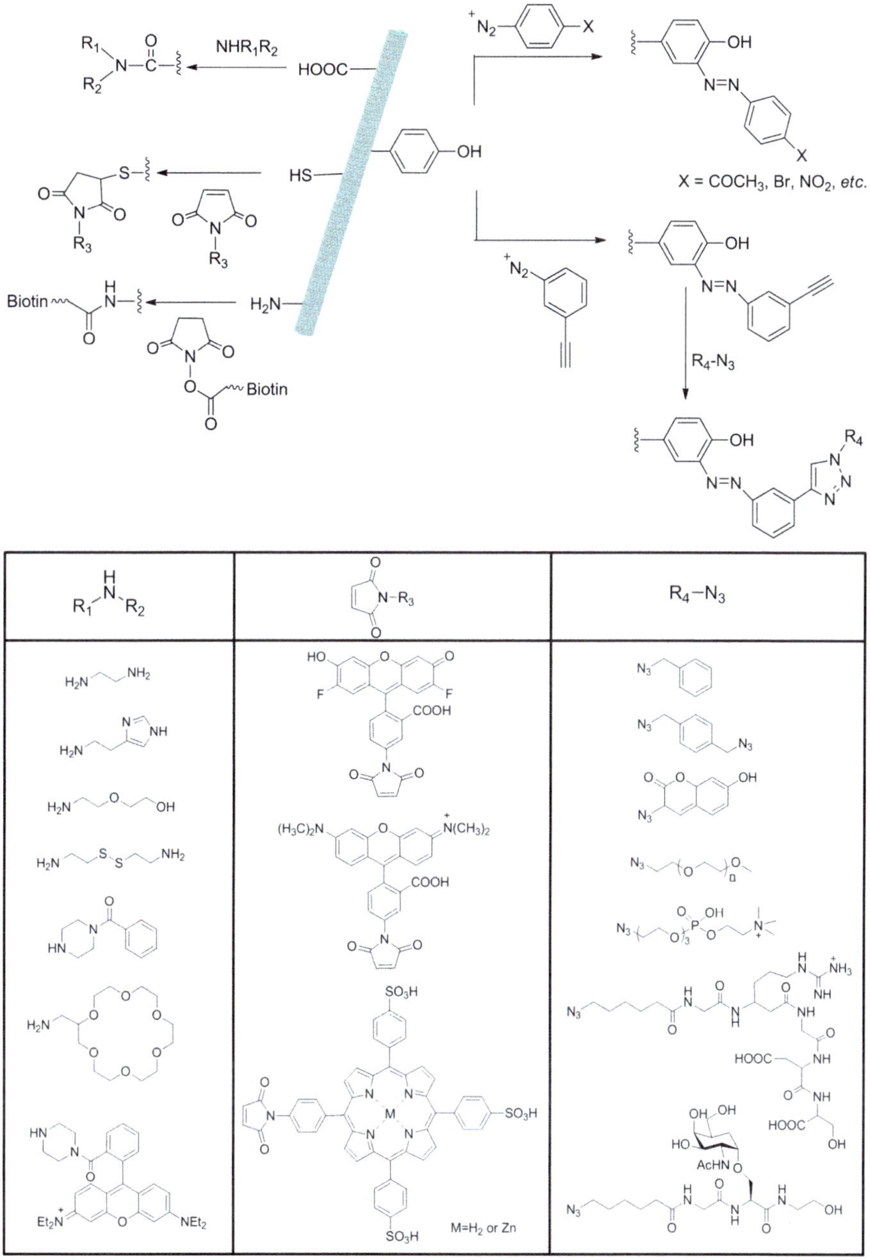

Figure 8.2   Bioconjugation on TMV protein capsid.[33,37,44]

the protonation can reduce the repulsion between Glu50 and Asp77,[50,51] hence long silicate-coated fibers can be synthesized mediated by the head-to-tail assembly of TMV particles.[49] Under neutral or alkaline pH conditions, metal ions such as Cd^{2+}, Pb^{2+} and Fe^{2+}/Fe^{3+} can bind with glutamic and aspartic acids on the external surface of TMV and consequently CdS-, PbS- or iron oxide-coated TMV nanotubes can be easily achieved in the presence of H_2S or NaOH.[49] When TMV was mixed with $[PtCl_6]^{2-}$ or $[AuCl_4]^-$, Pt- or Au-coated TMV nanotubes could be synthesized by treatment with a reducing agent.[52] Similarly, a silver salt could be specifically deposited on the inner surface and linear silver rods could be produced and trapped in the inner channel of TMV after a photochemical reduction.[52] Moreover, the insertion of two cysteine residues on the N-terminus of the coat protein greatly enhanced the affinities of the TMV template with Au, Ag, Pd and Pt clusters.[53,54]

CoPt, which has been applied in data storage, is one of the most attractive ferromagnetic bimetallic alloys.[55] Yamashita's group reported the synthesis of bimetallic alloy nanowires within the inner cavity of TMV using Co^{2+} and Pt^{2+} as starting materials.[56] Ultrasonication treatment was found to be essential for the entry of metal ions into the narrow internal channels of TMV particles. In addition, mutant TMV particles (*e.g.* S101K and E106K mutants) have been created to increase the nucleation sites, which consequently could promote the formation of CoPt nanowires.[57] The resulting CoPt and FePt alloy nanowires had uniform diameters and could potentially be used in sensing, imaging and high-density data storage.

Kern's group employed an electroless deposition method to produce high aspect ratio nanowires using the inner cavity of TMV as a template.[58] Dissolved in pure water, TMV particles were activated by treatment with $[PtCl_4]^{2-}$ and $[PdCl_4]^{2-}$ at pH 5. Although either $[PtCl_4]^{2-}$ or $[PdCl_4]^{2-}$ could bind to the negatively charged internal surface of TMV through electrostatic interactions, they could diffuse into the channel and form complexes with nitrogen-based ligands, a process which was independent of solution pH.[59] Upon activation, the mixture was dispersed in an electroless deposition bath containing dimethylamine borane (DMAB) and then Co(II) or Ni(II) was introduced at pH 6–8. Nickel or cobalt clusters would fill in the inner cavity of TMV, forming nanowires ~3 nm in diameter and several hundred nanometers in length, without destroying the integrity of the TMV shells. Employing the same strategy, Cu and CoPt can also be incorporated into the central channel of TMV.[60,61] Further studies revealed that the metal clusters can be selectively bonded both to the central channel and to the exterior surface of the TMV hollow tube, and the formation of nanowires depends critically on several factors:[32] the activation of the biomolecules, the choice of reductant, the deposition media and the ions to be deposited. For example, Ni and Co clusters could exclusively deposit on the outer surface when TMV in phosphate buffer was activated by Pd or Pt and gold from ascorbic acid bath had no spatial selectivity and could bind to both the exterior surface and the interior channel of the TMV tube.[32]

Atomic layer deposition (ALD), developed in the 1970s, has also been utilized to deposit Al_2O_3 or TiO_2 on the TMV surface.[62] In this approach, TMV solution was first dried on a solid substrate and subsequently exposed to gas phase of tetraisopropyltitanium or tetramethylaluminum in the ALD chamber. After chemisorption or physisorption of precursor molecules on the substrate, excess molecules in the gas phase was purged with Ar gas. The adsorbed molecules were hydrolyzed to form a layer of metal oxide on

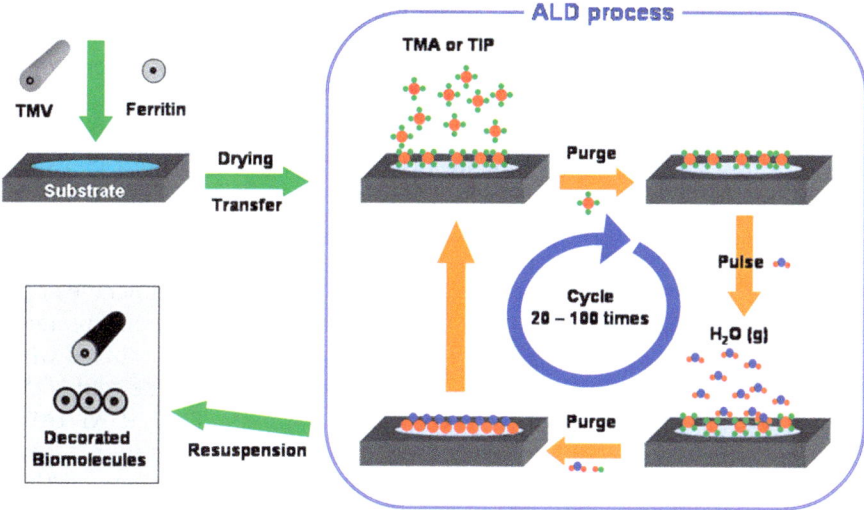

Figure 8.3 Overview of the ALD experimental process for the coating of TMV particles.[62] Reprinted with permission from ref. 62. Copyright 2006 American Chemical Society.

exposing the substrate to water molecules. This cycle could be repeated and the exposure time of the substrate to precursor molecules could be well controlled, thus making ALD an effective approach to manage the film thickness on the TMV surface (Figure 8.3). It should be mentioned with this method is that the metal oxide can deposit inside the hollow channel only when the virion cavity is accessible to the precursor molecules in gas phase.

Royston and co-workers introduced a silica coating strategy that could improve the stability of the TMV template and also its affinity for metal ions.[63,64] Enhanced stability of the TMV template was achieved by using an aniline polymerization step prior to silica coating.[65,66] The thick shell silica-coated TMV provides a highly stable and robust platform for further deposition of Pt, Pd, Au and Ag to generate multilayering of materials. This kind of multilayer material is potentially useful for designing conducting and optical devices.

8.4.2 Hybrid System Using TMV as Building Block

Self-assembly can be described as a process in which organized structures are formed from the spontaneous organization of the initial disordered nanoparticles or discrete components through local non-covalent interactions.[67,68] TMV particles are monodisperse and have a propensity to self-assemble into longer fibers in acidic media. In this section, we discuss the self-assembly of TMV into long nanofibers, nanofilms, spherical-like structures and microarrays.

8.4.2.1   Head-to-Tail Self-Assembly of TMV

In acidic conditions, wt TMV will automatically stack into long helical rods, attributed to the minimized repulsion of protonated carboxylic residues at the interface.[30] However, this complementary interaction is unstable and the stacking rods formed cannot be further employed. Niu and co-workers systematically studied the assembly process and devised a facile way to synthesize long nanofibers *via* the self-assembly of TMV assisted by aniline polymerization at nearly neutral pH.[65,69] Theoretically, at this pH, above the isoelectric point of TMV (~3.4), the amine groups in aniline molecules can accumulate on the negatively charged TMV surface through electrostatic attraction and hydrogen bonding. When aniline and TMV were mixed with ammonium persulfate (APS) as initiator, polymerization of aniline could take place on the surface of TMV. The synthesized polyaniline–TMV (PANI–TMV) fibers are long and tube-like, homogeneous in diameter, with high aspect ratio and excellent processability. Transmission electron microscope (TEM) images revealed that the pH of the reaction solution is a factor that can affect the morphologies of the composite fibers. At acidic pH, long fibers formed initially, but after 4 h of reaction, most of the virus particles had formed bundled structures, probably due to the accumulation of polyaniline on the outer surface of TMV leading to strong hydrophobic interactions between fibers. Further studies revealed that two essential elements contribute to the formation of long fibers: first, accumulation and polymerization of monomers on the TMV surface; and second, prolongation and stabilization of TMV helices.[70]

The 1D head-to-tail assembled long PANI–TMV fibers were not conductive, so they were subsequently treated with poly(sulfonated styrene) (PSS) (Figure 8.4).[66] PSS, with highly negative polyanions, can not only enhance the stability and solubility of the fibers, but also increase their conductivity. Scanning spreading resistance microscopy (SSRM) combined with atomic force microscopy (AFM) confirmed the conductivity of poly(sulfonated styrene)–polyaniline–TMV (PSS–PANI–TMV) nanofibers.[66]

8.4.2.2   Self-Assembly of TMV on a Solid Surface

Many practical applications require control of the arrangement of nanoparticles on a solid support over a broad range. The controlled alignment of virus particles, *e.g.* CCMV, CPMV, TYMV and bacteriophage M13, on various solid surfaces has been studied extensively.[71-74] For TMV, great progress has also been made in the preparation of ordered thin films.

For example, TMV can be directly coated on metal and hydroxyl-containing surfaces, such as gold, mica, glass and silicon wafers.[75] On a glass surface, divalent cations, such as Cd^{2+}, Pb^{2+}, Zn^{2+}, Ni^{2+} and Cu^{2+}, could promote the precipitation of TMV from solution and, after drying of the precipitate, long strips several centimeters in length separated by parallel cracks were observed.[76] In a chitosan-modified capillary tube, fluid flow

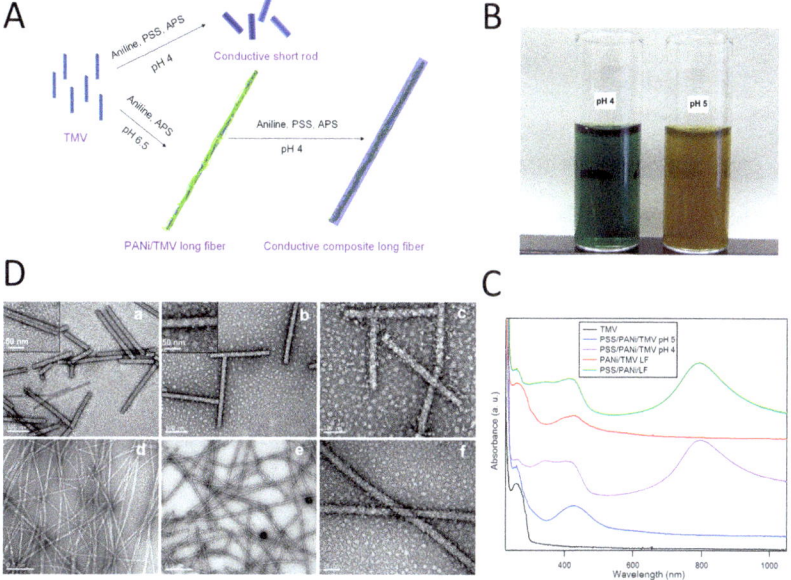

Figure 8.4   (A) Scheme of the preparation of conductive PANI–TMV; (B) PSS–PANI–TMV composite fiber in aqueous solution; (C) TEM images of (a) wt TMV; (b) and (c) PSS–PANI–TMV nanowires from the pH 5 and pH 4 reaction, respectively; (d) PANI–TMV; (e) PSS–PANI–TMV; (f) enlarged image of PSS–PANI–TMV; (D) UV/Vis absorbance of native TMV, PANI–TMV and PSS–PANI–TMV long fiber (LF).[66] Adapted with permission from ref. 66. Copyright 2007 American Chemical Society.

assembly of negatively charged TMV can generate a large-scale alignment that has potential to be utilized in tissue engineering and electronics fields.[77]

Microcontact printing (μCP) is a widely utilized technique to achieve patterned surfaces and its general principle is to transfer a solution of desired substance (ink) onto a target surface utilizing a micro-structured elastomeric stamp.[78–82] Wrinkles on a polydimethlysiloxane (PDMS) substrate could be generated *via* stretching and plasma treatment. The wavelength and amplitude of the wrinkles can be controlled with good accuracy by manipulation of the applied strain and the oxide layer. Horn and co-workers demonstrated that large areas with highly uniform TMV alignment can be created in the grooves of wrinkled PDMS substrates using a simple spin-coating procedure [Figure 8.5(A)].[82,83] Also, the spinning speed and TMV concentration have an influence on the formation of defined TMV arrays. With alignment of TMV particles in the grooves of wrinkled PDMS, striped arrays were then transferred onto a flat silicon wafer through μCP. Highly ordered TMV lines, several micrometers in length and only a few nanometers in width, could be obtained under ideal working conditions.[82]

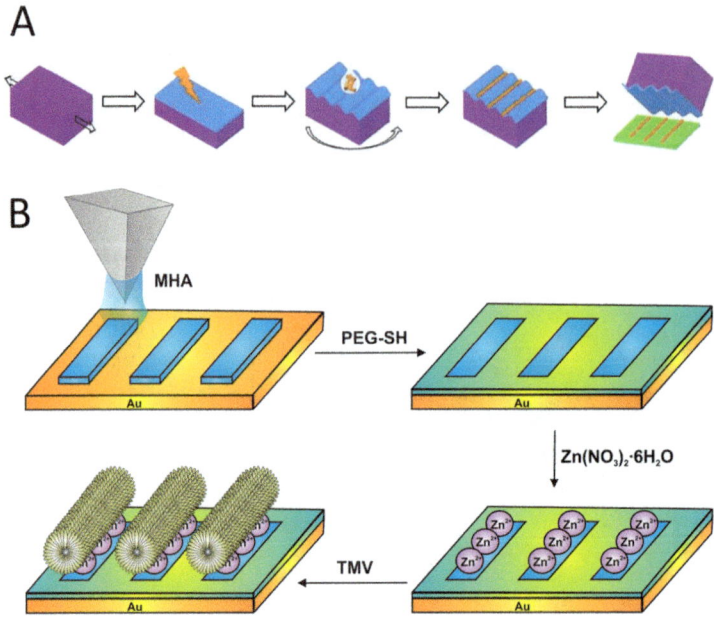

Figure 8.5   (A) Scheme of TMV alignment through microcontact printing: fabrication of a wrinkled substrate on a PDMS substrate *via* stretching and
plasma treatment, spin coating of TMV solution onto the substrate,
TMV particle alignment in the grooves and the final printing step.[82]
(B) Scheme of TMV alignment through dip-pen nanolithography:
first, the MHA template was generated by DPN, followed by treatment
of PEG-SH, affording a self-assembled monolayer; after exposure to
Zn(NO$_3$)$_2$ solution, a single layer of metal ions was deposited; TMV particles were anchored *via* coordination interaction.[84] Adapted from ref.
82 and 84 with permission from Wiley-VCH, Copyright 2010 and 2005,
respectively.

Mirkin's group described the combination of dip-pen nanolithography
(DPN) with coordination chemistry to control the positioning and orientation of TMV particles in the context of large arrays [Figure 8.5(B)].[84]
Initially, chemical templates of 16-thiohexadecanoic acid (MHA) were
generated on a gold thin film by DPN. By immersing the substrate in an
alkanethiol solution, the templates were passivated with a monolayer of
11-thioundecylpenta(ethylene glycol) (PEG-SH). This process can reduce
the opportunity for non-specific attachment of TMV to the spacing region.
The carboxylic acid groups from MHA can bind with Zn^{2+} when exposed to
Zn(NO$_3$)$_2$ solution. With the elimination of uncoordinated ions, TMV solution was brought into contact with a metalized substrate. At this stage, the
metal ions can help to bridge the MHA-modified patterned surface with
the carboxylate-rich surface of TMV. As a result, each individual MHA substrate can bind to one TMV particle with predetermined orientation in an
extended array.

8.4.2.3   Self-Assembly of TMV at an Interface

A liquid/liquid interface provides a confined platform for the preparation of thin films of metals or semiconductors. Virus particles assembled at a liquid/liquid interface based on a Pickering emulsion to form 2D crystal structures have been systematically studied.[72,85,86] In Pickering emulsions, the colloidal particles are positioned at the interface to decrease the interfacial energy and to stabilize emulsions.[72,87–89] In an optimized system, with a reduced particle exchange rate and increased adsorption time, highly ordered arrays can be produced at the interface.[85]

Russell *et al.* systematically studied the self-assembly of anisotropic TMV particles at an oil/water interface using tensiometry and small-angle X-ray and small angle neutron scattering.[73] They found that the orientation of TMV rods at a perfluorodecalin/water interface depends on the initial TMV concentration in the aqueous phase. At low TMV concentration, in order to mediate the greatest interfacial interactions per particle, TMV rods are oriented parallel to the interface. In comparison, at a high TMV concentration, increased dipole–dipole repulsion drives TMV particles to orient normal to the oil/water interface. Additionally, the ionic strength of the bulk solution is another important factor that dominates TMV assembly patterns. The repulsive interactions between TMV particles at the interface can be effectively changed by varying the ionic strength of the buffer solution, which will thus lead to different TMV assembly behavior.

In another study, Yang *et al.* modified a silicon wafer with a positively charged fluid lipid monolayer and then assembled TMV particles on it.[90] The charge density of the lipid monolayer is a key factor in controlling TMV assemblies. An ordered TMV assembly was observed by X-ray and AFM analysis, attributed to improved lateral mobility of the confined TMV particles on the modified plane compared with those on bare silicon substrates. However, the dominance of lipid charge diminished when the solution concentration of Ca^{2+} increased. Ca^{2+} ions in TMV solution reduced the average inter-particle spacing between TMV rods on the lipid monolayer from 42 to 20 nm. This behavior can probably be explained by the reduced repulsive inter-particle interactions in the presence of the polyvalent counter ion Ca^{2+} and the membrane-assisted attractions.

8.4.2.4   Assembly of TMV at a Three-Phase Contact Line

8.4.2.4.1   Convective Assembly.   The convective assembly process was first developed by Denkov, Velev and co-workers for the controlled deposition of micro- and nanoparticle coatings.[91–93] This is a shear-driven assembly process during which nearly monodisperse colloidal particles deposited in an orderly fashion on substrates to form thin films under the driving force from solvent evaporation [Figure 8.6(A)].[94] Through convective assembly, large-scale uniform silica, gold and ferritin films were formed successively.[95] Kuncicky *et al.* were the first to obtain large-scale ordered TMV assemblies by simply pulling a TMV-containing meniscus over the substrate.[94] They found that all the anisotropic TMV particles were arranged parallel to the receding direction of

the meniscus withdrawal and that the withdrawal speed and substrate wetta-
bility were two key parameters for controlling the film structure. On a hydrophilic substrate, a high withdrawal speed produced random and non-aligned
web-like fibers, whereas on a hydrophobic substrate, a high withdrawal
speed generated thin, highly branched fibers. The fibers formed on a partially hydrophobic substrate could be further processed and converted into
conducting nanowires. This is the first example of the formation of a 2D virus
coating on a multiple centimeters length scale with a facile process. Later,
Wargacki *et al.* systematically studied the impact of anisotropic particles and
substrate chemistry on the TMV convective assembly process.[96] The results
revealed that, compared with hydrophobic substrates, the hydrophilic substrates generated much flatter and more uniform films, and a high TMV concentration and a low assembly speed led to well-organized viral layers.

8.4.2.4.2 Evaporation Under a Confined Environment. The receding
meniscus of a TMV suspension was created in glass capillary tubes by Lin *et al.*
[Figure 8.6(B)].[97] The air/solution/substrate interface was driven towards
the middle part from both ends when the confined solution was evaporated
gradually under ambient conditions. TMV concentration, salt effects and the
properties of the capillary tube inner surface influenced the formation of
TMV patterns. For example, in the presence of salt, monolayer stripes with
same thickness formed and oriented parallel to the contact line to maximize
the interfacial coverage per particle, and when the concentration increased
but was lower than 0.5 mg mL^{-1}, multilayer stripes formed but they were still

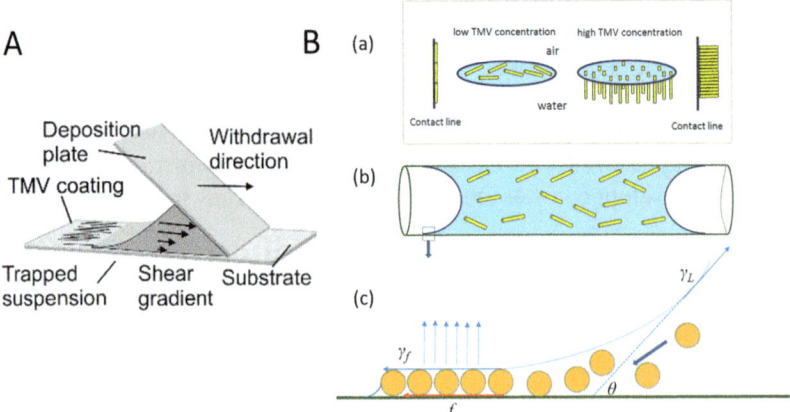

Figure 8.6 (A) Schematic representation of convective assembly.[94] (B) (a) Self-
assembly of TMV particles at the air/liquid interface and the contact
line: deposition parallel to the contact line at low concentration (left)
and perpendicular to the contact line at high concentration (right). (b)
Schematic illustration of TMV solution in a capillary tube and (c) its thin
meniscus formation at the contact line.[97] Adapted from ref. 94 and 97
with permission from Wiley-VCH, Copyright 2006 and 2010, respectively.

perpendicular to the long axis of the glass tube. On the other hand, when salt was removed, low concentrations of TMV particles randomly deposited on the substrate to form an irregular, loose multilayer structure as the repulsion between the negatively charged TMV particles could not be screened in pure water; accompanied by the increasing concentration, the dipole–dipole repulsion increased and finally drove the particle orientation perpendicular to the air/liquid interface. The patterned TMV structures in the capillary tube can direct smooth muscle cells which orient perpendicular to the long axis of the tube, and this orientation is essential for the function of blood vessels.

To create large-scale ordered structures on flat surfaces, which might have more potential applications, Lin *et al.* replaced the glass capillary tube with two horizontally stacked glass slides separated slightly from each other and developed a facile approach to form 2D TMV patterns on large surfaces.[98] In the new confined space, long-range ordered stripes of TMV patterns can be created and their height and width can be controlled *via* the TMV concentration. The large-scale patterned surfaces were successfully applied to regulate surface hydrophobicity and direct the growth of bone marrow stromal cells.[98]

8.4.2.5 Layer-by-Layer Assembly of TMV

Layer-by-layer (LBL) assembly is an extremely simple technique first developed by Decher and co-workers to construct highly tunable and ultra-thin films *via* electrostatic interactions on solid supports.[99] The films are generated by alternately depositing cationic and anionic materials with wash steps during the process. Functional molecules such as colloids, polymers, small organic molecules, inorganic clusters, viruses, proteins and nucleic acids can be incorporated into thin-film fabrication.[100–107] It has been demonstrated that viruses, such as CPMV and TYMV, can maintain their integrity upon LBL operation. By controlling the virus density and surface coverage, it is possible to modulate cell attachment, cell adhesion and cell spreading.[108,109] Steinmetz *et al.* studied the difference in LBL assembly between sphere-like CPMV and rod-shaped TMV.[110] The results showed that both of these viruses could be immobilized on polyelectrolyte assemblies, but the mechanisms were different because of their different shapes. CPMV could be incorporated into the architecture, whereas TMV particles preferred to float on the top owing to the inter-diffusion of polyelectrolytes. The same assembly behavior with TMV was also observed from M13.[111]

8.4.2.6 Three-Dimensional (3D) Assembly

Based on the assembly principle of the Pickering emulsion method, non-covalent interactions between TMV nanoparticles and poly(4-vinylpyridine) (P4VP) led to the formation of TMV–P4VP core–shell structures.[112] P4VP is a polymer that has shown great potential in self-assembly with other nanoparticles or polymers. TEM and field emission scanning electron microscopy (FESEM) were applied to analyze and characterize the formation of the TMV–P4VP core–shell particles. As a general approach, P4VP combined with 3%

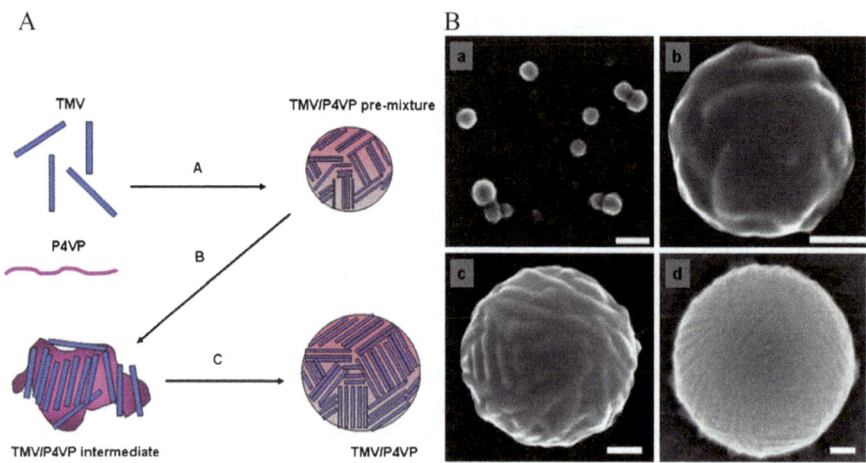

Figure 8.7   (A) Illustration of the assembly process of TMV and P4VP.[112] (B) Representative FESEM images of TMV–P4VP core–shell particles in pure water at (a) low magnification and (b–d) high magnification.[112] Adapted from ref. 112.

dimethylformamide (DMF) was mixed with aqueous TMV solution and aggregated into TMV-covered particles. After dialysis in water for 48 h, a TMV–P4VP intermediate formed. In this step, the amount of DMF declined and this led to a decrease in P4VP solubility and phase separation, which caused more TMV particles to move to surface to reduce the interfacial energy between P4VP and water. Finally, TMV particles were distorted to fit the curvature of the spherical particles and transformed into TMV–P4VP core–shell structures. Further studies confirmed that the coverage of TMV was only on the surface of P4VP balls (Figure 8.7). If the polymer is polystyrene (PS), TMV can also coat and even form liquid crystalline-like structures on the surface of PS beads.[113]

In another report, modulating the concentration of TEOS and aminopropyltriethoxysilane (APTES) at a modest ratio led to the formation of 3D spherical silica nanoparticles with nematic TMV particles radically anchored to the silica core.[114] In this process, the addition of TEOS–APTES destroyed the initial liquid crystalline state of TMV, so during the realignment state TMV rods anchored to the primary silica balls. Subsequently, the anchored rods deformed owing to the force of nematic state redevelopment. Finally, the whole length of the TMV rods broke and associated with silica particles to form core–shell spherical nanoparticles.

8.4.2.7   Development of TMV Microarrays

An important property of TMV is its ability to disassemble, which occurs upon cell entry. A few coat protein subunits will depart from the 5′ end, leaving exposed viral RNA connected to the remaining incomplete TMV.[115] This pendant genome sequence offers the possibility of fabricating DNA-containing devices through a nucleic acid hybridization-directed assembly

A

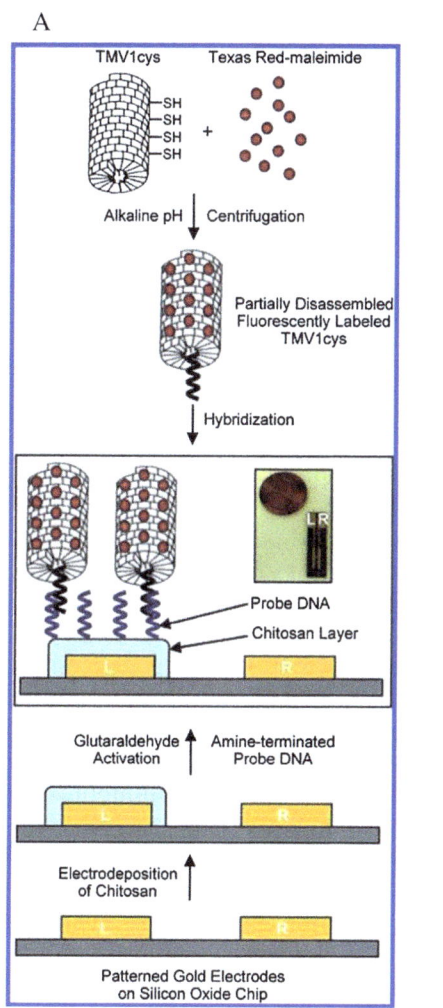

B

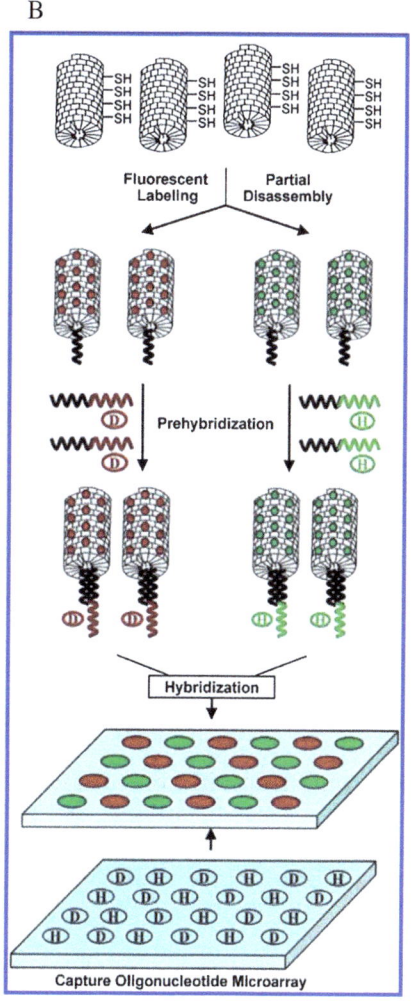

Figure 8.8 (A) Partially disassembled TMV1cys nanotemplates stand on a readily addressable site directed by DNA probe.[46] (B) Fluorescently labeled and partially disassembled TMV assembly onto multiple address on DNA oligonucleotide microarray platforms.[116] Adapted with permission from ref. 46 and 116. Copyright 2005 and 2007 American Chemical Society, respectively.

approach. The interaction between two complementary genome sequences shows unrivalled selectivity without the need for special chemical handling.

Culver's group labeled genetically engineered TMV1Cys virus particles with fluorescent dyes and partially disassembled them with alkaline solution to expose 5'-end RNA sequences and then hybridized these particles with the linker DNA. The DNA contains a complementary sequence to the virus 5' end and an amine group that can form covalent bonds with glutaraldehyde-activated chitosan on silicon chips [Figure 8.8(A)].[46] Thus, the linker DNA probe could anchor virus particles on silicon chips with a specific orientation. In another experiment, TMV1Cys particles were labeled with Cy5 or Cy3 and two different linker DNAs.[116] Two capture DNAs were engineered onto glass slides and they had complementary sequences to the linker DNAs. This facile approach realized two different dye-labeled viruses assembled on specific spots on aldehyde-functionalized glass slides [Figure 8.8(B)].[116] With a microfabrication procedure, doubly labeled TMV1Cys particles were incubated with stop-flow lithography-generated PEG-based encoded microparticles.[117] These microparticles consisted of three regions, a Rhodamine B-containing encoded region, a DNA capture region and a negative control area. These distinct areas were polymerized seamlessly by UV exposure through a photomask. The results demonstrated that the assembled TMV1Cys particles on the microparticles were of high density and integrity.[117]

The TMV1Cys particles can also stand directly on a gold solid substrate.[118] A cysteine residue was engineered at the N-terminus of TMV coat protein and exposed sufficiently to attach to the gold surface. Hence, through the gold–thiol interaction, TMV can assemble onto the gold patterned surfaces in a vertical orientation. Nickel and cobalt nanoparticles are further coated onto the assembled viruses through electroless deposition to make oriented high surface area materials, which could be used as electrodes.[47,118,119]

Wege and co-workers developed some elegant bottom-up approaches to assemble TMV on modified silicon wafers or gold cores.[120-122] For example, capture DNAs were first grafted to the gold surface. These DNAs have a thiol group on one end and another end complementary to the viral RNA 3′ end. Following hybridization with capture DNA, viral RNA can be immobilized. Beginning with the original assembly site on an RNA scaffold, TMV coat protein will aggregate helically around the RNA and finally form a star-like gold–virus hybrid architecture.[122]

8.5   Applications of TMV Hybrid Bionanomaterials

Owing to the highly ordered and robust architecture, inherent chemical and physical properties and genetic programmability, TMV was found to have a wide range of applications in nanoscale electronic devices, vaccine development, cell study and tissue engineering.

8.5.1   TMV-Based Nanodevices

Balandin's group found that in addition to acting as nanotemplates for silica or silicon inorganic semiconductor nanotubes, genetically programmed TMV could actually improve the low-temperature electron mobility and thermal conductivity of hybrid virus nanotubes because of the confined acoustic phonon redistribution effects.[123] Yang and co-workers reported a novel electronic memory effect established by the electroless deposition of Pt on TMV particles.[124] Not only did TMV function as the backbone, but the RNA core containing aromatic rings served as a charge donor and the protein capsid separating RNA and Pt played the role of a barrier to stabilize the charges. Charge transfer and charge traps in TMV–Pt systems could therefore allow conductance switching behavior and afford a memory effect.

TMV1Cys particles were applied to fabricate electrodes.[47,118,119] In an NiO–Zn battery system, a virus-coated electrode showed a higher capacity than non-TMV-modified electrodes.[118,119] A silicon virus-based anode in a lithium-ion battery performed much better than other reported silicon and carbon–silicon anodes because of the unique multilayer of the virus-based nanostructure.[47] It has a high capacity, low fading rate and excellent charge–discharge ability for a large number of operating cycles.

Yi and co-workers synthesized small, uniform and well-dispersed Pd crystalline nanoparticles along TMV1Cys without external reducing agents.[125] The generated nanoparticles can be integrated into a PEG-based hydrogel through a simple and robust replica molding approach. They demonstrated that these Pd–TMV–PEG hybrid microparticles had a higher catalytic activity than commercial Pd/C catalysts in reducing hexavalent chromate ion from industrial waste and drinking water.

Evans and co-workers reported a new method for the synthesis of Pt–TMV nanotubes with high active surface area using methanol, which served as both a reducing agent and solvent.[126] A thin, homogeneous and continuous layer of Pt could be deposited on the surface of TMV particles by modulating the water-to-methanol ratio and platinum salt concentration. The synthesized TMV nanoparticles with high aspect ratio and uniform anisotropy enhanced the available surface sites for the catalytic oxidation of methanol and offered potential application as an anode catalyst in methanol fuel cells.

8.5.2   TMV-Based Nanoparticles for Biomedical Research

It has been determined that both the C- and N-termini and even the intermediate area of TMV can fuse with foreign peptides without destroying the highly organized structure of TMV. Haynes *et al.* were the first to modify TMV coat protein gene with immunogenic poliovirus type 3 antigenic epitope at its 3′ end and express it in *Escherichia coli*.[41] The expressed fusion protein could reassemble to form virus-like rods under acidic conditions. Okada's group developed a new TMV RNA vector to produce both the intact coat protein and fusion proteins.[127] In this new vector, six nucleotides were inserted in the 3′ end just after the coat protein stop codon and immediately followed by the 5′ end of the foreign gene sequence. Angiotensin I-converting enzyme inhibitor peptide (ACE I) and epitopes from influenza virus hemagglutinin and from HIV-I envelope protein were all successfully synthesized using this new RNA vector. These proteins could replicate and assemble in tobacco plants. Further, 25 amino acids could be inserted into coat protein as genetic fusions;[128] an even larger protein, as long as a 133-mer, was successfully displayed on the C-terminus with a 15-amino acid linker.[43]

In the N-terminus, an additional lysine residue was inserted and then conjugated with streptavidin–GFP (green fluorescent protein) or streptavidin–L2 protein (AA 61–171, from canine oral papillomavirus) *via* a biotin linker.[129] The results showed that GFP-decorated TMV particles had augmented humoral IgG titers and the L2 protein fragment displayed on TMV was significantly more immunogenic. Another chimeric TMV consisting 10 or 15 amino acids from the spike protein of murine hepatitis virus (MHV) inserted between Ser154 and Gly155 of coat protein was propagated in tobacco plants.[42] The study indicated that intranasal and subcutaneous immunization of mice with the manipulated TMV particles containing MHV epitope elicited protective immunity against MHV infection.

There is increasing interest in applying hybrid biomaterials in tissue engineering to induce cellular processes, such as cell growth and cell differentiation. Bone marrow-derived stem cells, referred to bone marrow stromal cells or bone marrow-derived mesenchymal stem cells (BMSCs), can be differentiated into various cell types when induced in different chemical and environmental conditions owing to their multilineage differentiation potential.[130] Kaur *et al.* cultured BMSCs on TMV-coated 2D substrates and monitored their differentiation process into osteoblast-like cells.[131] They found that at studied time points of 7, 14 and 21 days, the calcium mineralization increased and osteo-specific gene (osteocalcin, osteopontin and osteonectin) expression was upregulated, and that the gene upregulation showed a maximum change in expression at 14 days, which was 7 days earlier compared with tissue culture plastic (Figure 8.9). As phosphates play a crucial role in cell mineralization and can promote the differentiation of BMSCs,[132,133] when TMV Tyr139 residue was tailored with a phosphate group (TMV–Phos), significantly higher upregulation of the osteo-specific genes was observed

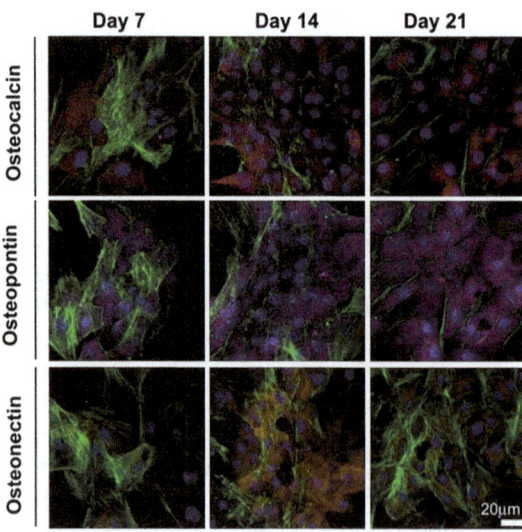

Figure 8.9   Gene expression analysis of cells grown on TMV-coated substrate by immune staining for 7, 14 and 21 days.[131] Reprinted from ref. 131, Copyright 2009, with permission from Elsevier.

compared with native TMV.[134] The 3D porous alginate hydrogels mixed with TMV can also enhance the attachment, proliferation and differentiation of BMSCs.[135]

Lee *et al.* designed fibronectin synergy mutants and collagen mimics, such as TMV–PHSRN3, TMV–RGD1 and TMV–P15, and used them for cell studies.[136] These mutants showed different effects on cell adhesion and attachment. Cells grown on TMV–RGD substrate formed filopodial extensions but attachment was much weaker, whereas those grown on TMV–P15 substrate displayed little spreading but much stronger attachment. Further studies found that when TMV–RGD1 was displayed on clean glass coated with polyelectrolyte on the topmost layer, it could promote BMSC cell differentiation into bone-like cells within just 2 days.[137] In addition, when it was co-spun with poly(vinyl alcohol) (PVA) to form fibers, it could efficiently enhance cell adhesion.[39]

8.6 Conclusion

Bionanoparticles are promising templates and building blocks for the fabrication of novel biohybrid materials and they have great potential in a variety of applications. In recent studies, TMV, a classic plant virus model, has been modified or assembled to synthesize inorganic and organic composite materials, such as nanowires, nanofilms, electrodes and light-harvesting systems. TMV has also shown potential in biomedical applications. Based on these studies, the hybrid materials area opens up a new path to fabricate more multifunctional nanodevices.

References

1. M. Fischlechner and E. Donath, *Angew. Chem., Int. Ed.*, 2007, **46**, 3184.
2. M. Young, W. Debbie, M. Uchida and T. Douglas, *Annu. Rev. Phytopathol.*, 2008, **46**, 361.
3. G. Jutz and A. Böker, *Polymer*, 2011, **52**, 211.
4. Z. Liu, J. Qiao, Z. Niu and Q. Wang, *Chem. Soc. Rev.*, 2012, **41**, 6178–6194.
5. S. K. Campos and M. A. Barry, *Curr. Gene Ther.*, 2007, 7, 189.
6. T. Ramqvist, K. Andreasson and T. Dalanis, *Expert Opin. Biol. Ther.*, 2007, 7, 997.
7. M. Manchester and P. Singh, *Adv. Drug Delivery Rev.*, 2006, **58**, 1505.
8. L. A. Lee and Q. Wang, *Nanomed.: Nanotechnol., Biol. Med.*, 2006, **2**, 137.
9. J. M. Polo and T. W. Dubensky, *Drug Discovery Today*, 2002, 7, 719.
10. M. C. Canizares, L. Nicholson and G. P. Lomonossoff, *Immunol. Cell Biol.*, 2005, **83**, 263.
11. Z. Niu, M. Bruckman, V. S. Kotakadi, J. He, T. Emrick, T. P. Russell, L. Yang and Q. Wang, *Chem. Commun.*, 2006, 3019.
12. C. Mao, D. J. Solis, B. D. Reiss, S. T. Kottmann, R. Y. Sweeney, A. Hayhurst, G. Georgiou, B. Iverson and A. M. Belcher, *Science*, 2004, **303**, 213.

13. K. T. Nam, D.-W. Kim, P. J. Yoo, C.-Y. Chiang, N. Meethong, P. T. Hammond, Y.-M. Chiang and A. M. Belcher, *Science*, 2006, **312**, 885.

14. C. E. Flynn, C. Mao, A. Hayhurst, J. L. Williams, G. Georgiou, B. Iverson and A. M. Belcher, *J. Mater. Chem.*, 2003, **13**, 2414.

15. C. Mao, C. E. Flynn, A. Hayhurst, R. Sweeney, J. Qi, G. Georgiou, B. Iverson and A. M. Belcher, *Proc. Natl. Acad. Sci. U. S. A.*, 2003, **100**, 6946.

16. B. D. Reiss, C. Mao, D. J. Solis, K. S. Ryan, T. Thomson and A. M. Belcher, *Nano Lett.*, 2004, **4**, 1127.

17. L. A. Lee and Q. Wang, *Nanomed.: Nanotechnol., Biol. Med.*, 2006, **2**, 137.

18. N. F. Steinmetz and D. J. Evans, *Org. Biomol. Chem.*, 2007, **5**, 2891.

19. L. A. Lee, Z. Niu and Q. Wang, *Nano Res.*, 2009, **2**, 349.

20. C. M. Soto and B. R. Ratna, *Curr. Opin. Biotechnol.*, 2010, **21**, 426.

21. L. A. Lee, H. G. Nguyen and Q. Wang, *Org. Biomol. Chem.*, 2011, **9**, 6189.

22. S. Y. Lee, J. S. Lim and M. T. Harris, *Biotechnol. Bioeng.*, 2011, **109**, 16.

23. P. van Rijn and A. Böker, *J. Mater. Chem.*, 2011, **21**, 16735.

24. W. M. Stanley, *Science*, 1935, **81**, 644.

25. P. J. G. Butler, *J. Gen. Virol.*, 1984, **65**, 253.

26. A. Klug, *Philos. Trans. R. Soc. London, Ser. B*, 1999, **354**, 531.

27. P. Goelet, G. Lomonossoff, P. Butler, M. Akam, M. Gait and J. Karn, *Proc. Natl. Acad. Sci. U. S. A.*, 1982, **79**, 5818.

28. K. Namba, R. Pattanayek and G. Stubbs, *J. Mol. Biol.*, 1989, **208**, 307.

29. B. Bhyravbhatla, S. J. Watowich and D. L. Caspar, *Biophys. J.*, 1998, **74**, 604.

30. A. Durham, J. Finch and A. Klug, *Nature*, 1971, **229**, 37.

31. R. Perham and T. Wilson, *Virology*, 1978, **84**, 293.

32. M. Knez, M. Sumser, A. M. Bittner, C. Wege, H. Jeske, T. P. Martin and K. Kern, *Adv. Funct. Mater.*, 2004, **14**, 116.

33. T. L. Schlick, Z. Ding, E. W. Kovacs and M. B. Francis, *J. Am. Chem. Soc.*, 2005, **127**, 3718.

34. M. A. Bruckman, G. Kaur, L. A. Lee, F. Xie, J. Sepulveda, R. Breitenkamp, X. Zhang, M. Joralemon, T. P. Russell, T. Emrick and Q. Wang, *ChemBioChem*, 2008, **9**, 519.

35. V. V. Rostovtsev, L. G. Green, V. V. Fokin and K. B. Sharpless, *Angew. Chem.*, 2002, **114**, 2708.

36. C. W. Tornøe, C. Christensen and M. Meldal, *J. Org. Chem.*, 2002, **67**, 3057.

37. J. L. Brennan, N. S. Hatzakis, T. R. Tshikhudo, N. Dirvianskyte, V. Razumas, S. Patkar, J. Vind, A. Svendsen, R. J. Nolte and A. E. Rowan, *Bioconjugate Chem.*, 2006, **17**, 1373.

38. Z. Yin, H. G. Nguyen, S. Chowdhury, P. Bentley, M. A. Bruckman, A. Miermont, J. C. Gildersleeve, Q. Wang and X. Huang, *Bioconjugate Chem.*, 2012, **23**, 1694.

39. L. Wu, J. Zang, L. A. Lee, Z. Niu, G. C. Horvatha, V. Braxtona, A. C. Wibowo, M. A. Bruckman, S. Ghoshroy and H.-C. zur Loye, *J. Mater. Chem.*, 2011, **21**, 8550.

40. W. O. Dawson, D. L. Beck, D. A. Knorr and G. L. Grantham, *Proc. Natl. Acad. Sci. U. S. A.*, 1986, **83**, 1832.

41. J. R. Haynes, J. Cunningham, A. von Seefried, M. Lennick, R. T. Garvin and S.-H. Shen, *Nat. Biotechnol.*, 1986, **4**, 637.
42. Y. Sugiyama, H. Hamamoto, S. Takemoto, Y. Watanabe and Y. Okada, *FEBS Lett.*, 1995, **359**, 247.
43. S. Werner, S. Marillonnet, G. Hause, V. Klimyuk and Y. Gleba, *Proc. Natl. Acad. Sci. U. S. A.*, 2006, **103**, 17678.
44. R. A. Miller, A. D. Presley and M. B. Francis, *J. Am. Chem. Soc.*, 2007, **129**, 3104.
45. M. Endo, M. Fujitsuka and T. Majima, *Chem.–Eur. J.*, 2007, **13**, 8660.
46. H. Yi, S. Nisar, S.-Y. Lee, M. A. Powers, W. E. Bentley, G. F. Payne, R. Ghodssi, G. W. Rubloff, M. T. Harris and J. N. Culver, *Nano Lett.*, 2005, **5**, 1931.
47. X. Chen, K. Gerasopoulos, J. Guo, A. Brown, C. Wang, R. Ghodssi and J. N. Culver, *ACS Nano*, 2010, **4**, 5366.
48. M. Demir and M. H. Stowell, *Nanotechnology*, 2002, **13**, 541.
49. W. Shenton, T. Douglas, M. Young, G. Stubbs and S. Mann, *Adv. Mater.*, 1999, **11**, 253.
50. J. N. Culver, W. O. Dawson, K. Plonk and G. Stubbs, *Virology*, 1995, **206**, 724.
51. B. Lu, G. Stubbs and J. N. Culver, *Virology*, 1996, **225**, 11.
52. E. Dujardin, C. Peet, G. Stubbs, J. N. Culver and S. Mann, *Nano Lett.*, 2003, **3**, 413.
53. S.-Y. Lee, E. Royston, J. N. Culver and M. T. Harris, *Nanotechnology*, 2005, **16**, S435.
54. S.-Y. Lee, J. Choi, E. Royston, D. B. Janes, J. N. Culver and M. T. Harris, *J. Nanosci. Nanotechnol.*, 2006, **6**, 974.
55. Z. Zhang, D. A. Blom, Z. Gai, J. R. Thompson, J. Shen and S. Dai, *J. Am. Chem. Soc.*, 2003, **125**, 7528.
56. R. Tsukamoto, M. Muraoka, M. Scki, H. Tabata and I. Yamashita, *Chem. Mater.*, 2007, **19**, 2389.
57. M. Kobayashi, M. Seki, H. Tabata, Y. Watanabe and I. Yamashita, *Nano Lett.*, 2010, **10**, 773.
58. M. Knez, A. M. Bittner, F. Boes, C. Wege, H. Jeske, E. Maiss and K. Kern, *Nano Lett.*, 2003, **3**, 1079.
59. M. Knez, M. Sumser, A. Bittner, C. Wege, H. Jeske, S. Kooi, M. Burghard and K. Kern, *J. Electroanal. Chem.*, 2002, **522**, 70.
60. S. Balci, A. Bittner, K. Hahn, C. Scheu, M. Knez, A. Kadri, C. Wege, H. Jeske and K. Kern, *Electrochim. Acta*, 2006, **51**, 6251.
61. S. Balci, K. Hahn, P. Kopold, A. Kadri, C. Wege, K. Kern and A. M. Bittner, *Nanotechnology*, 2012, **23**, 045603.
62. M. Knez, A. Kadri, C. Wege, U. Gösele, H. Jeske and K. Nielsch, *Nano Lett.*, 2006, **6**, 1172.
63. E. Royston, S.-Y. Lee, J. N. Culver and M. T. Harris, *J. Colloid Interface Sci.*, 2006, **298**, 706.
64. E. S. Royston, A. D. Brown, M. T. Harris and J. N. Culver, *J. Colloid Interface Sci.*, 2009, **332**, 402.

65. Z. Niu, M. Bruckman, V. S. Kotakadi, J. He, T. Emrick, T. P. Russell, L. Yang and Q. Wang, *Chem. Commun.*, 2006, 3019.

66. Z. Niu, J. Liu, L. A. Lee, M. A. Bruckman, D. Zhao, G. Koley and Q. Wang, *Nano Lett.*, 2007, **7**, 3729.

67. B. A. Grzybowski, C. E. Wilmer, J. Kim, K. P. Browne and K. J. Bishop, *Soft Matter*, 2009, **5**, 1110.

68. M. Grzelczak, J. Vermant, E. M. Furst and L. M. Liz-Marzán, *ACS Nano*, 2010, **4**, 3591.

69. Z. Niu, M. A. Bruckman, S. Li, L. A. Lee, B. Lee, S. V. Pingali, P. Thiyagarajan and Q. Wang, *Langmuir*, 2007, **23**, 6719.

70. M. A. Bruckman, Z. Niu, S. Li, L. A. Lee, T. Nelson, J. Lavigne, Q. Wang and K. Varazo, *NanoBiotechnology*, 2007, **3**, 31.

71. M. T. Klem, D. Willits, M. Young and T. Douglas, *J. Am. Chem. Soc.*, 2003, **125**, 10806.

72. J. T. Russell, Y. Lin, A. Böker, L. Su, P. Carl, H. Zettl, J. He, K. Sill, R. Tangirala and T. Emrick, *Angew. Chem., Int. Ed.*, 2005, **44**, 2420.

73. J. He, Z. Niu, R. Tangirala, J.-Y. Wang, X. Wei, G. Kaur, Q. Wang, G. N. Jutz, A. Böker, B. Lee, S. V. Pingali, P. Thiyagarajan, T. Emrick and T. P. Russell, *Langmuir*, 2009, **25**, 4979.

74. L. A. Lee, *Chem. Commun.*, 2008, 5185.

75. M. Knez, M. Sumser, A. Bittner, C. Wege, H. Jeske, D. Hoffmann, K. Kuhnke and K. Kern, *Langmuir*, 2004, **20**, 441.

76. A. Nedoluzhko and T. Douglas, *J. Inorg. Biochem.*, 2001, **84**, 233.

77. X. Zan, S. Feng, E. Balizan, Y. Lin and Q. Wang, *ACS Nano*, 2013, **7**, 8385.

78. J. L. Wilbur, A. Kumar, E. Kim and G. M. Whitesides, *Adv. Mater.*, 1994, **6**, 600.

79. R. J. Jackman, J. L. Wilbur and G. M. Whitesides, *Science*, 1995, **269**, 664.

80. M. Mrksich and G. M. Whitesides, *Trends Biotechnol.*, 1995, **13**, 228.

81. Y. Zhang, E. A. Matsumoto, A. Peter, P.-C. Lin, R. D. Kamien and S. Yang, *Nano Lett.*, 2008, **8**, 1192.

82. A. Horn, S. Hiltl, A. Fery and A. Böker, *Small*, 2010, **6**, 2122.

83. A. Horn, H. G. Schoberth, S. Hiltl, A. Chiche, Q. Wang, A. Schweikart, A. Fery and A. Böker, *Faraday Discuss.*, 2009, **143**, 143.

84. R. A. Vega, D. Maspoch, K. Salaita and C. A. Mirkin, *Angew. Chem.*, 2005, **117**, 6167.

85. G. Kaur, J. He, J. Xu, S. Pingali, G. n. Jutz, A. Böker, Z. Niu, T. Li, D. Rawlinson and T. Emrick, *Langmuir*, 2009, **25**, 5168.

86. S. Kewalramani, S. Wang, Y. Lin, H. G. Nguyen, Q. Wang, M. Fukuto and L. Yang, *Soft Matter*, 2011, **7**, 939.

87. S. U. Pickering, *J. Chem. Soc., Trans.*, 1907, **91**, 2001.

88. Y. Lin, H. Skaff, A. Böker, A. Dinsmore, T. Emrick and T. P. Russell, *J. Am. Chem. Soc.*, 2003, **125**, 12690.

89. Y. Lin, H. Skaff, T. Emrick, A. Dinsmore and T. Russell, *Science*, 2003, **299**, 226.

90. L. Yang, S. Wang, M. Fukuto, A. Checco, Z. Niu and Q. Wang, *Soft Matter*, 2009, **5**, 4951.

91. N. Denkov, O. Velev, P. Kralchevski, I. Ivanov, H. Yoshimura and K. Nagayama, *Langmuir*, 1992, **8**, 3183.

92. N. Denkov, O. Velev, P. Kralchevsky, I. Ivanov, H. Yoshimura and K. Nagayama, *Nature*, 1993, **361**, 26.

93. B. G. Prevo and O. D. Velev, *Langmuir*, 2004, **20**, 2099.

94. D. M. Kuncicky, R. R. Naik and O. D. Velev, *Small*, 2006, **2**, 1462.

95. B. G. Prevo, D. M. Kuncicky and O. D. Velev, *Colloids Surf., A*, 2007, **311**, 2.

96. S. P. Wargacki, B. Pate and R. A. Vaia, *Langmuir*, 2008, **24**, 5439.

97. Y. Lin, E. Balizan, L. A. Lee, Z. Niu and Q. Wang, *Angew. Chem., Int. Ed.*, 2010, **49**, 868.

98. Y. Lin, Z. Su, G. Xiao, E. Balizan, G. Kaur, Z. Niu and Q. Wang, *Langmuir*, 2010, **27**, 1398.

99. G. Decher and J. D. Hong, *Makromol. Chem., Macromol. Symp.*, 1991, **46**, 321.

100. J. Schmitt, G. Decher, W. J. Dressick, S. L. Brandow, R. E. Geer, R. Shashidhar and J. M. Calvert, *Adv. Mater.*, 1997, **9**, 61.

101. Y. Lvov, G. Decher and H. Moehwald, *Langmuir*, 1993, **9**, 481.

102. S. Srivastava and N. A. Kotov, *Acc. Chem. Res.*, 2008, **41**, 1831.

103. Y. Lvov, H. Haas, G. Decher, H. Moehwald, A. Mikhailov, B. Mtchedlishvily, E. Morgunova and B. Vainshtein, *Langmuir*, 1994, **10**, 4232.

104. G. Decher, B. Lehr, K. Lowack, Y. Lvov and J. Schmitt, *Biosens. Bioelectron.*, 1994, **9**, 677.

105. Y. Lvov, K. Ariga, I. Ichinose and T. Kunitake, *J. Am. Chem. Soc.*, 1995, **117**, 6117.

106. G. Decher, M. Eckle, J. Schmitt and B. Struth, *Curr. Opin. Colloid Interface Sci.*, 1998, **3**, 32.

107. Z. Tang, Y. Wang, P. Podsiadlo and N. A. Kotov, *Adv. Mater.*, 2006, **18**, 3203.

108. Y. Lin, Z. Su, Z. Niu, S. Li, G. Kaur, L. Lee and Q. Wang, *Acta Biomater.*, 2008, **4**, 838.

109. X. Zan, P. Sitasuwan, J. Powell, T. W. Dreher and Q. Wang, *Acta Biomater.*, 2012, **8**, 2978–2985.

110. N. F. Steinmetz, K. C. Findlay, T. R. Noel, R. Parker, G. P. Lomonossoff and D. J. Evans, *ChemBioChem*, 2008, **9**, 1662.

111. P. J. Yoo, K. T. Nam, J. Qi, S.-K. Lee, J. Park, A. M. Belcher and P. T. Hammond, *Nat. Mater.*, 2006, **5**, 234.

112. T. Li, L. Wu, N. Suthiwangcharoen, M. A. Bruckman, D. Cash, J. S. Hudson, S. Ghoshroy and Q. Wang, *Chem. Commun.*, 2009, 2869.

113. T. Li, Z. Niu, N. Suthiwangcharoen, R. Li, P. E. Prevelige and Q. Wang, *Sci. China: Chem.*, 2010, **53**, 71.

114. C. E. Fowler, W. Shenton, G. Stubbs and S. Mann, *Adv. Mater.*, 2001, **13**, 1266.

115. T. M. A. Wilson, *Virology*, 1984, **137**, 255.

116. H. Yi, G. W. Rubloff and J. N. Culver, *Langmuir*, 2007, **23**, 2663.

117. W. S. Tan, C. L. Lewis, N. E. Horelik, D. C. Pregibon, P. S. Doyle and H. Yi, *Langmuir*, 2008, **24**, 12483.

118. E. Royston, A. Ghosh, P. Kofinas, M. T. Harris and J. N. Culver, *Langmuir*, 2008, **24**, 906.

119. K. Gerasopoulos, M. McCarthy, E. Royston, J. N. Culver and R. Ghodssi, *J. Micromech. Microeng.*, 2008, **18**, 104003.

120. A. Mueller, F. J. Eber, C. Azucena, A. Petershans, A. M. Bittner, H. Gliemann, H. Jeske and C. Wege, *ACS Nano*, 2011, **5**, 4512.

121. C. Azucena, F. J. Eber, V. Trouillet, M. Hirtz, S. Heissler, M. Franzreb, H. Fuchs, C. Wege and H. Gliemann, *Langmuir*, 2012, **28**, 14867.

122. F. J. Eber, S. Eiben, H. Jeske and C. Wege, *Angew. Chem.*, 2013, **125**, 7344.

123. V. A. Fonoberov and A. A. Balandin, *Nano Lett.*, 2005, **5**, 1920.

124. R. J. Tseng, C. Tsai, L. Ma, J. Ouyang, C. S. Ozkan and Y. Yang, *Nat. Nanotechnol.*, 2006, **1**, 72.

125. C. Yang, C.-H. Choi, C.-S. Lee and H. Yi, *ACS Nano*, 2013, **7**, 5032–5044.

126. M. Ł. Górzny, A. S. Walton and S. D. Evans, *Adv. Funct. Mater.*, 2010, **20**, 1295.

127. H. Hamamoto, Y. Sugiyama, N. Nakagawa, E. Hashida, Y. Matsunaga, S. Takemoto, Y. Watanabe and Y. Okada, *Nat. Biotechnol.*, 1993, **11**, 930.

128. A. A. McCormick, T. A. Corbo, S. Wykoff-Clary, L. V. Nguyen, M. L. Smith, K. E. Palmer and G. P. Pogue, *Vaccine*, 2006, **24**, 6414.

129. M. L. Smith, J. A. Lindbo, S. Dillard-Telm, P. M. Brosio, A. B. Lasnik, A. A. McCormick, L. V. Nguyen and K. E. Palmer, *Virology*, 2006, **348**, 475.

130. A. I. Caplan, *J. Orthop. Res.*, 1991, **9**, 641.

131. G. Kaur, M. T. Valarmathi, J. D. Potts, E. Jabbari, T. Sabo-Attwood and Q. Wang, *Biomaterials*, 2010, **31**, 1732.

132. C. R. Nuttelman, D. S. Benoit, M. C. Tripodi and K. S. Anseth, *Biomaterials*, 2006, **27**, 1377.

133. P. Müller, U. Bulnheim, A. Diener, F. Lüthen, M. Teller, E. D. Klinkenberg, H. G. Neumann, B. Nebe, A. Liebold and G. Steinhoff, *J. Cell. Mol. Med.*, 2007, **12**, 281.

134. G. Kaur, C. Wang, J. Sun and Q. Wang, *Biomaterials*, 2010, **31**, 5813.

135. J. Luckanagul, L. A. Lee, Q. L. Nguyen, P. Sitasuwan, X. Yang, T. Shazly and Q. Wang, *Biomacromolecules*, 2012, **13**, 3949.

136. L. A. Lee, Q. L. Nguyen, L. Wu, G. Horvath, R. S. Nelson and Q. Wang, *Biomacromolecules*, 2012, **13**, 422.

137. L. A. Lee, S. M. Muhammad, Q. L. Nguyen, P. Sitasuwan, G. Horvath and Q. Wang, *Mol. Pharm.*, 2012, **9**, 2121.

CHAPTER 9

Virus-Based Systems for Functional Materials

MARTIJN VERWEGEN[a] AND JEROEN J. L. M. CORNELISSEN*[a]

[a]Laboratory for Biomolecular Nanotechnology, MESA+ Institute for Nanotechnology, University of Twente, P.O. Box 217, 7500 AE Enschede, The Netherlands
*E-mail: J.J.L.M.Cornelissen@utwente.nl

9.1   Introduction: Spherical Viruses

Virus-based bionanotechnology holds the promise of control over the structure, properties and functionality of materials at the nanometre scale. After all, viruses, and by extension virus-like particles (VLPs), represent some of the largest hierarchical protein constructs found in Nature. Their symmetrical architecture and their high degree of monodispersity, compared with other nanoparticles, make them unique as nanobuilding blocks. Furthermore, many of these particles seem to have specific and tuncable physical properties that can be utilized for their further function and manipulation.

Viruses and VLPs are therefore highly desirable nanobuilding blocks that could find applications ranging from nanocontainers, for studying reactions in confinement or drug delivery, to modular structural components, that allow for the creation of complex nanoarchitectures, and eventually functional materials. This chapter is intended to generate an understanding of how the structure, modification and organization of viruses enable them to

RSC Smart Materials No. 16
Bio-Synthetic Hybrid Materials and Bionanoparticles: A Biological Chemical Approach Towards Material Science
Edited by Alexander Böker and Patrick van Rijn
© The Royal Society of Chemistry 2015
Published by the Royal Society of Chemistry, www.rsc.org

be the key component in these potential functional materials, a field recently introduced as chemical virology. Ultimately, these functional virus-based materials could allow the construction of novel optical, electronic, catalytic, imaging and other nano-scale precision-based applications.

9.1.1 Basic Structure

Viruses come in many different shapes and morphologies, but for most chemical virology purposes they can be subdivided into rod-like and spherical viruses. In both cases a virus nanoparticle consists of an RNA or DNA core protected by a protein coat or capsid, which is held together by non-covalent interactions. Depending on the pH and ionic strength of the solution, these virus particles display a variety of swelling, maturation or other structural transformations. For example, the cowpea chlorotic mottle virus (CCMV), a typical icosahedral virus, is known to have pores that can be opened or closed based on pH and ionic strength, allowing for the influx of materials. Such structural changes are often associated with release mechanisms for the genome cargo carried by viruses.[1,2]

Key to understanding these natural nanoparticles is knowing their structure (see Figure 9.1). Common techniques for studying this area include X-ray diffraction (XRD) for the crystallographic structure and small-angle X-ray scattering (SAXS) and (cryo)electron microscopy (EM) to check dynamic structural changes during the various stages of maturation and assembly (see Figure 9.2). More recently, structural parameters have been studied using atomic force microscopy (AFM), revealing not only the surface topology but also physical properties, such as the shear and Young's modulus. Additional insights gained by nanoindentation have also revealed the importance and strength of structural components such as the individual subunits and the genetic material inside the capsid.[3,4]

Such techniques can also be used to probe the nature of the interaction between the RNA and coat protein, especially under conditions where pH and ionic strength are varied. For instance, Makino *et al.* used X-ray data to reveal several unknown protein segments and their interaction with the RNA strand inside the virus.[5] Small-angle neutron scattering (SANS) data obtained by Comellas-Aragones *et al.*[6] show the morphology of CCMV and CCMV capsids and confirm the pH and ionic strength-based swelling behaviour that had been observed by Speir *et al.*[7] using cryo-TEM, but also reveal that the RNA is bound close to the protein coat.

9.1.2 Virus Symmetry

Spherical virus protein capsids generally adopt an icosahedral symmetry, which was described by the Caspar and Klug triangulation number and corresponds to the number of subunits that the next symmetrical morphology will take. In essence, each set of three subunits is modelled as a triangle on a sphere. This forms a grid of triangles that is arranged into pentamers and

Figure 9.1 VIPER database-reconstructed images of some spherical viruses used in the functional materials that are discussed in this chapter.[2]

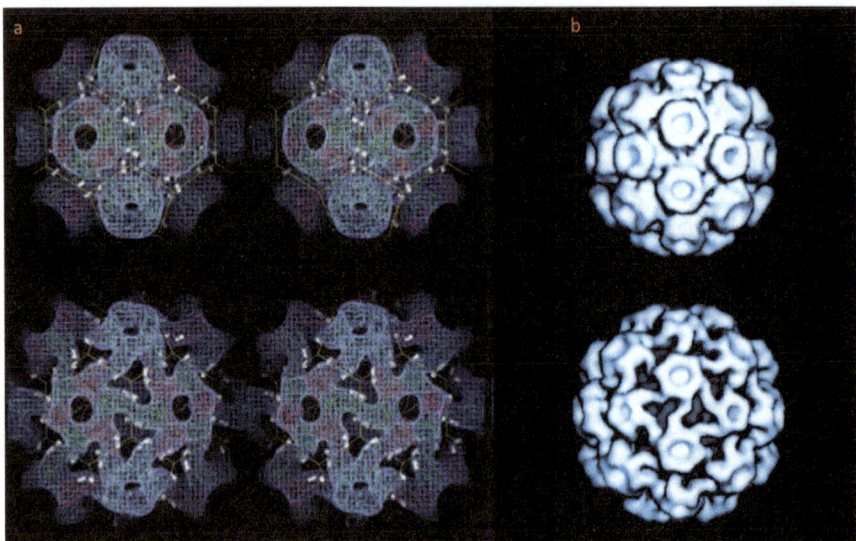

Figure 9.2   CCMV structures have been solved by both X-ray crystallography (a) and
cryo-TEM reconstruction (b), revealing a closed (top, pH <6.5) and swollen (bottom, pH > 6.5) morphology.[1,7] The symmetrical structure containing pentamers and hexamers is also clearly visible. Reprinted from
ref. 7, Copyright 1995, with permission from Elsevier.

hexamers around the spherical form. Symmetrical spheres occur at regular
intervals in the number of triangles, which conform to integer values called
triangulation numbers and follow $T = h^2 + hk + k^2$, where h and k represent
the distances between pentamers on the spherical grid. The smallest symmetrical assembly ($T = 1$) consist of 20 triangles or 60 proteins. For every sym-
metric assembly beyond this, more proteins are needed, thus the number of
proteins is $60T$ per capsid.[8,9]

The formation of an icosahedral symmetry cannot be inferred from free
energy minimization, and is therefore not necessarily a thermodynamic
process. To overcome this, Bruinsma *et al.* suggested additional structural
parameters based on the interaction of the sides of capsomers (hexamers
and pentamers) that make up the capsid shell.[10] Such interactions allow for
the formation of stable icosahedral forms, but also octahedral and cubic
capsid morphologies. Zandi *et al.* speculated that as icosahedral forms grow,
ruptures appear in the structure and thus other stable morphologies might
aid in the release of genomic cargo.[11]

Further molecular dynamics simulations and experiments have found that
additional stable capsid shells can be formed that do not obey the laws of
symmetry. This structural polymorphism was described in simulations by
Nguyen and Brooks.[12] As a basis of protein folding, they took the hexameric
and pentameric subunits found in Nature to be the first steps towards capsid
assembly and applied elementary kinetics to them. This reveals that a large
number of non-icosahedral, yet still symmetrical, assemblies can be formed

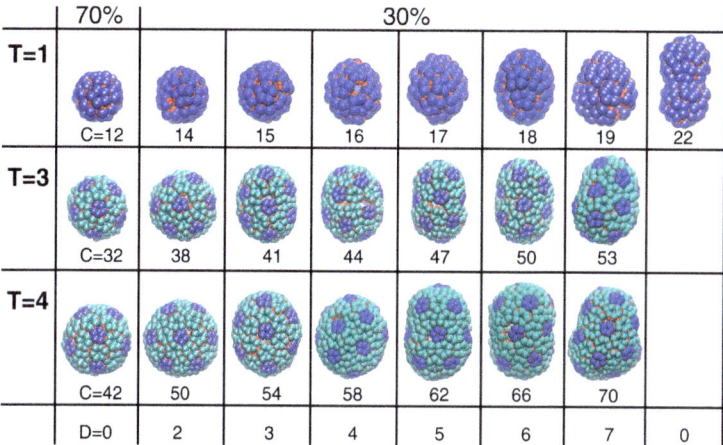

	70%				30%			
T=1	C=12	14	15	16	17	18	19	22
T=3	C=32	38	41	44	47	50	53	
T=4	C=42	50	54	58	62	66	70	
	D=0	2	3	4	5	6	7	0

Figure 9.3 Molecular dynamics predicts the formation of non-icosahedral assemblies under non-optimal conditions (43.5 µM of protein and 290 K). Still, around 70% of the proteins yield icosahedral T = 1, 3 or 4 capsids, the remainer forming non-icosahedral assemblies, with the relative yield decreasing as the number of capsomers (C) increases. Adapted with permission from ref. 12. Copyright 2008 American Chemical Society.

(Figure 9.3).[12] Additional simulations show this polymorphism generally results from hexameric dislocations. These dislocations increase the number of proteins in the coat by $6DT$, where D is the number of dislocations.[13]

Although Caspar–Klug symmetry and molecular dynamics reveal a great deal about the symmetry of viruses and what stable intermediates may form, the actual assembly process cannot be easily derived from this structure. Whereas some virus structures seem to assemble from single proteins or from dimers of proteins into trimetric or pentameric units, others need an origin of assembly (OAS) for the initial specific binding of the viral genome or form proto-capsids not fully adhering to Caspar–Klug symmetry before settling down into a final virus form.[14,15]

9.1.3 Virus Assembly

Electrostatic interactions between the protein coat and the RNA or DNA are one of the key mechanisms by which virus protein cages assemble and remain intact, a property that can be exploited for the encapsulation of polyelectrolytes and charged nanoparticles. In fact, there is a 1.6:1 charge balance for nearly all native viruses between coat protein charge and genome charge. Calculations by Belyi and Muthukumar[16] have shown that the virus genome is contained in a spherical shell within the cavity with a small gap between it and the protein shell, which is in good agreement with neutron scattering data.[6] More importantly, however, is the realization that the conformation adapted by a native virus is the one that results in the lowest free electrostatic energy.[16]

Still, electrostatic forces are not the only important contributor to viral assembly. Experimental results indicate that the native RNA of a virus is taken up far more efficiently than random cellular RNA fragments due to specific interactions induced by the packing sequence.[17] Castelnovo *et al.* presented theoretical models that explain these findings by entropy, showing that it likely plays an important role in virus assembly.[18] Qualitatively, this can be described by two processes. The first process, which preferentially selects viral RNA over cellular RNA, simply relies on the fact that viral RNA is larger than cellular RNA. Therefore, a given capsid morphology needs less viral RNA strands per particle compared with cellular RNA, thereby enhancing the entropy of the system. The second process, in which monodisperse polyanions can form stable small capsids at charge ratios that are not electrostatically favoured, is an entropic effect that favours multiple small capsids at the same protein concentration rather than a reduced number of larger capsids.

Probing the role of the virus sequence might therefore provide insights into the process of assembly, provided that such analysis is done uniquely for each virus. For example, in CCMV the deletion of several N-terminal residues does not affect the capsid assembly, although it did prevent encapsulation of the RNA. On the other hand, deletion of several C-terminal residues completely prevented any assembly from taking place.[19,20]

9.1.4 Capsid Assembly

Disassembly of a virus protein shell allows for the extraction of the coat proteins. This can be achieved for most viruses by changing the pH and ionic strength to trigger a structural change to open the capsid, followed by extraction and precipitation of the genetic material, for instance by precipitation with Ca^{2+}. For some viruses, these empty capsids can then be reassembled by changing the pH and ionic strength or by adding an anionic template, usually a polyelectrolyte or nanoparticle, to the coat proteins.[21]

A detailed experimental study by Bancroft[22] and later Lavelle *et al.*[23] on the pH- and ionic strength-dependent assembly of the CCMV capsid revealed substantial variety in the assembly. Depending on the conditions, the protein can fold itself into icosahedral capsids, multiwalled capsids, rods, or even dumbbell structures (Figure 9.4). Similar viruses, such as brome mosaic virus (BMV) and brown bean mottle virus (BBMV), only show spherical structures (BMV) or no aggregation without an RNA template (BBMV).[22,23] These observations can be readily explained by the electrostatic nature of the interaction. Furthermore, TEM images reveal many irregularly shaped capsids. Likely such structures, along with the dumbbell morphology, are an example of non-icosahedral assemblies as described by Nguyen and Brooks in their simulations.[12]

Metal binding sites, such as for Ca^{2+}, can also play a role in this capsid assembly. The addition of small amounts of divalent ions can electrostatically cross-link carboxylic acid groups and thus keep the capsid morphology intact or close pores. In red clover necrotic mosaic virus (RCNMV), divalent ions (Ca^{2+}) can be used to control the opening of its surface pores. This was

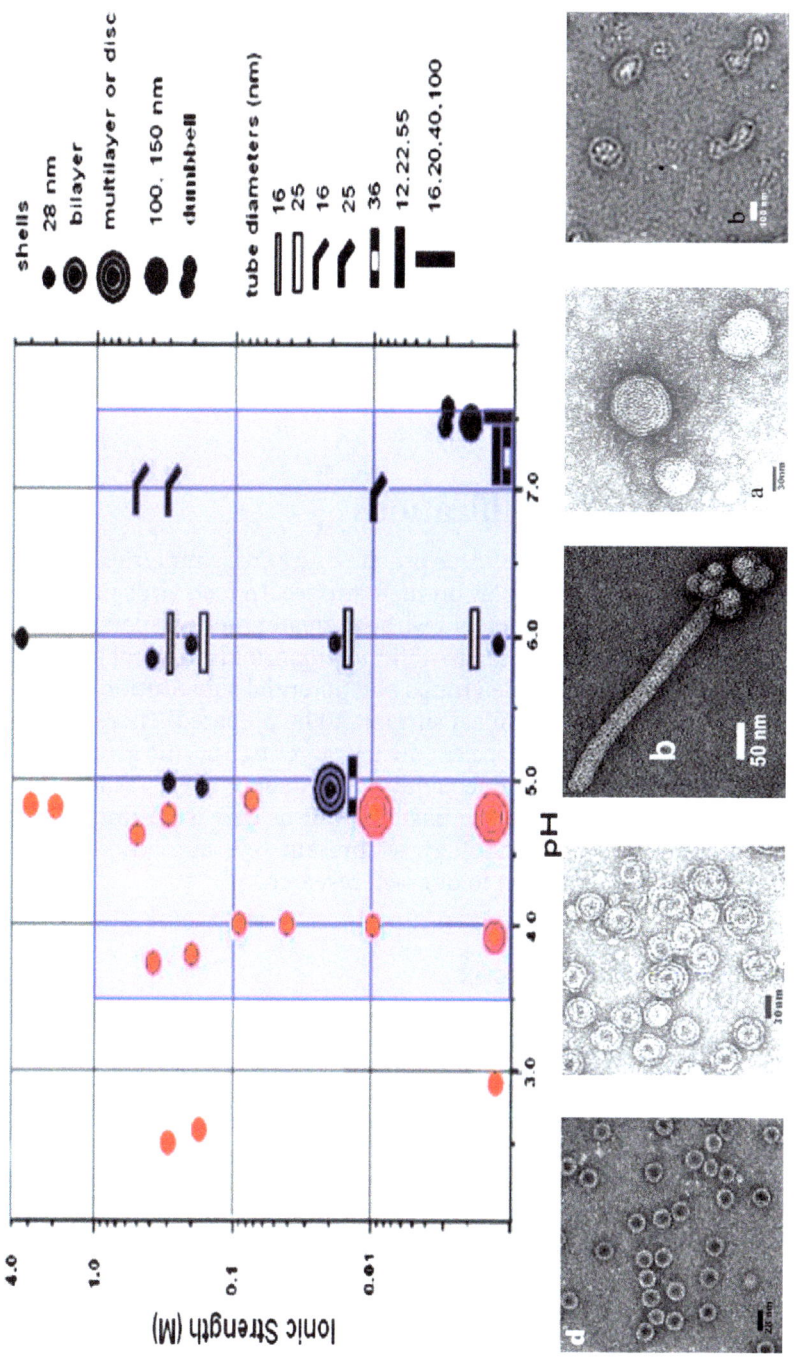

Figure 9.4 Top: the phase diagram of the CCMV protein assembly reveals a great variety of shapes and morphologies depending on the buffer conditions used. Bottom: TEM images of (left to right) single-walled, bilayered, tubular, multiwalled or disk, and dumb-bell morphologies. Adapted with permission from ref. 23. Copyright 2009 American Chemical Society.

used by Loo *et al.* to facilitate the controlled uptake and release of dye molecules.[24] Using EDTA to remove Ca^{2+} the pores can be opened, allowing dye molecules to be taken up, and subsequent addition of fresh Ca^{2+} trapped them inside. The release could be triggered by addition of EDTA or alternatively by disassembling the capsid shell by raising the pH to 10.

These metal binding sites can, however, also be used for other things. For instance, the metal binding capacity in CCMV was investigated by Basu *et al.*, showing that the metal binding sites in the protein structure are as predicted by its crystallographic structure.[25] The binding of metal was not dependent on the RNA or the positively charged N-terminus. In addition, Tb^{3+} was found to bind much more strongly than naturally occurring Ca^{2+}, showing that these capsids have the potential to bind potentially useful ions, *e.g.* for imaging. This was confirmed by Allen *et al.*, who used this to bind Gd^{3+} and confirmed that these capsids function as high-relaxivity magnetic resonance contrast agents.[26]

9.2 Synthesis and Modifications

Owing to the icosahedral structure, virus protein cages are symmetrical in the chemical functionalities they display on their surface. In fact, virus proteins have been compared to dendrimers, but with far greater monodispersity and easier synthesis. Moreover, their electrostatic and manifold chemical properties available from amino acid side groups and potential substitution procedures enable highly versatile chemical surfaces to be prepared. This surface can be easily modified using surface chemistry targeting specific groups, or by genetic modifications to introduce additional functionality.[27] This topic has been extensively reviewed in the past,[27–29] but as this is an important field within virus-based nanotechnology, we present several examples that we found most relevant in relation to our own research.

9.2.1 Chemical Modifications

Chemical modifications allow the protein to be equipped with a wide variety of functional groups. Amongst other things, this has permitted the formation of redox-active nanoparticles, templates for biomineralization, ligand attachment sites for high-contrast magnetic resonance imaging (MRI) agents, and charge-inversed VLPs.[30–32] The most common groups that are targeted are external amines from lysine residues and carboxylic acids (Figure 9.5), although by using naturally occurring pores or disassembled coat proteins, interior modification is also possible. Although thiol groups from cysteine residues are also a good target for chemical modification, this is generally after such a group has been genetically engineered into the coat protein, as they are usually not found on a native virus surface.

Determination of the amount of available chemical groups on a protein cage surface provides important information on the virus capacity to act as a chemical scaffold. An effective way to study this was reported by Barnhill *et al.*, who studied the addressability and reactivity of lysine and carboxylic acid groups on turnip yellow mosaic virus (TYMV).[33] To do this they used dyes with flexible linkers that contained reactive NHS esters for the lysine groups or could be coupled through EDC coupling in the case of carboxylic acids. The study showed 60 reactive lysines (one per monomeric subunit) and 180 carboxylic acids (three per subunit). More importantly, no self-quenching was observed, indicating that the reactive groups were evenly spaced at a good distance on the capsid. The reactivity of these groups is not always similar. For instance, CPMV has one easily addressable lysine that can be targeted using chemical modifications (Figure 9.6). Wang *et al.* explored this using

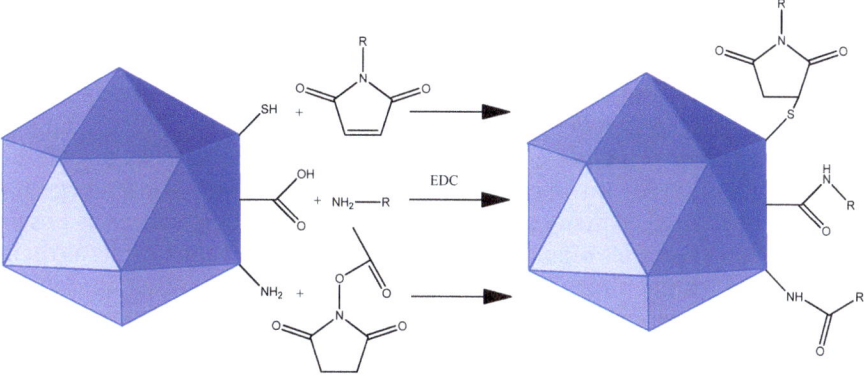

Figure 9.5   Cysteine thiol groups, lysine amine groups and carboxylic acid groups are common targets for chemical modification using maleimide, EDC [1-ethyl-3-(3-dimethylaminopropyl)carbodiimide] coupling or *N*-hydroxysuccinimide (NHS) esters.

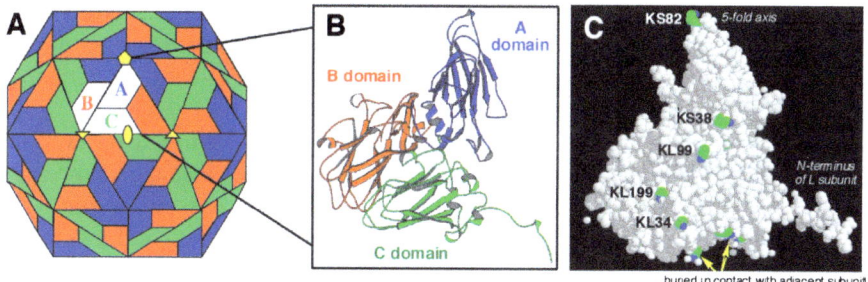

Figure 9.6   Exploring the capsid structure is key to understanding which residues might be available for chemical modification. In the CPMV structure presented (A, B), the amines of lysines in the small subunit (KS82 and KS38) and also of the three of the large subunit lysines (KL38, KL199 and KL199) seem solvent accessible, whereas the remaining lysine residues are buried in the structure (C).[34] Reprinted from ref. 34, copyright 2002, with permission from Elsevier.

acylation, protein digestion and mass spectrometry to determine the reactivity of the K38 reactive residue.[34] In addition, they explored the possibility of attaching dye labels and biotinylation. They showed that in addition to K38, up to four residues can be forced to react if a large excess of fluorescent dye [4000 equivalents per coat protein (CP)] is present. Similar work on other viruses also revealed that reactivity follows symmetry with a discrete amount of addressable groups being found on each capsid subunit.[33]

Chemically available amine groups occur on virtually all viral protein cages. The regular spacing of these groups essentially makes the virus ideal for applications that make use of such regularity and as such it makes a good substitute for dendrimeric structures, which after a certain degree of branching become increasingly difficult to synthesize.[35] Steinmetz *et al.* used this to create a redox-active VLP by decorating CPMV with ferrocene.[30] To achieve this, an NHS ester was coupled to ferrocene and subsequently reacted with the CPMV coat; 240 ± 10% redox active groups were found on the capsid, indicating that all available amines had reacted and therefore a good spacing of the redox centres had been achieved.

The dendrimer principle also enables the virus to be used as a scaffold for other species. For instance, the key to developing MRI contrast agents seems to be to load as many paramagnetic ions on a single carrier as possible. Virus nanoparticles offer an excellent scaffold for such materials as their symmetry, size and uniformity exceed those achievable with other attempted species, such as dendrimers. Additionally, they are by nature biocompatible. High relaxivity times have been achieved, for example, for MS2 by chemically attaching empty ligands capable of binding Gd^{3+} to lysine groups.[36]

The above-mentioned modifications are not only used to link functional groups, but can also affect the properties of the capsid. For instance, altering the charge on the exterior virus coat protein will change the interaction that it has with its environment. Aljabali *et al.* demonstrated this in their modification of CPMV by succinamate, converting amine groups into carboxylic acids (Figure 9.7).[37] This charge inversion enabled the protein coat to act as an efficient scaffold for the mineralization of iron oxide and cobalt nanoshells. Furthermore, these surfaces could be further modified with thiolated oligosaccharides.

Modification of a virus coat can cause it to dissociate into subunits if an energetically unfavourable state emerges. Artificial templates can, however, be presented to these coat proteins, such that they can reassemble the coat protein. CCMV modified with poly(ethylene glycol) (PEG) chains shows such dissociation into subunits. By presenting polystyrene sulfonate (PSS) as a template for the coat proteins, a smaller $T = 1$ capsid is formed (Figure 9.8).[38]

9.2.2   Genetic Modifications

One drawback to chemical modifications is that they rely on finding reaction conditions suitable for the capsid. Genetic modifications make use of natural mechanisms to introduce functionality to protein cages. A key issue

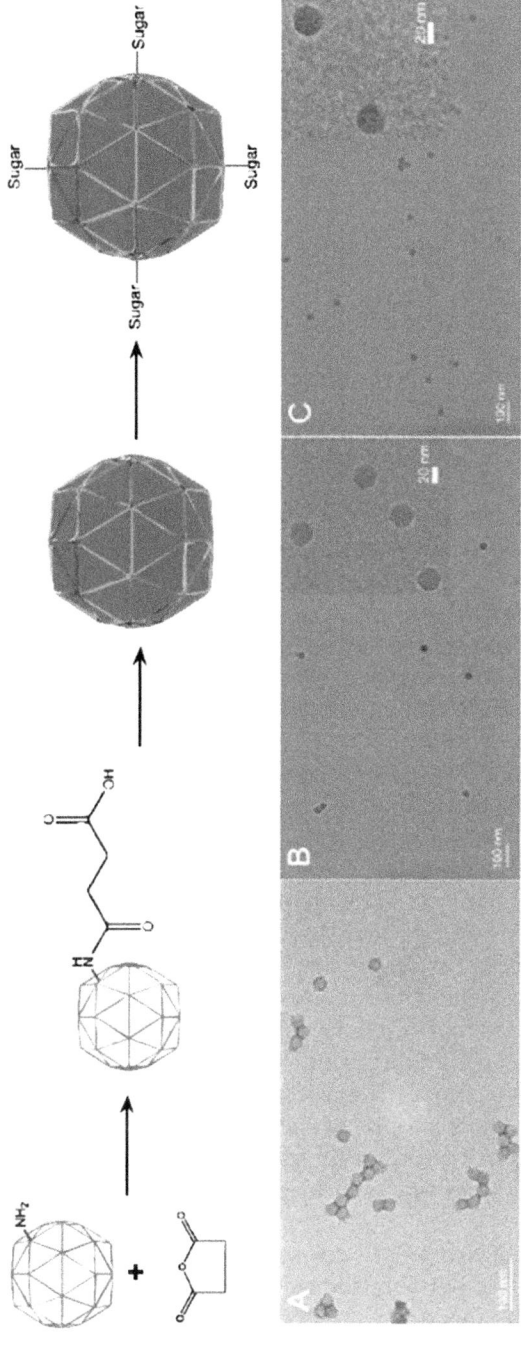

Figure 9.7 Top: synthetic pathway for the modification of CPMV with succinamate followed by mineralization of a nanoshell. Bottom: TEM images of (A) CPMV succinate, (B) CPMV with cobalt shell, and (C) CPMV with iron oxide shell.[37] Reprinted from ref. 37, copyright 2011, with permission from Wiley.

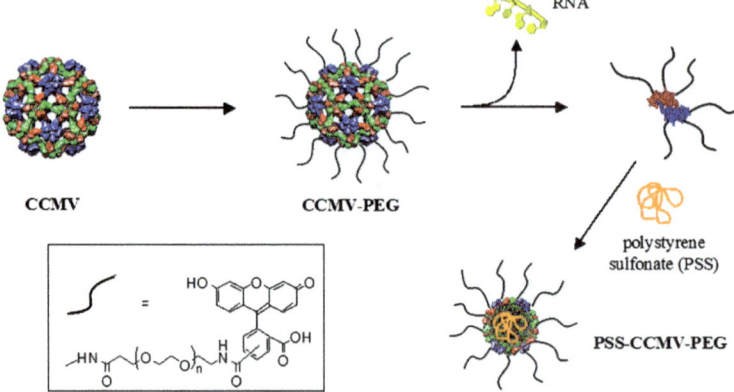

Figure 9.8   PEG chains on CCMV can render subunits unable to form capsid structures; however, the addition of an anionic template for the capsid
assembly can overcome this energetic barrier. Adapted with permission
from ref. 38. Copyright 2009 American Chemical Society.

here is to make a modification in the protein sequence without disrupting
its tertiary structure or ability to form a capsid shell. General strategies for
such modifications have used either point mutations on the surface of the
protein, modifications to the N-terminus or altering the composition of surface-exposed loops.[39] These modifications have allowed the charge alteration
of protein cages,[40,41] gold or nickel binding protein cages[42–45] and as anchoring points for further chemical modification.[32]

Introducing cysteine residues in the protein structure is perhaps the most
common modification.[39,46] This enables the protein cage to bind selectively
to gold and maleimide functionalized species, and has been used to anchor
cages to surfaces[47] and nanoparticles.[43] The symmetry of the capsid ensures
an even spacing of these modifications on the virus surface, which allows for
the organization gold nanoparticles when they are mixed with the modified
virus and bind to the surface cysteines. Using TEM, it was shown by Blum
et al. that gold particles of different size form a tight arrangement based on
the position of the cysteine thiol groups (Figure 9.9).[43] Taking this a step
further, the regular spacing of these gold nanoparticles make such VLPs
excellent candidates for the fabrication of conductive networks. Upon con-
necting the bound gold nanoparticles with $1,4\text{-}C_6H_4[\textit{trans}\text{-}(4\text{-}AcSC_6H_4C{\equiv}CPt{-}$
$(PBu_3)_2C{\equiv}C]_2$ (di-Pt) and oligophenylenevinylene (OPV), the formation of
distinctive conductive networks on the virus surface was revealed. The conductive properties of the networks are dependent on the size and spacing of
the nanoparticles.[48]

Genetically modified viruses can be used as anchoring points for a variety
of species as the location of the introduced surface group and also its nature
and orientation can be easily tuned. This allows for the use of the virus as a
scaffold to organize functional materials in a controlled manner. An example
of this can be found in CPMV with His-tags engineered on several positions

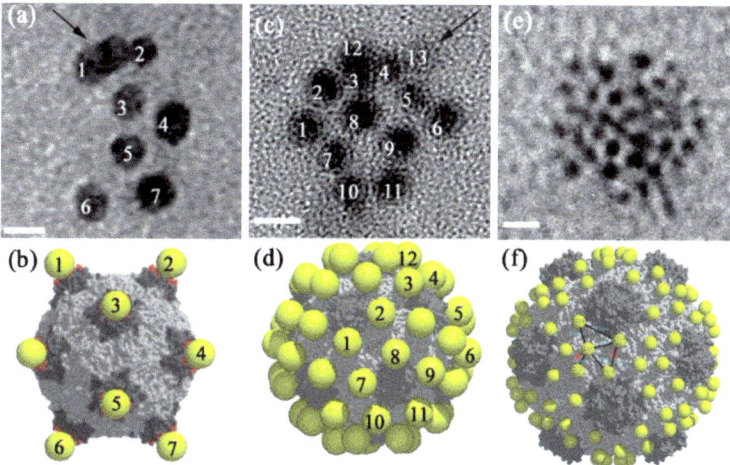

Figure 9.9 Three different cysteine mutants of CPMV (top to bottom: a & b, c & d and e & f) show good agreement between experimental TEM images of gold nanoparticle binding and theoretical predictions based on the capsid symmetry. Adapted with permission from ref. 43. Copyright 2004 American Chemical Society.

by Chatterji *et al.*[44] This modification enables CPMV to bind nickel in various positions on the capsid, each of which has different binding efficiencies and electrostatic properties. In this manner, protein shells with similar compositions and overall morphology can be given distinctly different properties, such as control over the electrostatics by altering the protonation of the histidine sequence.[44]

Introducing genetic modifications onto the surface also provides the ability to localize and orient the interactions on the coat proteins. Despite being composed of identical building blocks and having a high degree of symmetry in their structure, protein engineering can, in this way, be used to monofunctionalize a virus coat protein. This is achieved by coassembling modified and unmodified protein building blocks into a capsid, as was demonstrated by Li *et al.*[45] They engineered cysteine and histidine tags onto the surface of simian virus 40 (SV40). Coassembling these with native proteins and subsequently selecting the monofunctionalized VLPs with a nickel column yielded VLPs that could selectively bind a single gold nanoparticle using its surface-exposed cysteine group.

Recent studies have revealed that more complex modifications to virus capsids can take functionality even further. Not only combining genetic and chemical modifications but also coupling inorganic materials to them generates complex nanoarchitectures. Martinez-Moralez *et al.* used this technique to anchor iron oxide nanoparticles to the surface of the CPMV-T184C mutant.[49] To achieve this, they chemically functionalized the exposed cysteine group with an amine group and used carbodiimine chemistry to couple the carboxylic acid-coated iron oxide particles to the virus mutant. This

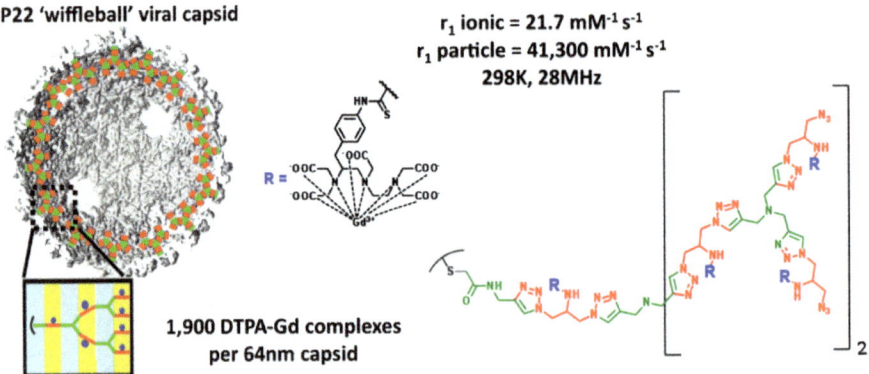

P22 'wiffleball' viral capsid

r_1 ionic = 21.7 mM^{-1}s^{-1}
r_1 particle = 41,300 mM^{-1}s^{-1}
298K, 28MHz

1,900 DTPA-Gd complexes
per 64nm capsid

Figure 9.10   The 'wiffleball' morphology of a mature bacteriophage P22 capsid allows for the growth of a branched oligomer due to diffusion through holes in its structure. Adapted with permission from ref. 32. Copyright 2009 American Chemical Society.

enabled them to obtain hybrids, linking multiple iron oxide nanoparticles to the virus surface. These hybrids showed an enhanced magnetic response due to dipole coupling between the regularly spaced iron oxide particles, showing the potential of such constructs for use as MRI contrast agents.

Qazi *et al.* used this approach by introducing a cysteine group to the capsid interior of bacteriophage P22 and coupling *N*-propargyl bromoacetamide to the capsid to act as the starting point for the synthesis of a branched oligomer (Figure 9.10).[32] This was possible due to P22's 'wiffleball' morphology that ensured sufficiently large holes for the reagents to flow through freely. The oligomeric components doubled the Gd^{3+} content with each branching step, thus quickly filling the capsid with the paramagnetic ions. A total of 1900 Gd^{3+} ions can be loaded into the capsid, ensuring a significant improvement in particle relaxivity compared with previous systems.

Protein engineering can be taken a step further in another direction. By creating a fusion protein comprising a capsid subunit and another (functional) protein, it becomes possible to encapsulate these secondary proteins within the confinement of a virus shell. This can be done by mixing the fusion protein with unmodified subunits and triggering capsid formation. Depending on the ratio, it would even allow different, albeit statistical, encapsulations of protein into the capsid. Patterson *et al.* demonstrated this for the encapsulation of CelB glycosidase inside P22.[50] Furthermore, these VLPs demonstrated conservation of enzyme activity, unlike previous encapsulations of enzymes, which showed an enhancement.

9.2.3   Biomineralization

Genetic manipulation also introduces biological functionality into a virus cage, which plays a role in biomineralization. The synthesis of monodisperse, well-defined nanoparticles and materials is a field in which biomineralization

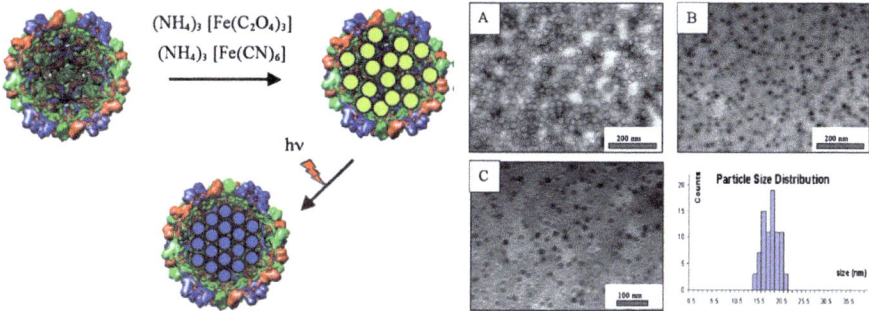

Figure 9.11 Left: synthetic pathway for the diffusion of monomers into CCMV and subsequent synthesis of Prussian blue particles inside the capsid. Right: uranyl acetate-stained (A) and unstained (B, C) TEM images reveal that the size distribution of the particles approximates the interior diameter of CCMV.[31]

offers an interesting perspective. Biological organisms have adapted a wide variety of means of controlling the crystallization of inorganic minerals on various length scales. Using empty virus protein cages as a scaffold for such crystallization is therefore no great leap. And whereas certain nanoparticles grow in native virus capsids, others can be fabricated using recombinant capsid proteins, with examples given below. This biomineralization inside the protein follows two main strategies: either a specific protein motif is engineered into the virus genome that facilitates the nucleation of a specific mineral or non-specific diffusion mechanics are used.

The interior of a virus protein cage is designed by Nature to interact with RNA or DNA. However, this environment is also ideally suited for the mineralization of various inorganic nanoparticles provided that the conditions for mineralization conform to those present in the protein cage. Biomineralization of this kind in viral capsids takes its original inspiration from the iron-binding properties of the ferritin protein capsid. This protein cage facilitates the reduction of iron salts to form iron oxide inside its cavity. Douglas and co-workers used CCMV capsids as a mimic of ferritin.[51-53] It shows a good affinity for the biomineralization of tungsten, vanadium and molybdenum salts. The key to this process seems to be the ability of virus capsids to have material diffuse through their pores and present a template for the crystal growth. Diffusion mechanics are greatly enhanced when negatively charged precursors are used, such as is the case for the Prussian Blue $\{NH_4Fe^{II}[Fe^{III}(CN)_6]\}$ nanoparticles synthesized inside CCMV reported by de la Escosura *et al.*, where the local apparent concentration of iron inside the capsid far exceeded the concentration of the precursor solution.[31] These VLPs are generally size and shape controlled by the templating virus capsid (Figure 9.11).

Diffusion does not always require charged precursors. An empty, unmodified CPMV protein mantle has been shown to facilitate the growth of iron oxide and cobalt. This simply uses the permeability of the protein coat to small ions. After incubation for 30 min (for $CoCl_2$) or overnight (for iron sulfates) a reducing agent can be added for 30–40 min, after which the crystals are formed inside the protein cage. The protein cage can then be modified, for instance by biotinylation, to allow for further interaction.[54]

Protein cages are highly robust structures that can be genetically engineered to alter the interior environment without losing their structure and external functionality. This was applied by Douglas *et al.* to CCMV to create a true mimic of ferritin in which an iron oxide particle was mineralized inside a genetically modified cavity.[40] To achieve this, nine basic residues on the N-terminus were substituted for glutamic acid (the subE mutant). This subE mutant therefore displayed an acidic rather than a basic interior cavity, which allowed the growth of monodisperse iron oxide nanoparticles. In this case, the CCMV cavity was modified to act similarly to some natural iron-binding ferritin protein cages.

By engineering specific protein sequences onto virus capsids, they can act as a scaffold for the mineralization of inorganic nanoparticles. These sequences are designed to act as a template for one specific compound. Steinmetz *et al.*[55] and Shah *et al.*[56] used a surface exposed loop on the small subunit of the CPMV coat protein as the insertion point for such mineralization sequences, which enabled them to grow thin shells of iron–platinum and silica. These particles are potentially of interest in magnetic and optical studies owing to the well-defined monodisperse template that the shells of these materials form. Furthermore, these crystallization processes can be performed in aqueous media without the need for organic solvents, allowing for green chemistry.[55,56]

9.3 Biohybrid Structures of Virus-Like Particles

As discussed, virus protein cages can be modified and altered in different ways to facilitate a variety of interactions. To build on this, and to introduce further interactions, biohybrid structures can be created out of artificial (synthetic) templates and capsid proteins to form a new class of nanomaterials. In this way, they combine the advantages of viruses, such as monodispersity, symmetry and ease of modification, with a wide variety of physical properties. Among other things, this has led to functional nanoreactors and plasmonic crystals and it has enabled the capsid size and morphology to be controlled.[57–60]

9.3.1 Electrostatic Loading

In 1969, Bancroft *et al.* showed that CCMV, CPMV and BBMV (broad bean mottle virus) display the ability to form a spherical shell when presented with a large variety of flexible charged polyanions.[61] Rigid polyanions, such as double-stranded DNA, can sometimes even trigger the assembly of tubes or other non-spherical assemblies. Therefore, although a virus coat protein is designed to encapsulate negatively charged RNA or DNA, it seems equally suited to encapsulate anionic polymers and organic molecules that mimic this genetic material. For instance, anionic PSS and poly(acrylic acid) (PAA) are encased in hibiscus chlorotic ringspot virus (HCRSV), whereas neutral dextran polymers of similar length do not. The VLPs formed showed similar electrostatic and size properties compared with native viruses.[62]

Controlling the morphology of a virus capsid with a polyanion permits the formation of new protein architectures induced by this the template. For instance, Caspar–Klug symmetry predicts that coat proteins of virus particles can form stable capsid shells of various sizes, which differ from their native conformation. This was investigated by introducing short anionic polymers to virus coat proteins. For example, Sikkema *et al.* found that PSS of 9.9 kDa could assemble CCMV coat protein into a stable 16–18 nm monodisperse $T = 1$ capsid.[59] Further work on the encapsulation of PSS in CCMV revealed that for higher mass polymers ($M_w > 400$ kDa) the capsid can form the $T = 2$ and even the native $T = 3$ morphology if the mass is above 2000 kDa.[63] As such, RNA viruses are shown to undergo electrostatic assembly upon addition of a charged polymer.

As discussed, for the encapsulation of genomic material, most RNA viruses require a charge ratio of 1.6 negative charges to compensate each positive charge. When Hu *et al.*[63] and Cadena-Nava *et al.*[64] used fluorescently labelled PSS to track the amount of PSS to be encapsulated, they found that CCMV can encapsulate an undercharge of 0.6 or 0.45 negative charges for each positive charge, whereas ≥400 kDa PSS displayed a 9:1 overcharge. These results indicate that other factors also contribute significantly to the encapsulation process. Theoretical simulations predict that the initial amount of polymer needed ($\langle \varphi \rangle *$) scales with the surface charge of the capsid interior (σ) and is indirectly proportional to the charge on the polymer (α) and the interior radius of the capsid (R), following $\langle \varphi \rangle * = 6\sigma/R\alpha$.[65] Alternatively, this could be explained in part by the entropy factor present in the encapsulation, as theoretically simulated by Castelnovo *et al.*[18]

Usually the formation of a protein shell follows Caspar–Klug symmetry; however, when an incompressible template is presented, the protein shell can be forced to assume additional symmetries due to the size and curvature of the template and naturally occurring kinetic traps. This was shown by assembling CCMV coat proteins around nanoemulsion droplets that were stabilized by anionic surfactants (Figure 9.12). Depending on the size and curvature of these droplets, the protein either followed Caspar–Klug symmetry or showed locally disordered states, such as hexagonal sheets of protein.[66] Similarly, a rigid template can force CCMV to be organized into tubular nanostructures. Mukherjee *et al.* showed that double-stranded DNA can cause such assembly.[67] This assembly is dependent on the ratio of CP to DNA base pairs (BP), with excess of CP forming short 17 nm diameter tubes (similar to the diameter of $T = 1$ capsids), while an excess of 10 BP per CP affords longer, but narrower, tube-like structures.

The principles applied to the encapsulation of polymers also hold true for other molecules. Anionic molecules can provide a template and cause the

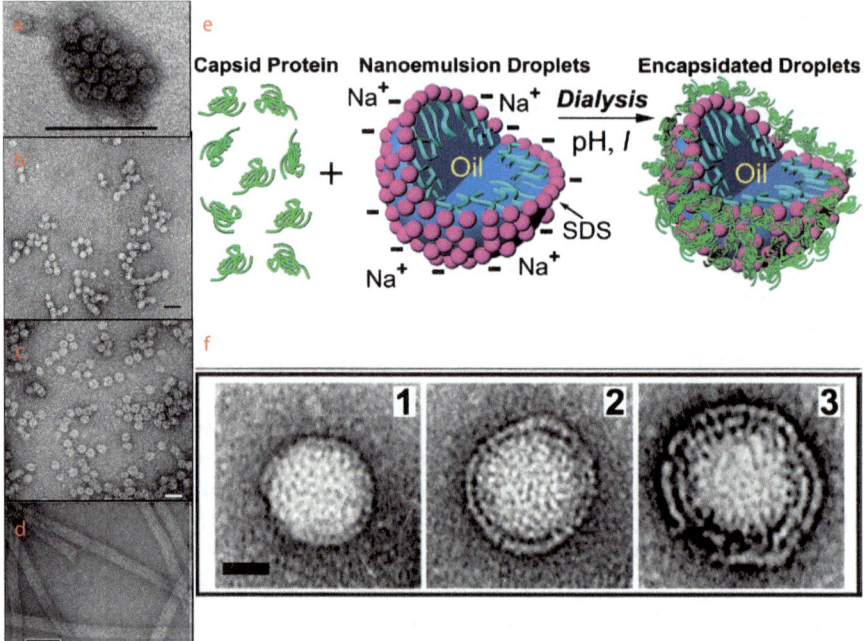

Figure 9.12   CCMV can be assembled into a variety of architectures based on the
template presented: (a) 16 nm T1 capsid using PSS;[59] (b) 22 nm T2
capsid using PSS;[63] (c) 27 nm T3 capsid using PSS;[63] and (d) 17 nm
diameter tubes using double-stranded DNA.[67] (e) Scheme showing the
encapsulation of nanoemulsion droplets in CCMV; (f) TEM images
of nanoemulsion droplets encapsulated by one, two or three protein
shells depending on buffer conditions.[66] (b, c) Reprinted from ref.
63, Copyright 2007, with permission from Elsevier. (d) Adapted with
permission from ref. 67. Copyright 2006 American Chemical Society.
(f) Adapted with permission from ref. 66. Copyright 2008 American
Chemical Society.

capsid to assemble. As such, the electrostatic loading of a virus shell is also
a strategy that can be employed to change the nature of the virus cavity. For
instance, micelles or DNA-amphiphiles can be introduced into the capsid. This
effectively creates a hydrophobic cavity that is stabilized by the virus shell.[68]

9.3.2   Nanoparticle Encapsulation

Electrostatic encapsulation is not limited only to organic molecules and polymers. The same can be applied to rigid inorganic nanoparticles, allowing
the creation of biohybrid structures that combine protein cages with inorganic physical properties and states, such as superparamagnetism, plasmon
absorption and similar confined electromagnetic states.[60,69]

Dragnea *et al.* first showed the tendency for citrate- or tannic acid-capped
gold nanoparticles to be trapped inside the virus capsid of BMV upon reassembling a capsid in the presence of these particles.[70] Small particles were shown
to be tightly bound inside the capsid and exhibited a change in spectroscopic

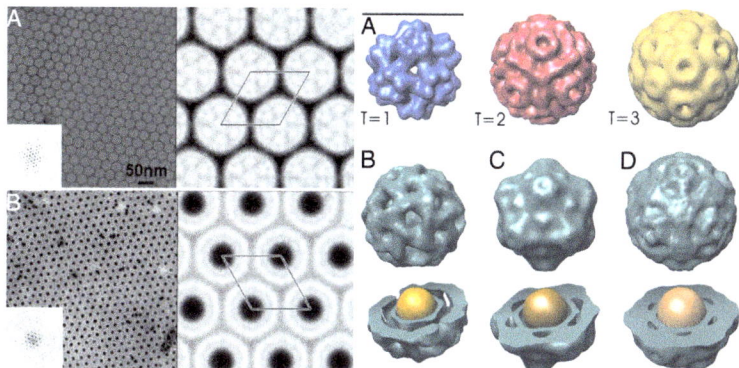

Figure 9.13 Encapsulation of AuNPs in CPMV leads to core-controlled polymorphism. Left: TEM images of (A) CPMV and (B) CPMV–AuNP. Right: (A) T = 1, T = 2 and T = 3 models from the VIPER database structures for CPMV and 3D cryo-TEM reconstructions showing (B) pseudo T = 1, (C) pseudo T = 2 and (D) pseudo T = 3 morphologies for different sized AuNP encapsulated in the capsid.[58] Reprinted with permission from ref. 58, copyright 2007 National Acadamy of Science. U.S.A.

properties. Although these surface ligands carry negative charge, the role of surface charge is more prominent when DNA linkers are attached to the gold particle prior to encapsulation.[71] By using a small DNA or RNA chain bound to a nanoparticle, it is possible to form a nucleation site for the coat protein shell. After incubation of this complex with the coat protein of red clover necrotic mottle virus, lowering the pH to trigger capsid formation is sufficient to enable this nucleation point to fully encapsulate the nanoparticle. The relative density can then be used to separate the full and empty protein shells.[72]

Encapsulation of a rigid gold nanoparticle core might therefore be stimulated by either the electrostatic interaction between the coat protein and the scaffold, or *via* a specific origin of asscmbly (OAS). This is done through a two-step encapsulation pathway. It involves first lowering the ionic strength in a neutral buffer to facilitate aggregation of proteins on the nanoparticle (NP) surface and second lowering the pH to trigger capsid shell formation. As such, only a fraction of the anionic AuNP is encapsulated. The efficiency of this encapsulation is often defined as $N_{AuVLP}/N_{AuTotal}$ and is normally only 2–3% for citrate ligands on gold. A flexible linker molecule can be attached to the nanoparticle surface to act as nucleation sites for capsid assembly, increasing the efficiency by up to 95%. Furthermore, by tuning the gold nanoparticle size, different core morphologies can be selected conforming to the T = 3, T = 1 and even metastable T = 2 morphologies (Figure 9.13). The encapsulation efficiency tends to vary with particle size as a competition between empty and filled capsid shells occurs. The resulting capsids are comparable to the native virus in that they show similar exterior characteristics and can be crystallized. These VLP crystals show interesting optical properties that differ from those of the free VLPs, likely due to electromagnetic interactions that are enhanced by the close proximity of the particles.[58,73]

Carboxylic acid-terminated triethylene glycol (TEG)-coated AuNPs can also be introduced into CCMV *via* the two-step encapsulation pathway. Similarly to BMV, the VLP size depends on the core size present. Surprisingly, it was found that N-terminus-deleted CCMV, which lacks many of the positively charged residues associated with electrostatic binding of RNA or polyelectrolytes, still showed an encapsulation efficiency of 72%. It was suggested that this could be of use in biomedicine by replacing such sequences with functional handles.[41] We consider that this also indicates that the (re)assembly of the virus coat protein is favoured by the presence of a template.

In addition to gold nanoparticles, BMV also has the potential to assemble around other inorganic cores. The key here is to manipulate the surface ligand selectively to create an anionic surface around the nanoparticle. For instance, oleic acid–iron oxide nanoparticles were modified by adding a carboxylic acid-terminated PEG chain attached to a phospholipid on their exterior surface (Figure 9.14). Subsequent assemblies showed the potential for core-controlled polymorphism even beyond the native $T = 3$ symmetry and displayed superparamagnetic behaviour in the particles. However, the encapsulation efficiency for the 10.5 and 8.5 nm particles was low (5% and 3%) compared with similar-sized gold nanoparticles. This likely results from the fairly spacious linker, which reduced the overall surface charge density due to folding of the PEG chain.[74] Anionic quantum dots can also be encapsulated inside BMV. Dixit *et al.* coated these particles with the same PEG and DNA ligands that were used to encapsulated gold nanoparticles.[75] Unlike the gold nanoparticles, however, multiple quantum dots were observed inside the virus shell.

Studies employing gold nanoparticles displaying a mixture of COOH- and OH-terminated PEG chains as surface ligands showed that the surface charge density plays an important role in the templated assembly of the BMV protein cage. Below a critical surface charge concentration little encapsulation is observed, whereas above it the efficiency increases drastically with increasing surface charge. VLP assembly is initiated at neutral pH where the nanoparticles are fully charged and subsequently brought to acidic pH to

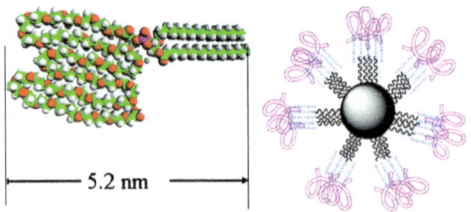

5.2 nm

Figure 9.14   Left: a carboxylic acid-terminated PEG chain is attached to a phospholipid (PL). Right: this lipid intercalates with oleic acid groups on the surface of an iron oxide nanoparticle to allow the encapsulation in BMV. COOH–PEG–PL is shown in pink (PEG) and blue (PL) with black oleic acid covering the iron oxide nanoparticle. Adapted with permission from ref. 74. Copyright 2007 American Chemical Society.

trigger shell formation. These results indicate the necessity to form a critical nucleus of coat proteins through electrostatic association, thus permitting further growth and encapsulation on decreasing the pH.[76]

Recent work by Tsvetkova *et al.* showed that both cooperative and non-cooperative absorption of protein subunits can promote the assembly of a virus capsid around a nanoparticle, but the pathway is dependent on the surrounding medium.[77] These studies indicated that the organization of BMV proteins around anionic AuNP shows such behaviour. At acidic pH, initially little capsid formation is observed, until a critical concentration is reached. Presumably, at that point, the coat protein has a sufficiently large nucleus to promote further cooperative assembly. At neutral pH, when the AuNP charge is far higher, the protein readily adsorbs on the nanoparticle, until it is saturated. This is likely due to a greater charge interaction with the nanoparticle surface, in that way excluding the need for a nucleus that facilitates cooperative encapsulation.

Gold nanoparticles can be coated with PEG or DNA to prepare them for encapsulation into simian virus 40 (SV40) capsids. DNA-coated particles of all sizes can be encapsulated, but neutral PEG-coated particles are only encapsulated if their core size is at least 15 nm. This size dependence is in contrast to the normal electrostatic interaction and seems to be dependent on the total size of the PEG–AuNP construct, not merely the gold core.[78] This furthermore suggests that a charged template is not always needed. Li *et al.* also showed that SV40 can encapsulate both anionic and cationic quantum dots with similar efficiencies.[79] This might be attributed to the difference in the charge landscape of its coat protein. Whereas the isoelectric points of the inner coat for most virus capsids that have been studied are well above the pH of the assembly conditions used, SV40 has an isoelectric point close to this pH. This could explain the ability to encapsulate particles regardless of charge, indicating that the type of virus protein coat has a significant effect on the encapsulation.

For encapsulation of anionic AuNP cores in animal alpha viruses, the encapsulation efficiency was never found to be above 62%.[80] However, in simpler viruses, such as red clover necrotic mosaic virus (RCNMV), capsid formation is also sometimes induced by more than electrostatics. In this case, an OAS is required, comprised of a small RNA or DNA sequence found within the viral genome, which pre-organizes some of the monomers, which subsequently triggers the further capsid formation. Loo *et al.* showed that by attaching this sequence to a variety of nanoparticles, such as Au, CdSe and $CoFe_2O_4$, capsid formation around the nanoparticle will occur.[81] This encapsulation process is limited by the size of the nanoparticle, as particles in excess of the natural cavity cannot be encapsulated.

Gold nanoparticles encapsulated in a viral coat protein present a stable scaffold with a large plasmon response. These physical properties can be exploited for their plasmon properties, as the virus shell provides sufficient spacing to prevent quenching effects. Instead, a dye that is positioned on the surface of such a particle is expected to show enhanced fluorescence as the dipole is enlarged due to interaction with the electric field of the gold plasmon. Capehart *et al.* used such gold nanoparticle-based VLPs and attached a fluorophore with a DNA linker to the surface and showed that the fluorescence enhancement depended on the linker length and thus the spacing of the fluorophore to the gold nanoparticle (Figure 9.15).[82] Similar results have been achieved with other nanostructures.[83–85] However, the virus capsids allow for a scaffold that is easily formed around a variety of nanoparticles and yet maintains similar surface functionality and chemical addressability.

9.3.2.1 Simulations

Simulations by Hagan confirmed that strong core–shell interactions can lead to core-controlled polymorphism.[86] However, these simulations predict that the stability of the assembly can be undermined if the curvature of the core is incompatible with the subunit–subunit interaction that results in the lowest free energy. Therefore, although core-controlled polymorphism is allowed, it leads to the formation of metastable particles. Additionally, chemisorption

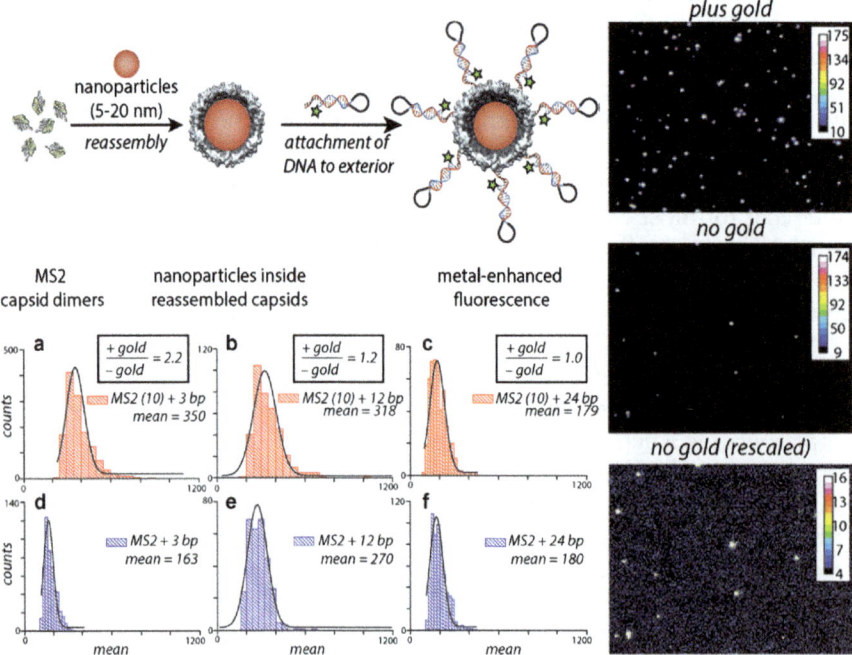

Figure 9.15   Top left: encapsulation of NP inside and attachment of fluorescent dye to the exterior of MS2. Right: total internal reflection microscopy (TIRF) of CPMV–dye with and without a gold core. Bottom left: TIRF intensity histograms for MS2 with AuNP (red) and without AuNP (blue) with dye at 3 (a, d), 12 (b, e) or 24 bp (c, f) distance spacing showing the relative increase in fluorescence. Adapted with permission from ref. 82. Copyright 2013 American Chemical Society.

of subunits could lead to kinetically trapped states that prevent proper shell formation. More flexible cores overcome the curvature issues, but kinetic traps might occur more frequently.

As shown by these simulations, the ability of protein subunits to self-assemble around a rigid electrostatic core is dominated primarily by the surface charge density of the nanoparticle. After a certain threshold value has been reached, the encapsulation efficiency is nearly 100%, although before that encapsulation is stunted. The theory is incomplete in that it does not account for the metastable phases that seem to occur in experimental results before the threshold, which explains a gradual increase in efficiency with increase in surface charge. As the early stages of the assembly are dominated by electrostatic adsorption on the nanoparticle surface, an excess of the coat protein is needed to allow for rapid encapsulation as otherwise the process requires desorption of coat protein before the capsid shells can be completed.[87]

A minimal charge density on the surface of a nanoparticle is required in order to trigger the assembly of a coat protein into a core-controlled sized VLP. Siber *et al.*'s simulations (Figure 9.16) showed not only that the core diameter of the inorganic nanoparticle core affects the size of the final assembly, but also that the surface charge density plays an important role. In particular, the flexible N-terminus that contains the positive charge, such as found in CCMV, can act as a spacer for sufficiently charged particles, in which the energy landscape will favour a different size assembly. Depending on the particle diameter and surface charge, multiple transitions between smaller and larger VLP assemblies are observed.[88]

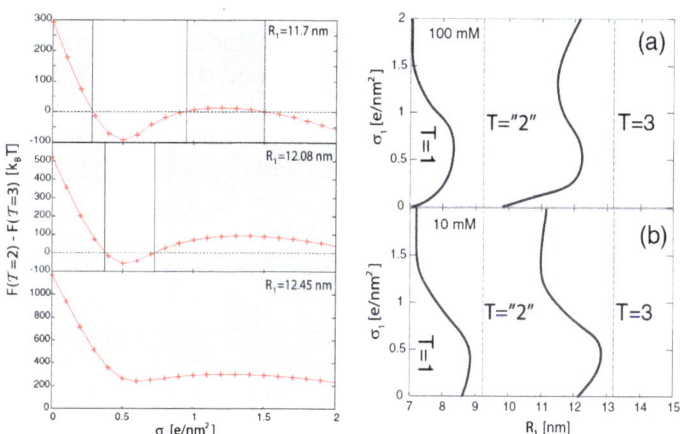

Figure 9.16   Left: the difference between the $T = 2$ and $T = 3$ capsid free surface energies for different core radii as a function of core surface charge density, showing either $T = 2$ (white) or $T = 3$ (grey) as being the energetically favoured assembly. Right: the energetically most favoured capsid morphology as a function of both charge (σ_1) and core radius (R_1) at (a) 100 mM and (b) 10 mM of monovalent salt.[88] Reprinted with permission from ref. 88, copyright 2010 American Physical Society.

Anionic nanoparticles are typically stabilized by weakly acidic groups. Additionally, virus proteins tend to be stable and assemble around physiological pH or slightly acidic pH (5–7). As such, most encapsulation experiments have been performed around the pK_a of the acidic groups, which in the case of small particles can lead to significant charge variations between the particles, an effect that is less pronounced for larger nanoparticles. Simulations by Lin *et al.*[89] showed that this accounts for the gradual increase in encapsulation efficiency beyond the critical charge density, rather than an immediate sharp rise as was predicted by Hagan.[87]

9.3.3   Outer Surface Electrostatics

Although the inner surface might readily and controllably promote the encapsulation of a wide variety of species, the outer surface also carries the potential for electrostatic interaction. Similarly to genetic and chemical modifications, the outer charged shell is dependent on the locally available groups and is highly symmetrical. In effect, this creates a surface which, more than the inner surface, carries patches of charge in a highly symmetrical pattern. Similarly to the inner surface, these patches can be used to bind polyelectrolytes or nanoparticle.

Binding polymers to the surface of a virus can be used to transform the virus particle into a template for the synthesis of an inorganic shell. For instance, Evans and co-workers electrostatically bound polyallylamine hydrochloride to the surface of CPMV to promote the absorption of gold nanoparticles.[90] These particles could subsequently be incubated with gold salts, which after reduction formed a gold shell around the virus particle.

Owing to the capsid symmetry, the electrostatic interactions of protein cages with small metallic nanoparticles allowed the creation of organized nanostructures. Not only does the symmetry of the protein cage result in a symmetrical distribution of charges, but often these charges are concentrated in patches on the surface, which, depending on pH and ionic strength, allow for multivalent electrostatic interactions at specific sites. Kale *et al.* used this to organize 5 nm CdS quantum dots on the surface of P22 bacteriophage capsids.[91] In this case, structures with hexagonal and pentagonal organization of quantum dots on the 60 nm capsid were obtained at pH 4, below the isoelectric point of 4.97, where the subunits appear to pop out of the capsids. The organization was lost at around pH 9.

Both internal and external functionalization of the capsids can be combined in the same structure. For instance, a protein mantle can be formed around a nanoparticle and subsequently coated with different nanoparticles, adhering to the outer shell to create complex architectures. Owing to the supramolecular nature of the interactions in these architectures, they can be easily manipulated using pH and ionic strength. SV40 capsids have been used to demonstrate this for gold nanoparticles adhering to amine-rich spots on the surface of an SV40 VLP containing a variety of different nanoparticles inside the VLP (Figure 9.17).[92]

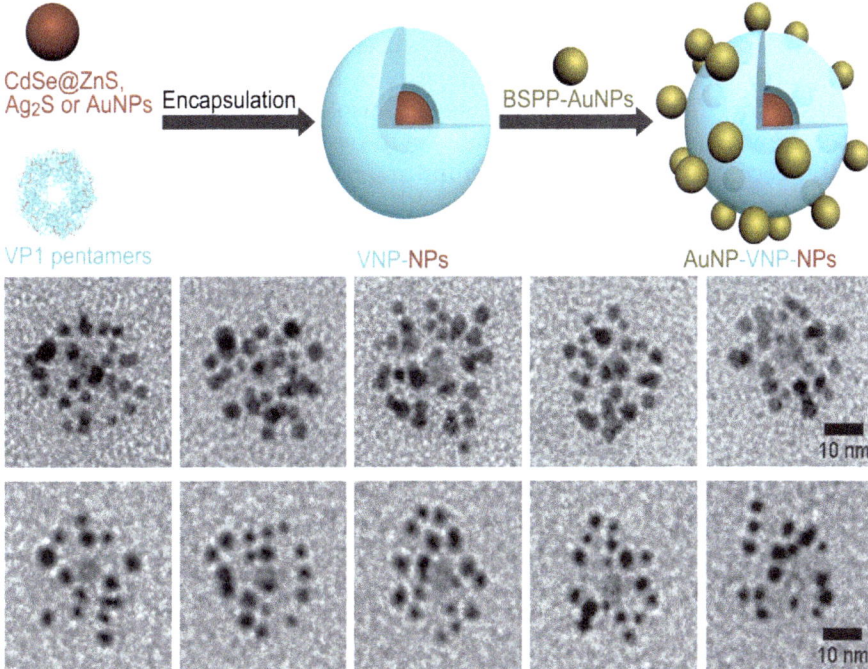

Figure 9.17   Top: scheme for combining internal and external electrostatic assembly for the creation of complex architectures. Bottom: TEM images of AuNP assembling around an SV40 capsid containing a CdSe/ZnS quantum dot.[92] Reprinted with permission from ref. 92, copyright 2012 John Wiley and Sons.

9.4   Organization of Viruses

As shown above, the organization of particles or molecules on viruses allows the fabrication of organized nanostructures on a single VLP. This can be taken a step further, for example, by linking multiple VLPs to a single surface, by layer-by-layer assembly or by non-covalent clustering. These architectures range from simple monolayers to layered structures and even complex hierarchical 3D crystals. These virus-based nanomaterial assemblies can be separated into two broad groups based on the particle interaction: either as covalent and biomolecular assembly, or as electrostatic assembly.

9.4.1   Covalent and Biomolecular Assembly

Using chemical linkers to direct viruses to assemble onto surfaces in 2D assemblies or layer-by-layer into 3D assemblies allows strong binding and precise control over the composition of the layers that are formed. Either the virus is crystallized first and subsequently stabilized using chemical cross-linkers or, to promote a specific interaction, a recombinant VLP can be used, containing, for instance, cysteine residues or His-tag loops. Two approaches to control this organization exist: (i) top down (pre)patterning of a surface or scaffold, or (ii) dynamic self-assembly at an interface or by inter-particle interaction. The latter generally shows a greater degree of control over the structure's size and order. We attribute this to the symmetry of the virus, which in dynamic self-assembly becomes an important structural and packaging parameter.

Under well-chosen conditions, virus particles can be crystallized; however, these crystals often need to be stabilized using chemical cross-linking. For example, Russell and co-workers showed that CPMV particles can be organized on the interface in a mixture of perfluorodecalin or chloroform and water.[93,94] The resulting close packing of virus nanoparticles was cross-linked using glutaraldehyde. No disruption of the virus nanoparticle integrity was observed; however, the resulting membranes would crumple upon the removal of the solvents. These cross-linked systems can be used as porous scaffolds for the synthesis of nanocomposites. To demonstrate this, Falkner *et al.* incubated such a crystal with a precursor solution containing palladium ions and subsequently platinum ions, which were catalytically reduced to generate metal deposits inside the virus scaffold.[95]

Genetically modified CPMV that presents a cysteine group on the exterior surface of the capsid can be attached to a gold surface that has been modified with maleimide groups, as shown by the formation of a layer of cysteine-labelled CPMV. Using a top-down strategy, this layer can subsequently be patterned, using e-beam lithography, leaving 30 nm wide lines of CPMV on the surface, in effect showing a pseudo-1D pattern of the viruses.[47] As an alternative strategy, the surface itself can be pre-patterned. To do this, Cheung *et al.* further modified the CPMV cysteine with six His-tags.[96] This enabled them to form a surface pattern with lines of Ni-NTA, allowing binding of the modified CPMV to the surface on these lines. In both cases, no higher order organization within the lines could be observed, or in the 2D patterns that were made from them.

Using a self-assembly approach may thus make it easier to organize viruses on a surface, by controlling and limiting the number of interactions. Klem *et al.* used A163C genetically modified CCMV virus capsids and self-assembled them onto Au surfaces.[42] They found that capsids that have a limited number of surface-exposed thiols show a greater degree of organization than capsids that have all of the 180 thiol groups exposed.

Moving towards a 3D pattern, an interlayer can help with the organization of virus particles without the need to modify viruses separately. Steinmetz *et al.* used biotinylated CPMV for the creation of alternating layers of CPMV linked by streptavidin molecules.[97] Before assembly into layers, different dyes were coupled to CPMV to monitor the process. The layer-by-layer process was thus demonstrated to create unique and separate layers that showed no intermixing (Figure 9.18). To enhance the biotin–streptavidin–CPMV assembly further, the effect of the density and spacer length of the biotinylation on CPMV was studied. Depending on these factors, the order within each layer can be affected. More importantly, shifts in the frequency

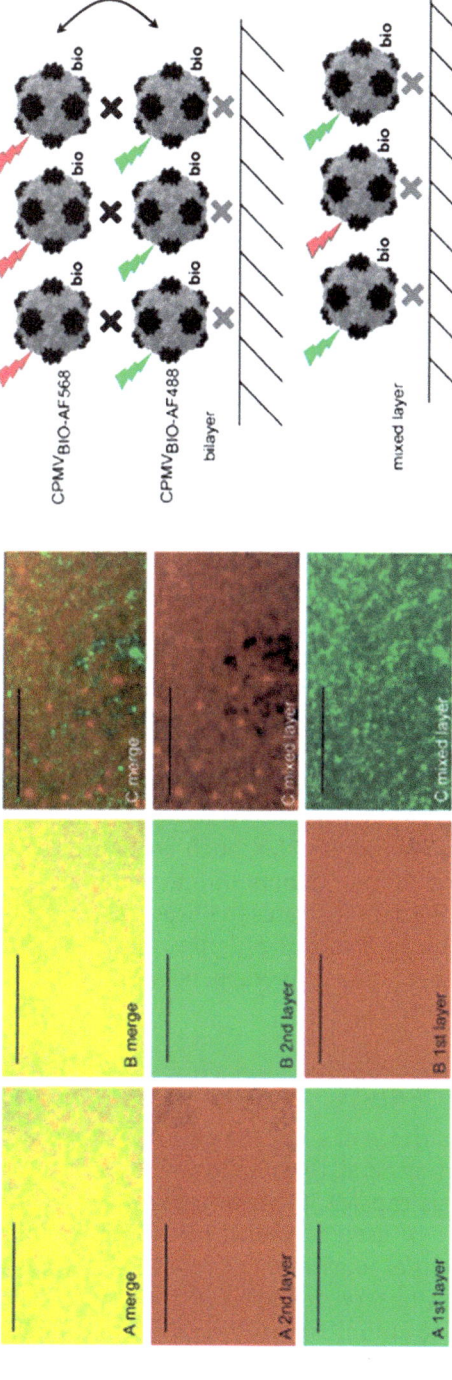

Figure 9.18 Left: fluorescence microscopy images of biotin–streptavidin–CPMV bilayers of (A) AF488-labelled CPMV layer followed by AF568-labelled CPMV layer, (B) AF568-labelled CPMV layer followed by AF488-labelled CPMV layer and (C) both AF-568 and AF-488 dye labels attached to CPMV in each layer. The top images show the merged images of the red and green filtered images shown below. Right: scheme showing the biotin–streptavidin–CPMV–dye assembly. Adapted with permission from ref. 97. Copyright 2006 American Chemical Society.

in quartz crystal microbalance measurements show that this also affects the mechanical properties of the array.[98]

Interestingly, the aggregation of coat proteins can also be achieved simply using coat protein alone, if suitable conditions and modifications are chosen. Porta *et al.* found that expression of foreign peptides, genetically engineered into an exposed surface loop, resulted in the aggregation of protein cages within infected plant tissue.[99] This behaviour was only observed in certain chimeric particles, and never for the wild-type virus.

The organization of viruses and VLPs is attractive as such particles are naturally designed to interact with biological cells. A large research effort is ongoing towards the use of these particles as drug-delivery vesicles; however, this is beyond the scope of this chapter.[100] A similar interaction principle can be used for other purposes when combined with surface growth. In a study by Lin *et al.*, CPMV was used to promote cell adhesion to a surface.[101] To achieve this, poly(diallyldimethylammonium chloride) (PDDA) was used to adhere CPMV to an Ag-coated wafer by alternating PDDA and CPMV in a layer-by-layer fashion. The coverage of CPMV in this case is dependent simply on both time and pH, favouring a pH near the isoelectric point. Cells grown on this layer show good adhesion and are still biologically active.

Ultimately, the goal we envision is to generate a large 3D architecture of viruses and integrate functional components into this structure. In work by Cigler *et al.*, this was achieved by using a DNA programmed assembly technique to form binary crystals of Qβ phage capsids and gold nanoparticles.[102] To this end, opposing strands of DNA were grafted on the capsid, using surface modification techniques similar to those described above, and the gold nanoparticle, using a thiol ligand exchange. Upon combination of these two particles of DNA, base pairing allows for the formation of NaTl lattice crystal structures for particles with similar radii. As an open structure with regularly spaced plasmonic gold nanoparticles, such structures can be vital in the design of the next generation of nanophotonic materials. Furthermore, we feel that this example also demonstrates the broad potential of viruses as a structural nanobuilding block in forming highly organized structures, as it is likely that the virus symmetry and monodispersity improved the degree of crystallinity found by the authors.

9.4.2 Electrostatic Assembly

As described, the exterior surface of a virus is decorated with symmetrically placed patches of charge. In general, to compensate for the cationic N-terminus found inside most virus capsids, the exterior tends to be anionic. This enables viruses and VLPs to be used as a substitute for other anionic species in many systems. The potential to assemble viruses in alternating multilayers of opposing charge was observed when PSS could be replaced with by carnation mottle virus as the anionic species in an alternating PSS and polyallylamine (PAA) multilayer.[103]

Electrostatic interactions can have a profound influence on the ability of viruses to adsorb onto and later assemble on surfaces. Little to no adsorption was observed by Suci *et al.* for CCMV on bare silica or Formvar surfaces.[104] After the surface charge had been changed by coating the surface with poly-lysine, a change in the coverage is observed, with greater positive charge inducing greater coverage. Variations in the pH after surface modification can be further used to alter the surface interaction. The electrostatic adsorption of CCMV on a charged surface, such as cysteamine on gold or amines on silica, can thus be used to preorganize a 2D layer of virus nanoparticles (see above).

Not all viruses show similar behaviour, and a markedly different behaviour between spherical and rod-like species is observed. In experiments by Yoo *et al.*, a layer of polyelectrolytes was grown and subsequently either rod-like (TMV, M13) or spherical viruses (CCMV, CPMV) were deposited on these layers.[105] The spherical particles show rapid surface absorption compared with the rod-like particles, but both eventually showed close packing of the particles. To achieve further growth, alternating layers of polyelectrolytes and virus nanoparticles were deposited. Spherical viruses are integrated successfully in this structure, although rod-like viruses are excluded from the layers and are found to float to the top of the polyelectrolyte layer.[106]

Going beyond surface-bound techniques might allow control over crystallization directly in solution. To achieve this, the electrostatic interaction of CCMV with a variety of positively charged polymers was studied by Kostiainen *et al.*[107] Amine-functionalized Newkombe-type dendrons, polyamidoamine (PAMAM) and poly-λ-lysine have all shown the ability to induce aggregation of CCMV. An increased charge on the macromolecules allows for a more efficient clustering, with four or fewer positive charges showing no clustering at all. Both the salt concentration and pH have a significant effect on this type of clustering. Furthermore, the clustering was also shown to work with Prussian Blue-containing CCMV-based VLPs, showing the general applicability of this method to generate nanomaterial arrays.

The assembly of viruses and VLPs into large clusters by itself may make for an interesting nanoarray, but dynamic systems that allow control over this clustering can also be created. Kostiainen *et al.* created such dynamic systems that allow the reversible assembly and disassembly of densely packed and highly ordered arrays to be studied (Figure 9.19).[108] These systems are based on same the Newkombe-type dendrimers as before, but now containing a photolabile group. After clustering, UV irradiation can be used to break up the dendrimers, thus dropping the overall charge to four or fewer positive charges per polyelectrolyte, and disrupting the array. Similarly, temperature-responsive block copolymers can be electrostatically bound to CCMV. Upon applying an external stimulus by raising the temperature, the mantle of these polymers collapses and causes the CCMV–polymer complex to cluster. A subsequent lowering of the temperature releases the virus–polymer complexes back into solution.[109]

This clustering behaviour is not limited to polymeric soft materials. Using the electrostatic interaction between anionic CCMV protein cages and cationic gold nanoparticles, Kostiainen *et al.* were able to fabricate complex architectures.[110] These architectures resulted from the naturally occurring

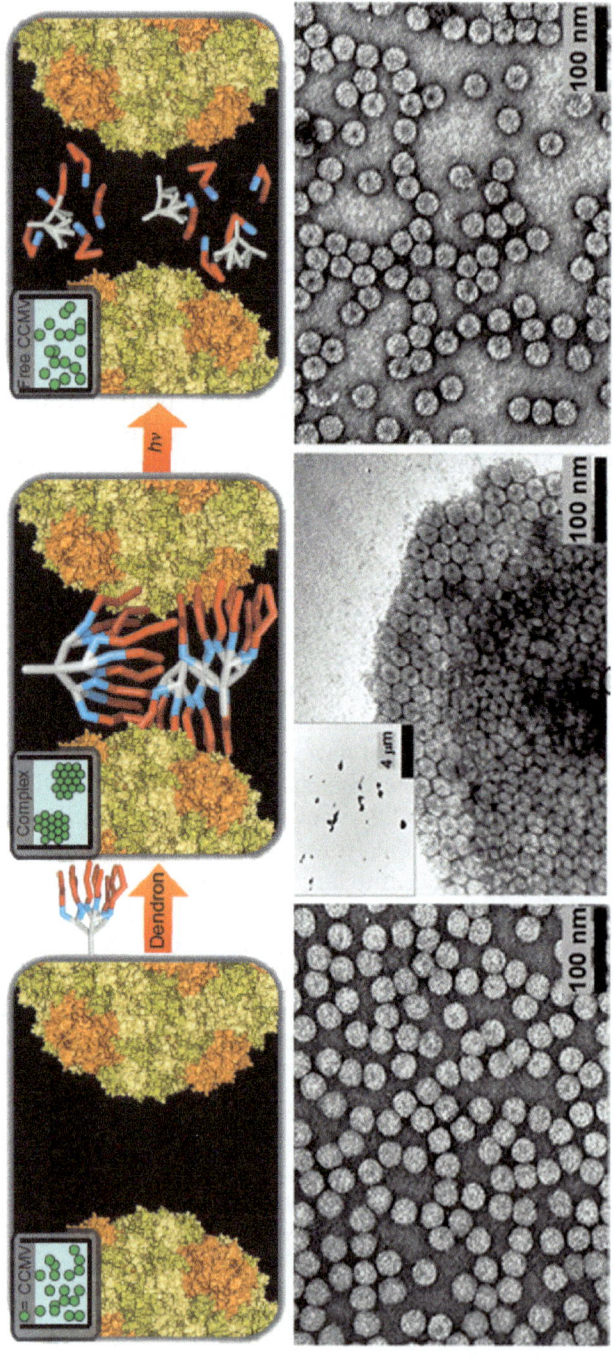

Figure 9.19 (Top) schematic representation and (bottom) TEM images of (left) free CCMV before any clustering agent is added, (centre) clustered CCMV due to the addition of spermine (red)-terminated Newkome-type dendrimers containing a photolabile group (blue) and (right) CCMV released from clustering by UV irradiation to disrupt the photolabile group. Reprinted from ref. 108.

patches of negative charge found on the CCMV surface. The spacing and position of these 60 patches are defined by the structure of the protein cage, and they are found near the naturally occurring pores on the threefold axis of the icosahedral structure. Allowing space for electrostatic repulsion between the gold nanoparticles, and thus leaving an empty patch between each nanoparticle binding site, it was found that 24 of these patches could be used to bind the gold nanoparticles.

In addition to proper clustering conditions, other factors could explain why Kostiainen *et al.* found clusters rather than single VLPs decorated with gold nanoparticles. De *et al.* demonstrated that nanoparticle and protein interactions are partially dominated by the relative size of these particles.[111] If proteins are much larger than the nanoparticles, the protein complexes multiple nanoparticles, and the reverse applies if the nanoparticles are much larger than the protein. If they are of similar size, however, the particles are capable of forming extended networks and form complex self-assembled structures.

VLP clustering might even lead to an energetically favoured state due to the patchy nature of the surface. For viral capsids, owing to the symmetry in the protein shell, these surface charge variations are especially pronounced. Božič and Podgornik showed that these variations can lead to a reduction in the interaction energy between the shells of two equally charged virus particles.[112] Furthermore, if the particles allow sufficient orientation variation to minimize the interaction energy, this can even lead to a short-range attractive force between the particles. If multiple symmetrical axes can be defined on a particle, such as on larger VLPs with increased Caspar–Klug symmetries, these effects are even more apparent. Additionally, such symmetrical concerns can, of course, influence the interaction of virus shells with other charged species, such as polyelectrolytes or multivalent ions.

9.5 Conclusion

Viruses and virus-like nanoparticles offer a highly adaptable scaffold for nanotechnology. They combine important natural characteristics, such as symmetry, chemical addressability and versatility and genetic modifiability. Furthermore, through manipulation and templating they can be tuned in size and shape. These properties can be utilized to build a wide variety of functional systems, from paramagnetic contrast agents and mineralization templates to nanoreactors and containers. More importantly, the nature of a virus capsid allows different modifications to be combined into complex systems and, when needed, arrays of these materials can be made through a variety of chemical and electrostatic clustering techniques.

It is in recent decades that VLPs have been studied for use in nanotechnological applications, and the research interest in this field is still far from abating, as there are both new challenges to be addressed and steps to be taken to improve upon the existing work. These challenges range from the structural scaling towards larger, organized clusters of these particles to seeking control over interactions between materials confined in or in the proximity of these protein cages.

There are, however, natural limitations to these particles, as they are and remain based on soft protein matter. Although generally more robust than other biological or soft materials, their biological nature limits the range of conditions, such as solvents, pH and temperature, in which they will thrive. The recent discovery of other, non-viral, protein cages, such as encapsulins from bacteria, proves that these limits could be stretched significantly if needed. Other opportunities lie in, for instance, combined hard–soft materials such as core–shell systems where proteins' monodispersity and symmetry form the template for materials with radically different properties.

Ultimately, we feel that although an extensive library of different VLPs and techniques for organization exists, so far few research papers have reported studies of the properties of VLPs in materials and the effect of the protein cage upon the resulting physical properties. This, however, will be a next crucial step in the future of this field, as it holds the promise of modular nanomaterials with a uniform VLP building block—an endeavour that will allow us to create nanostructured materials with a wide variety of new physical properties. Key to the success of this endeavour will be a clear understanding of the properties, parameters and limitations that govern these structures combined with exhaustive synthetic tools for their creation.

References

1. L. O. Liepold, *et al.*, Structural transitions in cowpea chlorotic mottle virus (CCMV), *Phys. Biol.*, 2005, 2(4), S166.
2. M. Carrillo-Tripp, *et al.*, VIPERdb2: an enhanced and web API enabled relational database for structural virology, *Nucleic Acids Res.*, 2009, 37(suppl 1), D436–D442.
3. K. K. Lee and J. E. Johnson, Complementary approaches to structure determination of icosahedral viruses, *Curr. Opin. Struct. Biol.*, 2003, 13(5), 558–569.
4. T. S. Baker and R. H. Cheng, A model-based approach for determining orientations of biological macromolecules imaged by cryoelectron microscopy, *J. Struct. Biol.*, 1996, 116(1), 120–130.
5. D. L. Makino, *et al.*, Investigation of RNA structure in satellite panicum mosaic virus, *Virology*, 2006, 351(2), 420–431.
6. M. Comellas-Aragones, *et al.*, Solution scattering studies on a virus capsid protein as a building block for nanoscale assemblies, *Soft Matter*, 2011, 7(24), 11380–11391.
7. J. A. Speir, *et al.*, Structures of the native and swollen forms of cowpea chlorotic mottle virus determined by X-ray crystallography and cryo-electron microscopy, *Structure*, 1995, 3(1), 63–78.
8. D. L. D. Caspar and A. Klug, Physical principles in the construction of regular viruses, *Cold Spring Harbor Symp. Quant. Biol.*, 1962, 27, 1–24.

9. D. L. D. Caspar and A. Klug, Structure and assembly of regular virus particles, in *Viruses, Nucleic Acids and Cancer. 17th Annual Symposium on Fundamental Cancer Research, University of Texas*, Williams & Wilkins, Baltimore, 1963, pp. 27–39.

10. R. F. Bruinsma, *et al.*, Viral self-assembly as a thermodynamic process, *Phys. Rev. Lett.*, 2003, **90**(24), 248101.

11. R. Zandi, *et al.*, Origin of icosahedral symmetry in viruses, *Proc. Natl. Acad. Sci. U. S. A.*, 2004, **101**(44), 15556–15560.

12. H. D. Nguyen and C. L. Brooks, Generalized structural polymorphism in self-assembled viral particles, *Nano Lett.*, 2008, **8**(12), 4574–4581.

13. H. D. Nguyen, V. S. Reddy and C. L. Brooks, Invariant polymorphism in virus capsid assembly, *J. Am. Chem. Soc.*, 2009, **131**(7), 2606–2614.

14. J. E. Johnson and J. A. Speir, Quasi-equivalent viruses: a paradigm for protein assemblies, *J. Mol. Biol.*, 1997, **269**(5), 665–675.

15. A. Zlotnick and S. J. Stray, How does your virus grow? Understanding and interfering with virus assembly, *Trends Biotechnol.*, 2003, **21**(12), 536–542.

16. V. A. Belyi and M. Muthukumar, Electrostatic origin of the genome packing in viruses, *Proc. Natl. Acad. Sci. U. S. A.*, 2006, **103**(46), 17174–17178.

17. B. R. Linger, *et al.*, Sindbis virus nucleocapsid assembly: RNA folding promotes capsid protein dimerization, *RNA*, 2004, **10**(1), 128–138.

18. M. Castelnovo, D. Muriaux and C. Faivre-Moskalenko, Entropic control of particle sizes during viral self-assembly, *New J. Phys.*, 2013, **15**(3), 035028.

19. J. M. Johnson, *et al.*, Interaction with capsid protein alters RNA structure and the pathway for *in vitro* assembly of cowpea chlorotic mottle virus, *J. Mol. Biol.*, 2004, **335**(2), 455–464.

20. A. Zlotnick, *et al.*, Mechanism of capsid assembly for an icosahedral plant virus, *Virology*, 2000, **277**(2), 450–456.

21. X. Zhao, *et al.*, *In vitro* assembly of cowpea chlorotic mottle virus from coat protein expressed in *Escherichia coli* and *in vitro*-transcribed viral cDNA, *Virology*, 1995, **207**(2), 486–494.

22. J. B. Bancroft, C. E. Bracker and G. W. Wagner, Structures derived from cowpea chlorotic mottle and brome mosaic virus protein, *Virology*, 1969, **38**(2), 324–335.

23. L. Lavelle, *et al.*, Phase diagram of self-assembled viral capsid protein polymorphs, *J. Phys. Chem. B*, 2009, **113**(12), 3813–3819.

24. L. Loo, *et al.*, Infusion of dye molecules into red clover necrotic mosaic virus, *Chem. Commun.*, 2008, 88–90.

25. G. Basu, *et al.*, Metal binding to cowpea chlorotic mottle virus using terbium(III) fluorescence, *J. Biol. Inorg. Chem.*, 2003, **8**(7), 721–725.

26. M. Allen, *et al.*, Paramagnetic viral nanoparticles as potential high-relaxivity magnetic resonance contrast agents, *Magn. Reson. Med.*, 2005, 807–812.

27. L. A. Lee, H. G. Nguyen and Q. Wang, Altering the landscape of viruses and bionanoparticles, *Org. Biomol. Chem.*, 2011, **9**(18), 6189–6195.

28. E. Strable and M. G. Finn, Chemical modification of viruses and virus-like particles, in *Viruses and Nanotechnology*, ed. M. Manchester and N. Steinmetz, Springer, Berlin, 2009, pp. 1–21.

29. M. Young, *et al.*, Plant viruses as biotemplates for materials and their use in nanotechnology, *Annu. Rev. Phytopathol.*, 2008, **46**(1), 361–384.

30. N. F. Steinmetz, G. P. Lomonossoff and D. J. Evans, Decoration of cowpea mosaic virus with multiple, redox-active, organometallic complexes, *Small*, 2006, **2**(4), 530–533.

31. A. de la Escosura, *et al.*, Viral capsids as templates for the production of monodisperse Prussian blue nanoparticles, *Chem. Commun.*, 2008, 1542–1544.

32. S. Qazi, *et al.*, P22 viral capsids as nanocomposite high-relaxivity MRI contrast agents, *Mol. Pharm.*, 2012, **10**(1), 11–17.

33. H. N. Barnhill, *et al.*, Turnip yellow mosaic virus as a chemoaddressable bionanoparticle, *Bioconjugate Chem.*, 2007, **18**(3), 852–859.

34. Q. Wang, *et al.*, Natural supramolecular building blocks: wild-type cowpea mosaic virus, *Chem. Biol.*, 2002, **9**(7), 805–811.

35. Q. Wang, *et al.*, Icosahedral virus particles as addressable nanoscale building blocks, *Angew. Chem., Int. Ed.*, 2002, **41**(3), 459–462.

36. E. A. Anderson, *et al.*, Viral nanoparticles donning a paramagnetic coat: conjugation of MRI contrast agents to the MS2 capsid, *Nano Lett.*, 2006, **6**(6), 1160–1164.

37. A. A. A. Aljabali, *et al.*, Charge modified cowpea mosaic virus particles for templated mineralization, *Adv. Funct. Mater.*, 2011, **21**(21), 4137–4142.

38. M. Comellas-Aragonès, *et al.*, Controlled integration of polymers into viral capsids, *Biomacromolecules*, 2009, **10**(11), 3141–3147.

39. Q. Wang, *et al.*, Natural supramolecular building blocks: cysteine-added mutants of cowpea mosaic virus, *Chem. Biol.*, 2002, **9**(7), 813–819.

40. T. Douglas, *et al.*, Protein engineering of a viral cage for constrained nanomaterials synthesis, *Adv. Mater.*, 2002, **14**(6), 415–418.

41. S. E. Aniagyei, *et al.*, Synergistic effects of mutations and nanoparticle templating in the self-assembly of cowpea chlorotic mottle virus capsids, *Nano Lett.*, 2009, **9**(1), 393–398.

42. M. T. Klem, *et al.*, 2-D array formation of genetically engineered viral cages on Au surfaces and imaging by atomic force microscopy, *J. Am. Chem. Soc.*, 2003, **125**(36), 10806–10807.

43. A. S. Blum, *et al.*, Cowpea mosaic virus as a scaffold for 3-D patterning of gold nanoparticles, *Nano Lett.*, 2004, **4**(5), 867–870.

44. A. Chatterji, *et al.*, A virus-based nanoblock with tunable electrostatic properties, *Nano Lett.*, 2005, **5**(4), 597–602.

45. F. Li, *et al.*, Monofunctionalization of protein nanocages, *J. Am. Chem. Soc.*, 2011, **133**(50), 20040–20043.

46. J. M. Hooker, E. W. Kovacs and M. B. Francis, Interior surface modification of bacteriophage MS2, *J. Am. Chem. Soc.*, 2004, **126**(12), 3718–3719.

47. C. L. Cheung, *et al.*, Fabrication of assembled virus nanostructures on templates of chemoselective linkers formed by scanning probe nanolithography, *J. Am. Chem. Soc.*, 2003, **125**(23), 6848–6849.

48. A. S. Blum, *et al.*, An engineered virus as a scaffold for three-dimensional self-assembly on the nanoscale, *Small*, 2005, **1**(7), 702–706.

49. A. A. Martinez-Morales, *et al.*, Synthesis and characterization of iron oxide derivatized mutant cowpea mosaic virus hybrid nanoparticles, *Adv. Mater.*, 2008, **20**(24), 4816–4820.

50. D. P. Patterson, *et al.*, Virus-like particle nanoreactors: programmed encapsulation of the thermostable CelB glycosidase inside the P22 capsid, *Soft Matter*, 2012, **8**(39), 10158–10166.

51. T. Douglas and M. Young, Virus particles as templates for materials synthesis, *Adv. Mater.*, 1999, **11**(8), 679–681.

52. M. Allen, *et al.*, Protein cage constrained synthesis of ferrimagnetic iron oxide nanoparticles, *Adv. Mater.*, 2002, **14**, 1562–1565.

53. T. Douglas and M. Young, Host–guest encapsulation of materials by assembled virus protein cages, *Nature*, 1998, **393**(6681), 152–155.

54. A. A. A. Aljabali, *et al.*, Cowpea mosaic virus unmodified empty virus like particles loaded with metal and metal oxide, *Small*, 2010, **6**(7), 818–821.

55. S. N. Shah, *et al.*, Environmentally benign synthesis of virus-templated, monodisperse, iron-platinum nanoparticles, *Dalton Trans.*, 2009, (40), 8479–8480.

56. N. F. Steinmetz, *et al.*, Virus-templated silica nanoparticles, *Small*, 2009, **5**(7), 813–816.

57. M. Comellas-Aragones, *et al.*, A virus-based single-enzyme nanoreactor, *Nat. Nanotechnol.*, 2007, **2**(10), 635–639.

58. J. Sun, *et al.*, Core-controlled polymorphism in virus-like particles, *Proc. Natl. Acad. Sci. U. S. A.*, 2007, **104**(4), 1354–1359.

59. F. D. Sikkema, *et al.*, Monodisperse polymer-virus hybrid nanoparticles, *Org. Biomol. Chem.*, 2007, **5**(1), 54–57.

60. Z. Liu, *et al.*, Natural supramolecular building blocks: from virus coat proteins to viral nanoparticles, *Chem. Soc. Rev.*, 2012, **41**(18), 6178–6194.

61. J. B. Bancroft, E. Hiebert and C. E. Bracker, The effects of various polyanions on shell formation of some spherical viruses, *Virology*, 1969, **39**(4), 924–930.

62. Y. Ren, S.-M. Wong and L.-Y. Lim, *In vitro*-reassembled plant virus-like particles for loading of polyacids, *J. Gen. Virol.*, 2006, **87**(9), 2749–2754.

63. Y. F. Hu, *et al.*, Packaging of a polymer by a viral capsid: the interplay between polymer length and capsid size, *Biophys. J.*, 2008, **94**(4), 1428–1436.

64. R. D. Cadena-Nava, *et al.*, Exploiting fluorescent polymers to probe the self-assembly of virus-like particles, *J. Phys. Chem. B*, 2011, **115**(10), 2386–2391.

65. P. van der Schoot and R. Bruinsma, Electrostatics and the assembly of an RNA virus, *Phys. Rev. E*, 2005, **71**(6), 061928.

66. C. B. Chang, *et al.*, Curvature dependence of viral protein structures on encapsidated nanoemulsion droplets, *ACS Nano*, 2008, **2**(2), 281–286.

67. S. Mukherjee, *et al.*, Redirecting the coat protein of a spherical virus to assemble into tubular nanostructures, *J. Am. Chem. Soc.*, 2006, **128**(8), 2538–2539.

68. M. Kwak, *et al.*, Virus-like particles templated by DNA micelles: a general method for loading virus nanocarriers, *J. Am. Chem. Soc.*, 2010, **132**(23), 7834–7835.

69. S. E. Aniagyei, *et al.*, Self-assembly approaches to nanomaterial encapsulation in viral protein cages, *J. Mater. Chem.*, 2008, **18**(32), 3763–3774.

70. B. Dragnea, *et al.*, Gold nanoparticles as spectroscopic enhancers for *in vitro* studies on single viruses, *J. Am. Chem. Soc.*, 2003, **125**(21), 6374–6375.

71. C. Chen, *et al.*, Packaging of gold particles in viral capsids, *J. Nanosci. Nanotechnol.*, 2005, **5**(12), 2029–2033.

72. L. Loo, *et al.*, Controlled encapsidation of gold nanoparticles by a viral protein shell, *J. Am. Chem. Soc.*, 2006, **128**(14), 4502–4503.

73. C. Chen, *et al.*, Nanoparticle-templated assembly of viral protein cages, *Nano Lett.*, 2006, **6**(4), 611–615.

74. X. Huang, *et al.*, Self-assembled virus-like particles with magnetic cores, *Nano Lett.*, 2007, **7**(8), 2407–2416.

75. S. K. Dixit, *et al.*, Quantum dot encapsulation in viral capsids, *Nano Lett.*, 2006, **6**(9), 1993–1999.

76. M. C. Daniel, *et al.*, Role of surface charge density in nanoparticle-templated assembly of bromovirus protein cages, *ACS Nano*, 2010, **4**(7), 3853–3860.

77. I. Tsvetkova, *et al.*, Pathway switching in templated virus-like particle assembly, *Soft Matter*, 2012, **8**(17), 4571–4577.

78. T. J. Wang, *et al.*, Encapsulation of gold nanoparticles by simian virus 40 capsids, *Nanoscale*, 2011, **3**(10), 4275–4282.

79. F. Li, *et al.*, Viral coat proteins as flexible nano-building-blocks for nanoparticle encapsulation, *Small*, 2010, **6**(20), 2301–2308.

80. N. L. Goicochea, *et al.*, Core-like particles of an enveloped animal virus can self-assemble efficiently on artificial templates, *Nano Lett.*, 2007, **7**(8), 2281–2290.

81. L. Loo, *et al.*, Encapsidation of nanoparticles by red clover necrotic mosaic virus, *J. Am. Chem. Soc.*, 2007, **129**(36), 11111–11117.

82. S. L. Capehart, *et al.*, Controlled integration of gold nanoparticles and organic fluorophores using synthetically modified M52 viral capsids, *J. Am. Chem. Soc.*, 2013, **135**(8), 3011–3016.

83. G. Schneider, *et al.*, Distance-dependent fluorescence quenching on gold nanoparticles ensheathed with layer-by-layer assembled polyelectrolytes, *Nano Lett.*, 2006, **6**(3), 530–536.

84. R. Bardhan, *et al.*, Fluorescence enhancement by Au nanostructures: nanoshells and nanorods, *ACS Nano*, 2009, **3**(3), 744–752.

85. G. P. Acuna, *et al.*, Fluorescence enhancement at docking sites of DNA-directed self-assembled nanoantennas, *Science*, 2012, **338**(6106), 506–510.

86. M. F. Hagan, Controlling viral capsid assembly with templating, *Phys. Rev. E*, 2008, **77**(5), 051904.

87. M. F. Hagan, A theory for viral capsid assembly around electrostatic cores, *J. Chem. Phys.*, 2009, **130**(11), 114902.

88. A. Siber, R. Zandi and R. Podgornik, Thermodynamics of nanospheres encapsulated in virus capsids, *Phys. Rev. E*, 2010, **81**(5 Pt. 1), 051919.

89. H. K. Lin, P. van der Schoot and R. Zandi, Impact of charge variation on the encapsulation of nanoparticles by virus coat proteins, *Phys. Biol.*, 2012, **9**(6), 066004.

90. A. A. A. Aljabali, G. P. Lomonossoff and D. J. Evans, CPMV-polyelectrolyte-templated gold nanoparticles, *Biomacromolecules*, 2011, **12**(7), 2723–2728.

91. A. Kale, *et al.*, Directed self-assembly of CdS quantum dots on bacteriophage P22 coat protein templates, *Nanotechnology*, 2013, **24**(4), 045603.

92. F. Li, *et al.*, Three-dimensional gold nanoparticle clusters with tunable cores templated by a viral protein scaffold, *Small*, 2012, **8**(24), 3832–3838.

93. J. T. Russell, *et al.*, Self-assembly and cross-linking of bionanoparticles at liquid–liquid interfaces, *Angew. Chem., Int. Ed.*, 2005, **44**(16), 2420–2426.

94. A. Boker, *et al.*, Self-assembly of nanoparticles at interfaces, *Soft Matter*, 2007, **3**(10), 1231–1248.

95. J. C. Falkner, *et al.*, Virus crystals as nanocomposite scaffolds, *J. Am. Chem. Soc.*, 2005, **127**(15), 5274–5275.

96. C. L. Cheung, *et al.*, Physical controls on directed virus assembly at nanoscale chemical templates, *J. Am. Chem. Soc.*, 2006, **128**(33), 10801–10807.

97. N. F. Steinmetz, *et al.*, Plant viral capsids as nanobuilding blocks: construction of arrays on solid supports, *Langmuir*, 2006, **22**(24), 10032–10037.

98. N. F. Steinmetz, *et al.*, Assembly of multilayer arrays of viral nanoparticles *via* biospecific recognition: a quartz crystal microbalance with dissipation monitoring study, *Biomacromolecules*, 2008, **9**(2), 456–462.

99. C. Porta, *et al.*, Cowpea mosaic virus-based chimaeras: effects of inserted peptides on the phenotype, host range, and transmissibility of the modified viruses, *Virology*, 2003, **310**(1), 50–63.

100. Y. Ma, R. J. M. Nolte and J. J. L. M. Cornelissen, Virus-based nanocarriers for drug delivery, *Adv. Drug Delivery Rev.*, 2012, **64**(9), 811–825.

101. Y. Lin, *et al.*, Layer-by-layer assembly of viral capsid for cell adhesion, *Acta Biomater.*, 2008, **4**(4), 838–843.

102. P. Cigler, *et al.*, DNA-controlled assembly of a NaTl lattice structure from gold nanoparticles and protein nanoparticles, *Nat. Mater.*, 2010, **9**(11), 918–922.

103. Y. Lvov, *et al.*, Successive deposition of alternate layers of polyelectrolytes and a charged virus, *Langmuir*, 1994, **10**(11), 4232–4236.

104. P. A. Suci, *et al.*, Influence of electrostatic interactions on the surface adsorption of a viral protein cage, *Langmuir*, 2005, **21**(19), 8686–8693.

105. P. J. Yoo, *et al.*, Spontaneous assembly of viruses on multilayered polymer surfaces, *Nat. Mater.*, 2006, **5**(3), 234–240.
106. N. F. Steinmetz, *et al.*, Layer-by-layer assembly of viral nanoparticles and polyelectrolytes: the film architecture is different for spheres *versus* rods, *ChemBioChem*, 2008, **9**(10), 1662–1670.
107. M. A. Kostiainen, *et al.*, Electrostatic self-assembly of virus-polymer complexes, *J. Mater. Chem.*, 2011, **21**(7), 2112–2117.
108. M. A. Kostiainen, *et al.*, Self-assembly and optically triggered disassembly of hierarchical dendron–virus complexes, *Nat. Chem.*, 2010, **2**(5), 394–399.
109. M. A. Kostiainen, *et al.*, Temperature-switchable assembly of supramolecular virus–polymer complexes, *Adv. Funct. Mater.*, 2011, **21**(11), 2012–2019.
110. M. A. Kostiainen, *et al.*, Electrostatic assembly of binary nanoparticle superlattices using protein cages, *Nat. Nanotechnol.*, 2013, **8**(1), 52–56.
111. M. De, *et al.*, Size and geometry dependent protein-nanoparticle self-assembly, *Chem. Commun.*, 2009, (16), 2157–2159.
112. A. L. Božič and R. Podgornik, Symmetry effects in electrostatic interactions between two arbitrarily charged spherical shells in the Debye-Hückel approximation, *J. Chem. Phys.*, 2013, **138**(7), 074902.

CHAPTER 10

Photoluminescent Hybrid Inorganic–Protein Nanostructures for Imaging and Sensing In Vivo and In Vitro

ANDREI V. ZVYAGIN*[a,b,c], VARUN K. A. SREENIVASAN[a],
EWA M. GOLDYS[d], VLADISLAV Ya. PANCHENKO[b],
AND SERGEY M. DEYEV[c,e]

[a]Department of Physics and Astronomy, Macquarie University, Sydney, NSW 2109, Australia; [b]Institute of Laser and Information Technologies, Russian Academy of Sciences, 40700 Shatura, Moscow Region, Russia; [c]Laboratory of Optical Theranostics, Nizhny Novgorod State University, Russia; [d]ARC Centre of Excellence for Nanoscale Biophotonics, Macquarie University, Sydney, NSW 2109, Australia; [e]Shemyakin and Ovchinnikov Institute of Bioorganic Chemistry, Russian Academy of Sciences, 117997 Moscow, Russia
*E-mail: andrei.zvyagin@mq.edu.au

10.1 Introduction

Nanotechnology and biophotonics provide the life sciences with unique methods for investigating a wide of range of biological systems at unprecedented levels of precision and control. The behaviour of cells is determined by their nanoscale constituents, including organelles and other nanoscale

RSC Smart Materials No. 16
Bio-Synthetic Hybrid Materials and Bionanoparticles: A Biological Chemical Approach Towards Material Science
Edited by Alexander Böker and Patrick van Rijn
Published by the Royal Society of Chemistry, www.rsc.org

molecular complexes such as ribosomes, receptors and macromolecules. The interaction of these constituents with engineered nanostructures allows their visualization and control.

Optical imaging assisted by molecular-specific labelling with photoluminescent nanoparticles (NPs) is one of the most significant applications of these nanomaterials. This direct, minimally invasive approach makes it possible to investigate the cellular morphology and various processes in living cells and/or tissues in their full biological context. Their detection sensitivity is high, reaching the regime of single molecules. Single-molecule imaging unobscured by ensemble averaging is a significant tool for understanding complex biological systems. Single-molecule imaging is required, for example, when studying molecular trafficking events associated with cell signalling *via* the activation of membrane receptors, and in studies of individual receptors and ligands. These studies place stringent demands on the fluorophore performance: fluorescence intermittency (blinking) and irreversible light-induced transitions to dark states (photobleaching) pose significant limitations. Dark-state transitions are particularly undesired in single-molecule studies and also in super-resolution microscopy [for example, stimulated emission depletion microscopy (STED)], which require high illumination intensities. Typical organic dye fluorophores survive only about one million excitation–emission cycles before undergoing an irreversible transition to the dark state (photobleaching), and this means that the continuously emitting fluorophore persists for only ~3 ms (assuming the fluorescence lifetime $\tau = 3$ ns and saturation excitation conditions). Longer wavelength fluorophores such as Cy5 and Cy7, Alexa Fluor 750 and CF dyes, whose excitation–emission wavelengths fall into the biological tissue transparency window (wavelength range 700–1300 nm) (Figure 10.1(a)) are especially prone to photobleaching.

Owing to the controllable chemical surface properties, NPs are also widely employed in applications such as targeted delivery.[1] This is realized by designing and assembling nanoparticulate biocomplexes whose basic structure is presented schematically in Figure 10.2. Such a biocomplex includes a solid, preferably, physically stable and chemically inert NP core, which is surface functionalized, for example, by a polymer surface coating (coat), which decorates the surface with biocompatible chemical moieties. These moieties serve as anchors for biomolecules fulfilling specific functions, including targeting (tar) specific molecular sites in cells or tissues with high affinity, and conjugating to therapeutic molecules, *e.g.* drugs (ther). It is often desirable to render NP biocomplexes highly visible over the non-specific autofluorescence background so that the effects and fate of the nanoparticles can be visualized. This is realized by means of contrast agents (CA), which are either attached to the NP core or, more conveniently, the NP core is designed to function as the CA. In targeted delivery applications, cell toxicity is an important consideration and many nanomaterials, including nanodiamonds, have been shown to be non-toxic or have limited cytotoxicity.[2]

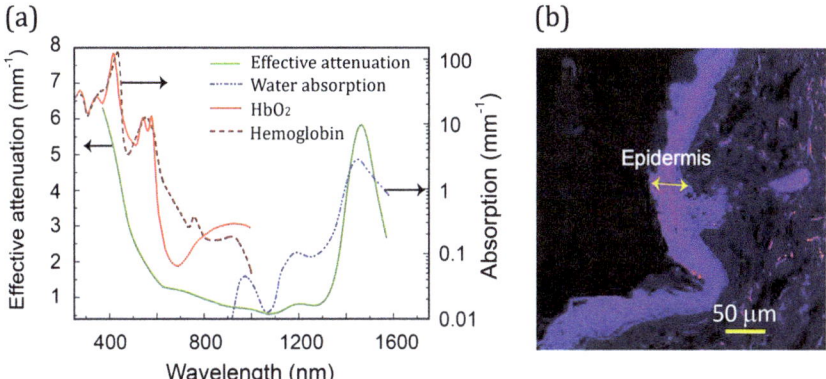

Figure 10.1 (a) Optical attenuation spectrum of living skin tissue (solid green line) dominated by water (blue dot-dashed line), haemoglobin (brown dashed line), oxyhaemoglobin (HbO$_2$, red solid line) and proteins (not shown), with the scattering effect also taken into account. The biological tissue transparency window ranges from 700 to 1300 nm. (b) Autofluorescence image of human skin under excitation at 405 nm. The viable epidermis layer is colour coded purple and marked with an arrow, with cell nuclei visible as dark ovals. The dermis, visualized primarily *via* collagen and elastin bundles, is situated right next to the epidermis extending to the right. Reproduced from ref. 6.

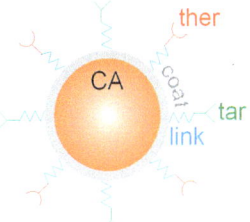

Figure 10.2 Schematic diagram of the photoluminescent nanoparticle biocomplex, featuring the key components: CA, contrast agent; ther, therapeutic vector; tar, targeting vector; link, molecular linker; coat, surface-functionalized coating.

In this chapter, we review several types of photoluminescent nanoparticles in the context of optical cellular/tissue imaging: semiconductor quantum dots (QDs), now an established commercial product, which provide a benchmark for comparisons; fluorescent nanodiamonds (FNDs), nanorubies and upconversion nanoparticles (UCNPs). The rationale behind the choice of these NPs is the diversity of the photoluminescence properties of these important classes of nanoemitters. A QD represents a semiconductor bandgap emitter whose emission properties are governed by the quantum confinement effect. An FND is a nanodiamond host that contains several isolated colour-centre defects. A nanoruby is a nanocrystal of aluminium oxide doped with a number of chromium, atom-like, emitters. A UCNP represents a nanocrystal doped with two types of rare earth ions. The UCNP photoluminescence is due to the (non-linear) excitation transfer processes between strongly coupled ion emitters. FNDs and UCNPs are largely photostable;[3] however, blinking has been reported in FNDs, and it seems to be related to surface conditions and environment.[4]

The UCNPs and nanorubies offer an improved signal-to-noise ratio, and thus high detection contrast, in a time-gated imaging modality.[5] These nanoparticles show good chemical inertness, which is essential for preserving their integrity in often aggressive biological environments, such as lysosomes, and for minimizing their potential cytotoxicity.

Surface moieties are crucial for interfacing NPs with biological cells. Nanoparticle conjugation with biomolecules of various biological functionalities (bioconjugation) permits easy design of the NP biocomplexes for interaction with biological systems. Very promising self-assembling approaches based on high-affinity molecular linkers, such as avidin:biotin and barnase:barstar, provide the crucial flexibility for designing and assembling nanoparticulate biocomplexes of increasing complexity. The bioconjugated NP–biomolecule complex is capable of performing targeted functions by, for example, activating a communication protocol with a cell that can result in its entry into the cell. Several examples of this specific cellular internalization have been observed.[7]

This chapter is organized as follows. We first introduce the photoluminescent nanoparticles first discovered, quantum dots (QDs), and describe their basic material, optical properties and surface chemistry. Using the example of QDs, we describe the key principles of bioconjugating nanoparticles to biological molecules. We also present representative applications of QDs for photoluminescence-assisted imaging of biological systems. At the end of Section 10.2, we describe several applications of QDs, including cellular imaging of non-specific and specific interactions of nanoparticle-based biocomplexes with cells. With this backdrop, in Sections 10.3–10.5 we introduce the remaining three relatively new photoluminescent nanomaterials and discuss their physical, optical and chemical properties in relation to biological imaging. In the final Conclusion section, we summarize current progress in photoluminescent molecular probe technology.

10.2   Quantum Dots

10.2.1   Photoluminescence: Mechanisms and Characteristics

QDs are basically very small pieces of semiconducting materials, such as CdSe, with sizes ranging from a few to several hundred nanometres in diameter. Their photoluminescence is a result of quantum confinement, where the conduction and valence bands of the QD material form an electron and hole quantum well, while the crystal boundary forms its edges. They display the well-known size-tuneable emission spectra, and exhibit exceptionally high quantum yields with additional design elements, such as a capping layer, made from a higher bandgap material.

Size-tuneable QD photoluminescence is governed by the nanoscale quantum confinement, which modifies the electronic density of states.[8] In a first approximation, the energy levels can be predicted by using a simple quantum mechanical model of a particle-in-a-box. This model predicts that the energy level separation is increased as the QD size is decreased. The optical absorption and photoluminescence emission properties are also dominated by the position of the quantum-confined energy levels, as shown in Figure 10.3. Upon photoexcitation, the electron–hole pairs form excitons, with radii within the nanometre range (~5 nm); these excitonic effects also contribute to the observed emission energies.[9] The emission bands are narrow and spectrally tuneable by varying the QD size (Figure 10.3); however, this can only be observed when the QD sample is monodisperse, *i.e.* characterized by a narrow colloidal size distribution.

In terms of their photoluminescence properties, QDs excel with very high quantum yields reaching values greater than 50%.[10] However, accurate measurements of the absorption coefficients of QDs are difficult owing to a variety of factors, such as the unknown colloidal concentration after the synthesis, the assumption of the bulk crystal parameters, spherical approximation, and difficulty of removing the synthesis precursors.[11] The reported peak absorption cross-sections for CdSe QDs at their excitonic absorption band are of the order of 10^4–10^6 L mol^{-1} cm^{-1} depending on the QD size.[12,13] For comparison, the absorption cross-section of Rhodamine 6G is 10^5 L mol^{-1} cm^{-1}.[12]

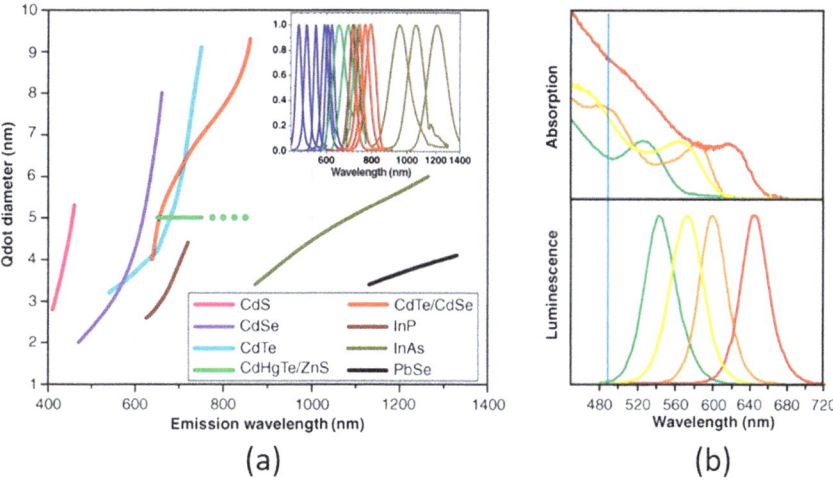

Figure 10.3   (a) Photoluminescence emission characteristics of QDs with different compositions and sizes. (b) Photoluminescence excitation (top) and emission (bottom) spectra of four CdSe/ZnS QDs of varying sizes. From ref. 14. Reprinted with permission from AAAS. (b) Adapted with permission from ref. 15. Copyright 2001 American Chemical Society.

10.2.2   Synthesis and Functionalization of QDs

The synthesis of quantum-confined CdSe (S, Te) QDs was demonstrated in 1993.[16] The method was based on a high-temperature procedure, where rapid nucleation of a supersaturated solution containing Cd and Se ions was followed by controlled crystallite growth around the nuclei, forming uniformly sized QDs. Most of the current QD synthesis methods follow this principle, because it offers independent control over the size and size dispersion, and the possibility of extracting QDs of varying sizes at progressive time points, as they grow by Oswald ripening.[8,17] Other methods of QD synthesis have also been reported.[18,19]

The quantum yield of the first-generation QDs reported by Murray *et al.*[16] was low (~10%) owing to dangling surface bonds, stoichiometric variations and other intrinsic defects.[20] The second-generation design of core–shell QDs successfully improved the quantum yield by capping the nanoparticle core with a lattice-matched semiconductor crystal layer, characterized by a higher energy bandgap compared with the core. The lattice match ensures the absence of crystal deformation at the core/shell interface. The capping layer also reduces the density of surface states, thereby enhancing the quantum yield to >50%.[10,20] The wide bandgap shell is optically transparent to optical wavelengths resonant with the core bandgap, and also creates an additional energy barrier for the excitons. The shell also prevents direct environmental exposure of the core materials, which usually contain toxic metals, such as Cd; CdSe/ZnS is one of the most popular choices for the core–shell QDs.[9] A comprehensive review of the synthesis and growth procedure can be found elsewhere.[21,22]

Biological applications of QDs demand their chemical and colloidal stability in water containing a mixture of different salts.[10,20] This is impossible for as-produced QDs synthesized using high-temperature methods, because their surface is populated with hydrophobic compounds, such as tri-*n*-octylphosphine (oxide) [TOP(O)].[21] Hence achieving water solubility requires additional surface treatments, referred to as surface functionalization. Two broad types of functionalization principles are (i) cap exchange, where the TOP(O) molecules on the QD surface are replaced with other biocompatible molecules bearing a hydrophilic group (*e.g.*, carboxyl, amine or hydroxyl),[23] and (ii) molecular encapsulation, where an additional layer of amphiphilic molecules form a full/quasi-micellar structure, exposing their hydrophilic groups to the aqueous exterior.[24] The stability of both types of functionalization is usually improved by cross-linking/polymerizing the exchange or encapsulation groups, for example by silanization.[15,18]

Several groups have explored aqueous routes for QD synthesis, which directly yield nanocrystals forming stable colloids in water.[25-27] Zhu *et al.*[26] reported an aqueous synthesis route for preparing mercaptosuccinic acid-stabilized CdTe–CdS core–shell QDs, which were characterized by quantum yields as high as 84%. Encapsulation of QDs in a hydrophilic polymer, *e.g.* latex, that renders these complexes stable in water has also been demonstrated.[28] Perhaps the most intriguing way to make QDs is their production in organisms, referred to as "biosynthesis." Such methods, demonstrated in yeast[29] and earthworms,[30] take advantage of endogenous detoxification mechanisms in these organisms. "Green" synthesis routes to non-cytotoxic heavy metal-free QDs are also under active investigation.[27,31] For example, Manzoor *et al.*[31] described the synthesis of non-cytotoxic metal ion-doped ZnS QDs *via* colloidal precipitation using sodium sulfide.[32] A two-stage aqueous synthesis to produce Cd-free core–shell QDs has also been reported.[33]

10.2.3 Bioconjugation of Nanoparticles

In order to interface nanoparticles, such as QDs, with biological systems, their surface needs to be conjugated to antibodies,[34,35] ligands,[31,36] other proteins or peptides,[23] drugs[37] or nucleic acids.[37–39]

Most of the NP surface functionalization procedures include the formation of oxygen-containing surface functional groups, such as carboxyls, hydroxyls, ketones and aldehydes. Many of these groups permit direct covalent attachment of biomolecules. Carboxyl (COOH) is a common group used for the assembly of proteins by fusing a COOH terminus of one protein with an amino terminus (NH_2) of another protein, forming a strong covalent amide bond. This bioconjugation strategy is also applicable to other types of photoluminescent nanoparticles described below, especially FNDs, thanks to the abundance of COOH groups on the acid-cleaned FND surface. However, this is not straightforward because of the colloidal instability of nanodiamonds in saline solutions, typically present in cells. Also, amide-bonding reactions can alter the functionality of biomolecules, and need adjustments specific to each biomolecule. Therefore, a universal bioconjugation method is highly desirable.

To this end, versatile biofunctionalization approaches based on high-affinity molecular pairs have been developed.[40,41] One such high-affinity molecular pair is streptavidin:biotin (Sav:B). A QD biofunctionalization procedure based on Sav:B is as follows. First, streptavidin molecules are covalently immobilized on the QD surface to form a QD–Sav subunit. This is done by activating carboxylic groups on the QD surface introduced by TOP(O) cap-exchange/encapsulation with chemicals such as mercaptoacetic acid. These activated carboxylic groups are reacted with the exposed amine groups in a streptavidin molecule to form stable amide bonds.[23] Commercially available streptavidin-coated QDs typically offer 6–8 immobilized streptavidin molecules per QD. The second subunit, B–X, where X is a targeting moiety such as an antibody or DNA probe, is produced by standard biochemistry methods.[34] Further, the covalently stabilized subunit QD–Sav strongly binds to the B–X to form a QD–Sav:B–X complex, which is realized by simply mixing the two subunit solutions. Another novel nanoparticle bioconjugation platform was described by Deyev *et al.*[41] This approach is based on a high-affinity protein pair barstar:barnase (Bs:Bn), a functional analogue of Sav:B, which also allows the flexible design of biofunctional complexes of QD–Bs:Bn–X.[35,41] Similarly to QD–Sav, the QD–Bs complex can be formed by engaging carboxyl groups on the QD surface and amine groups in barstar. The Bn–X subunit, where X is a protein, peptide or another biomolecule, can be synthesized either *via* standard biochemical procedures or using recombinant methods. Together, the two affinity pairs provide universal platforms for nanoparticle biofunctionalization, because X can be any type of molecule or even another nanoparticle.

In comparison, the Bn:Bs platform has some important merits to offer over the Sav:B pair.[42] Unlike streptavidin and biotin, barstar and barnase are prokaryotic monomeric proteins that are water soluble, stable in their native form and tolerate temperatures up to 50 °C and pH between 2 and 12. This is possible because their tertiary structures are stabilized exclusively by hydrogen, as opposed to disulfide, bonds. Bn being a protein devoid of cysteine residues, the Bn–X sub-unit can be produced using recombinant methods, which is often not possible with other proteins because of cross-linking *via* disulfide bonds. Such an approach is also not possible with B–X, since biotin is not a protein. Interestingly, the prokaryotic origin of Bs:Bn proteins also makes them inherently less susceptible to non-specific binding to mammalian tissues in comparison with streptavidin and biotin, which are both of eukaryotic origin.[43] Therefore, the Bs:Bn platform also offers increased signal-to-noise ratios in nanoparticle-based imaging and sensing applications. It is worth mentioning that the Sav:B pair has its own advantages over Bs:Bn, including the unique 1 : 4 binding ratio, where each streptavidin molecule can bind to up to four molecules of biotin, creating multivalent QD–Sav:B–X complexes. On the other hand, the 1 : 1 binding ratio of Bs:Bn allows switching of the subunit design to QD–Bn and Bs–X to produce QD–Bn:Bs–X. This flexibility is, however, not available with the Sav:B owing to the possibility of cross-linking to form (QD–B)$_4$:Sav–X, which can lead to irreversible agglomeration.

Recently, we studied the specific and non-specific binding characteristics of barstar:barnase and other affinity molecules such as streptavidin:biotin and antibody-based systems. Aghayeva *et al.* compared the influence of a number of chemical conditions, such as denaturing agents (urea, guanidinium hydrochloride), high salt concentrations and extreme pH conditions, on the binding and dissociation of colloidal supraparticular constructs prepared using these affinity molecules.[44] Barstar:barnase-based assemblies were virtually unaffected well beyond physiological regimes, but their binding was susceptible to extreme conditions. The effect of the harsh conditions on the association/dissociation kinetics was attributed to van der Waals forces and hydrophobic interactions that persisted despite the denaturation. Overall, under these conditions, the performance of the barstar:barnase system was similar to that of streptavidin:biotin. In another study, by Sreenivasan *et al.*,[42] the barstar:barnase and streptavidin:biotin systems were compared with respect to their susceptibility to non-specific binding to solid hydrophilic and hydrophobic surfaces in the presence and absence of serum proteins. It was found that the barstar:barnase system is less prone to non-specific binding, and can provide better sensitivity.[42]

Many reports on the biofunctionalization of the QDs and FNDs using these two universal platforms can be found in the literature. One example is the construction of FND-based complexes: FND–Bs:Bn–nanogold and FND–Bs:Bn–EGFP. The synthesized complexes revealed their functional stability and structural integrity[45], for example Figure 10.4 (inset TEM image). Most recently, we have demonstrated the implementation of the same protocol to obtain bioconjugates of mini-antibody and upconversion nanoparticles, UCNP–Bs:Bn–4D5, where 4D5 is a recombinant mini-antibody to the HER2/ neu receptor, type epithelial growth factor receptor.[46]

A more recent approach of QD bioconjugation is based on "click" chemistry methods, which provide single-step procedures with high yields under mild reaction conditions. Copper(I)-catalysed azide–terminal alkyne cycloaddition is the most widely used click reaction;[47] however, its application for QD bioconjugation is challenging for various reasons, including the toxicity of Cu and quenching of CdSe QD photoluminescence in the presence of Cu.[48] This limitation was overcome by Agard *et al.*, demonstrating a strain-promoted azide–alkyne cycloaddition (SPAAC) reaction[49] to functionalize azide-terminated QDs and cyclooctyne-modified transferrin molecules in the absence of any catalysts. The transferrin-functionalized QDs were shown to be internalized efficiently by HeLa cells and transported through early endosomes, reaching late endosomes during a 2 h period. However, the large size of the cyclooctyne molecule could potentially result in reduced coverage of the QD surface area, lowering the number of the terminal groups. Zuilhof and co-workers reported alternative click

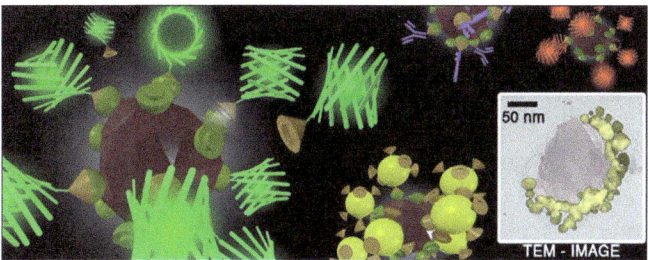

Figure 10.4 Diagram of a universal platform for bioconjugation of macromolecules and nanoparticles to nanovehicle surfaces – in this example, a nanodiamond. Covalently grafted to the diamond surface, molecules of barstar (green caps) are able to bind selectively to molecules of barnase (brown cones), which in turn are preconjugated with macromolecules such as green fluorescent protein (fragment), antibodies (fragment at the top) or nanoparticles, for example, nanogold (fragment). False colour-coded transmission electron microscopy image shows the assembly of a nanodiamond–nanogold complex, called "nano-diadem" (inset in the lower right corner). Reproduced from ref. 45.

chemistry-based approaches for efficient QD biofunctionalization based on thiol–ene and thio–yne methods.[47,48] This reaction depends on photo- or thermal initiator species, excitable by UV irradiation or elevated temperatures within several minutes to hours. The authors were able to demonstrate facile termination of biocompatible Si-based QDs with several biocompatible moieties, including carboxyl groups, thioacetic acid, mercaptoethanol and ethylene glycol. The carboxyl group-terminated QDs were further biofunctionalized with amine-terminated DNA, demonstrating the robustness of this technique.

10.2.4   QDs as Photoluminescent Molecular Probes

Large action cross-section (product of absorption cross-section and quantum yield), photostability, large Stokes shift and narrow spectra (spectral width ~50 nm) make QDs a suitable choice for a variety of biological applications.[10] A broad excitation spectrum and narrow and tuneable emission spectrum are unique to QDs and are very useful for multispectral labelling applications. QDs also exhibit a relatively long photoluminescence lifetime of over 20 ns in comparison with intrinsic and extrinsic organic molecules with fluorescent lifetimes of typically ~3 ns.[43,50] QDs exhibit much greater two-photon action cross-sections than organic fluorophores [>10 000 GM (Göppert-Mayer unit) compared with 1–300 GM for the latter],[51] making them attractive for non-linear optical imaging applications. In the light of these properties, QDs have found broad applications in cellular imaging. Here, we briefly discuss several specific aspects associated with the use of QDs, and NPs in general, for labelling cells.

A cell has to absorb molecules from the extracellular medium to sustain its life cycle and regulate its sensitivity to molecular signal transmitters. The absorption mechanisms belong to two categories: specific internalization that requires the cell to recruit molecules actively into the cytoplasm, and non-specific internalization, which is essentially a random process in which the cell exercises no "active" control, *i.e.* it shows poor material selectivity. Receptor-mediated endocytosis is an important example of a specific internalization mechanism that facilitates the import of selected extracellular molecules and mediates cellular regulation. This process is initiated upon activation of plasma membrane receptors by receptor-specific ligands. As a result, only biomolecular complexes grafted with specific ligands can activate corresponding receptors and gain access into the cell, and these are primarily discussed hereafter. In the process of endocytosis, the plasma membrane is engulfed inwards from the specialized membrane microdomains, forming either clathrin- or caveolin-coated pits. This specific cellular uptake mechanism has been a subject of intense research, driven by the demand to elucidate cellular molecular trafficking and potential applications in targeted drug delivery.

When the highest possible target delivery specificity is required, we need to understand non-specific internalization of nanoparticles, so that it can be minimized. Nanoparticle parameters were found to affect non-specific internalization, especially their size, surface charge and surface functional groups, as shown in Figure 10.5.[52] The surface charge considerably affects the nanoparticle uptake in the cells and should be taken into account, especially when conducting experiments with cell cultures. Despite the fact that the mechanism of the cellular uptake is defined as non-specific, it might be mediated by membrane receptors and realized either through direct interaction between charged particles and a receptor or through preliminary attachment of free proteins to the NP surface. Surface modification of nanoparticles with hydrophilic polymers, such as poly(ethylene glycol) (PEG), inhibits the nanoparticle uptake, making it ideally suitable as negative controls in various cellular models, where the terminal amino groups appear to suppress the internalization effects significantly, as shown in Figure 10.5 (columns, N-pQD$_e$, C-pQD$_e$). It is expected that PEG-free amino- or carboxyl-surface functionalization of NPs will facilitate intensive uptake of the particles in the cell, where the terminal charges play a secondary role, albeit the positively-charged nanoparticles are absorbed less effectively, and represent a good choice for experiments focusing on specific labelling and activation of cells.

In order to investigate molecular trafficking using NP bioconjugates, the biomolecular part of the NP module must activate a biological process. In the case of cell communications, it is often a ligand attached to the NP that activates cell receptors, thereby launching a cascade of biomolecular events that may be regarded as a process of communication of the cell and the nanoparticle. As a result, the particle is translocated to the cell cytoplasm, where it participates in other post-endocytosis events. As an example of a ligand that activates such a communication protocol within a cell, we consider the peptide somatostatin (SST). Among the multitude of physiological

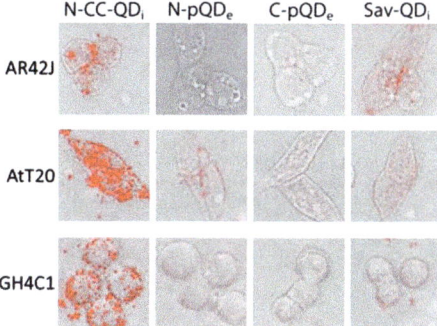

Figure 10.5 Images showing the level of the QD uptake in three immortalized cell cultures, AR42J, AtT20, GH4C1, depending on the surface functional groups. Direct attachment of amino groups (left-most column, N-CC-QD$_i$) shows the highest level of internalization, while a PEG linker (amino-terminal, N-pQD$_e$ and carboxyl-terminal, C-pQD$_e$) heavily suppresses internalization of nanoparticles. Subscripts 'i' and 'e' represent manufacturers Invitrogen and eBiosciences. Reproduced from ref. 52. © IOP Publishing. Reprinted with permission of IOP Publishing Limited.

functions in which SST participates, some are of high clinical relevance. The biological functions of SST are initiated upon its binding to transmembrane G-protein-coupled receptors of six subtypes, including sst_{2A}, which are differentially distributed in organs such as the brain, spinal cord and pancreas, and overexpressed in neuroendocrine tumour cells. The physiological effects of SST, evoked after ligand binding and receptor activation, are mediated through many signalling mechanisms, including adenylyl cyclase inhibition, K^+ channel opening and, in some cases, activation of phospholipase C leading to opening of intracellular Ca^{2+} stores.

The development of a somatostatin analogue with a photoluminescent nanoparticle contrast agent represented by a QD has recently been reported.[7] A biotinylated analogue of SST (SST-2B, where 2 stands for a diethylene glycol spacer between SST and biotin) was conjugated to a streptavidin-coated QD (Sav–QD), forming an SST-2B:Sav–QD complex. However, this preformed SST-2B:Sav–QD complex was incapable of binding/activating the somatostatin receptors. In order to address this problem, an *in situ* two-step conjugation strategy was introduced to overcome the Sav-induced inhibitory effect, an approach that allowed specific targeting of somatostatin receptors. Using this strategy, the receptor-mediated endocytosis of the nanoparticle-labelled SST was imaged (see Figure 10.6). Despite the original inactivity, SST-2B and similar biotinylated SST ligands can be made functional again using the demonstrated two-step approach. The application of such probes to targeted delivery into the cells that express somatostatin receptors, for example, in brain regions responsible for the regulation of blood pressure, will be a significant next step.

Despite a variety of merits and broad application scope in cellular imaging, the toxicity of the QDs largely precludes their application in many *in vivo* studies and investigative and practical medicine. The biological toxicity of

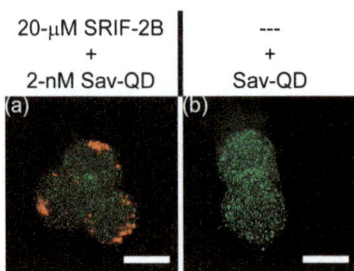

Figure 10.6   Fluorescence confocal images of cells treated with (a) 20 μM SST-2B, also known as SRIF-2B, or (b) blank negative control, followed by 2 nM Sav–QD at low temperature. The cells, when restored to 37 °C, internalized the preformed receptor–ligand complexes. The QD photoluminescence and cell morphology contrast due to the laser backscattering are colour coded red and green, respectively. Scale bar, 10 μm. Reprinted from ref. 7 with permission from Future Medicine Limited.

QDs is caused by the escape and dissolution of heavy metals, such as highly toxic Cd, from the QD core.[53] Several methods to overcome this have been developed, including the use of additional capping layers, and also replacing toxic elements in the core by non-toxic ones as described earlier.[27,31,54] Ye *et al.*[55] reported that phospholipid micelle-encapsulated CdSe/CdS/ZnS QDs appeared to be non-toxic in living animals over a relatively long period of 90 days. Other reports have suggested that the internalization of QDs by cells is the primary reason for this cytotoxicity, suggesting its minimization by special surface functionalization (*e.g.* with PEG) that inhibits the cellular uptake.[56] However, such approaches are not feasible under all experimental conditions and consequently the cytotoxicity of QDs continues to be an issue hindering applications of QDs in the life sciences.

10.3 Fluorescent Nanodiamonds (FNDs)

10.3.1 Nanodiamond (ND)

A fluorescent nanodiamond (FND) is a diamond nanocrystal that hosts colour centres. A diamond crystalline matrix is the most rigid material on Earth, made of carbon atoms with a binding energy of ~7.4 eV per atom. The diamond core is chemically inert, and the colour centres are well separated from environmental perturbations. Even high-energy cosmic radiation inflicts little damage on this crystalline material. At the same time, the diamond surface is chemically reactive, especially when the surface is well developed, as in the case of surface-functionalized NDs. For example, the most popular acid treatment[57] or high-temperature annealing in air[58] results in the formation of oxygen-containing surface moieties on the diamond surface that facilitate interfacing with photonic devices and macromolecules. NDs can be produced by two methods: (1) high-temperature, high-pressure (HTHP) synthetic growth followed by ball-milling produces NDs of high crystal quality 4 nm and larger in size[59] and (2) detonation of explosives in an inert atmosphere followed by disintegration yields remarkably monodisperse 5 nm NDs,[60] referred to as detonation nanodiamonds or ultradisperse nanodiamonds. Subsequent acid treatment and/or annealing in air removes the surface layer of amorphous carbon, replacing it with a variety of oxygen-containing groups, such as carboxyl groups.[58]

The very high refractive index of the ND core (2.4 in the visible spectral range) in comparison with most cellular constituents (1.34 for cell cytoplasm to 1.5 for lipids) allows easy visualization of NDs amidst the translucent background of a cell by conventional optical microscopy. Figure 10.7 shows NDs internalized in cells *via* micellar transport.[61] The ND clusters encapsulated in endosomes are seen as dominant scatterers even stronger than the intrinsic scattering signals coming from, *e.g.*, mitochondria, while the nuclei remain dim. However, this scattering contrast is insufficient by far for achieving the detection sensitivity at the single nanoparticle level due to the cell scattering

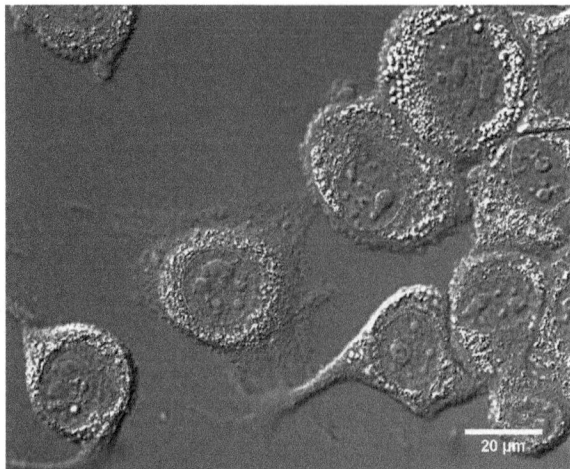

Figure 10.7   Differential interference contrast image of nanodiamond particle
uptake into 3T3 cells. The nanodiamonds exhibit high contrast against
the cell background due to strong elastic scattering and are observable
as bright rims around the dim cell nuclei, which are impermeable for
the NDs. Adapted from ref. 61, Copyright 2007, with permission from
Elsevier.

background. FNDs in spectrally selective imaging modality will improve the
imaging contrast and, hence, detection sensitivity.

10.3.2   Nitrogen-Vacancy Centre

The wide energy bandgap of diamond (5.5 eV) can accommodate up to ~500
distinctive colour centres, but nitrogen-vacancy (NV) and silicon-vacancy[62]
centres are among the brightest, with emission bands falling into the biological tissue transparency window (*cf.* Figure 10.1(a)). The NV centre, in particular, exhibits several exceptional properties that have attracted considerable
interest,[63] and this NV centre in an ND host will be discussed here, referred to
as FND. The NV centre in diamond is formed by replacing one carbon atom
with a nitrogen atom and a vacancy at a location adjacent to the nitrogen
atom, as shown schematically in the inset in Figure 10.8. The production
process usually involves irradiation of diamond samples with high-energy
electron or light-ion beams forming vacancy defects. In order to initiate
migration of these vacancy defects in a diamond (nano)crystal, the sample
is annealed at high temperature (~800 °C) in high vacuum or an oxygen-free
atmosphere. This leads to trapping of vacancies at the substitution nitrogen
sites that are abundant in the majority of synthetic diamonds. The NV centre
can remain neutral (NV^0) or acquire an additional electron from the neighbouring donor defects, becoming a negatively charged centre (NV$^-$). Both
centres are fluorescent, with the spectra shown in Figure 10.8, and exhibit

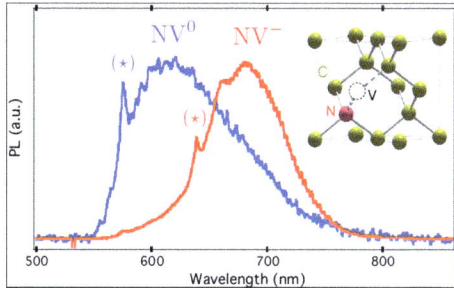

Figure 10.8 Photoluminescence spectra normalized to their respective maximum value of single NV^- (red curve) and NV^0 (blue curve) colour centres in diamonds. The zero-phonon lines (denoted*) of NV^- and NV^0 emissions are located at wavelengths 637 and 575 nm, respectively. The inset shows the atomic structure of the NV colour centre, consisting of a substitutional nitrogen atom (N) associated with a vacancy (V) in an adjacent lattice site of the diamond crystalline matrix. Reprinted with permission from ref. 64. Copyright 2010 American Physical Society.

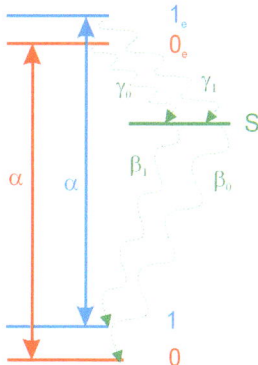

Figure 10.9 Simplified five-level schematic diagram of the electronic levels of the NV colour centre appropriate for room temperature. The centre energy levels due to its magnetic spin projections 0 and ±1 are indicated in red and blue, respectively, and the singlet energy level is indicated in green. The double-headed arrows indicate the radiative transitions. The solid blue and red arrows display the spin conserving transitions. The green wavy arrows indicate non-radiative decay that involves E-vibrations.

a characteristic sharp spectral feature, known as zero-phonon lines. NV^- has interesting physical, optical and magneto-optical properties.

The NV^- centre action cross-section is large, $\eta\sigma_a \approx 3 \times 10^{-17}$ cm^2 (σ_a, absorption cross-section; η, quantum yield),[65] with the quantum yield reaching ~80% under favourable excitation conditions.[66] The energy diagram of the NV^- centre (Figure 10.9) shows that the estimate of η can be obtained from the branching ratio of the radiative and non-radiative transitions. The non-radiative decay from the upper excited states to the intermediate singlet state S (green) accounts for the NV⁻ optical spin polarization property, which is useful in quantum science for spin magnetometry, and can be applied to probe the magnetic and spin properties of biological matter locally at nanometre length scales.

Magnetometry applications rely on the high sensitivity of spin-polarized NV centres to (local) magnetic fields, as shown later in Figure 10.12 and briefly described below. The spin-state perturbation is read out optically by means of the optically detected magnetic resonance and reports on the local value and orientation of the perturbing magnetic field vector. The ultimate sensitivity is estimated to be at the level of a single spin in the nanoscale region.[67]

The virtually unlimited photostability of the NV centre is another property important in biomedical imaging applications. An NV centre represents a solution to the problem of the poor photostability of organic fluorescent dyes that undergo irreversible conversion to a low-η or non-fluorescent state (or photobleaching). Photoluminescence intermittency, or blinking, is reported in virtually all single emitters,[68] especially affecting semiconductor quantum dot emission.[69] Although the NV⁻ centres has been also reported to exhibit blinking behaviour, this can be readily avoided under controllable conditions.[70]

10.3.3 Fluorescent Nanodiamonds for Cellular Imaging Applications

The first observation of FND imaging in cells was reported by Fu et al.,[71] where the high-contrast imaging and virtually unlimited photostability of 35 nm FNDs were reported. The authors demonstrated non-specific internalization of FNDs in cells (Figure 10.10), later followed by the demonstration of several applications of single FND imaging and tracking in the cellular environment.[72]

Our own observation of the non-specific internalization of a single fluorescent nanodiamond-sized (~100 nm) nanoparticle in Chinese hamster ovary (CHO) cells confirmed the feasibility of single fluorescent nanoparticle imaging, with the scattering background largely suppressed,[45] as shown in Figure 10.11. Combined with the reported minute cytotoxicity of the nanodiamonds,[2,73] the demonstration of single-particle imaging sparked considerable interest in the applications of FND as a bioprobe in live systems. These high expectations were underpinned by the nanodiamond surface properties that made this material amenable to bioconjugation, enabling the realization of targeted delivery in cells and specific tissue sites. Indeed, a number of interesting demonstrations of the specific internalization in cells of FND bioconjugated with targeting biomolecules have been demonstrated (see, for example, the comprehensive review by Schrand et al.[74]). It is worth noting that FND bioconjugates can serve not only as molecular bioprobes, but also as biomolecule delivery vehicles.[75]

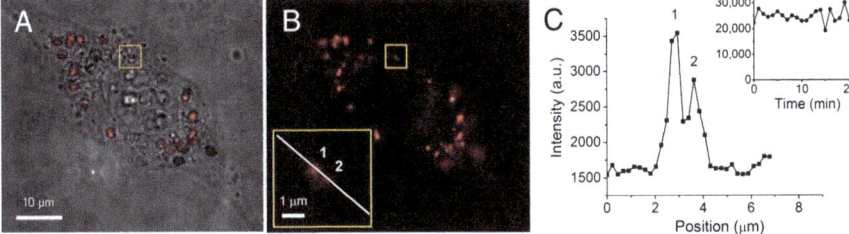

Figure 10.10 Observation of single FNDs in a HeLa cell. (A) Bright-field image of a HeLa cell after uptake of 35 nm FNDs. Most of the taken-up FNDs are seen to distribute in the cytoplasm. (B) Epi-luminescence image of a single HeLa cell after the FND uptake. An enlarged view of the photo-luminescence spots (denoted by "1" and "2") with diffraction-limited sizes (FWHM ≈ 500 nm) is shown in the inset. (C) Intensity profile of the photoluminescence image along the line drawn in the inset in (B). (C, inset) Integrated photoluminescence intensity (after subtraction of the signals from cell autofluorescence and background fluorescence from the microscope slides) as a function of time for particle "1". The signal integration time was 0.1 s. No sign of photobleaching was detected after continuous excitation of the particle for 20 min. Reproduced with permission from ref. 71. Copyright 2007 National Academy of Sciences.

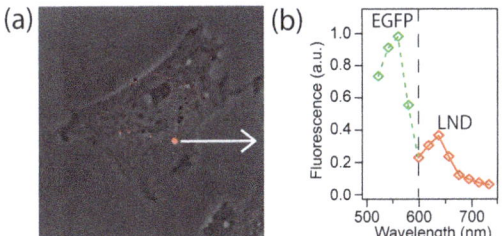

Figure 10.11 Observation of a single FND bioconjugate in a CHO cell. (a) Laser scanning confocal microscopy of a 100 nm fluorescent nanodiamond bioconjugated to green fluorescent protein (EGFP), shown as an red spot, an arrow from which points to (b), the combined fluorescence spectra of the FND bioconjugate. Reproduced from ref. 45.

The optical spin-selective properties of FND have also found new applications in cellular imaging applications. In particular, monitoring the decoherence rates in response to the changes in local environment may provide new information about intracellular processes.[76] The reported experiments demonstrate the viability of controlled single-spin probes for nanomagnetometry in biological systems, opening up a host of new possibilities for quantum-based imaging in the life sciences. Figure 10.12 shows optically detected magnetic resonance signals from single NV centres in different ND crystals spatially localized in the cell cytoplasm. These signals appeared to be capable of reporting on the intracellular environment.

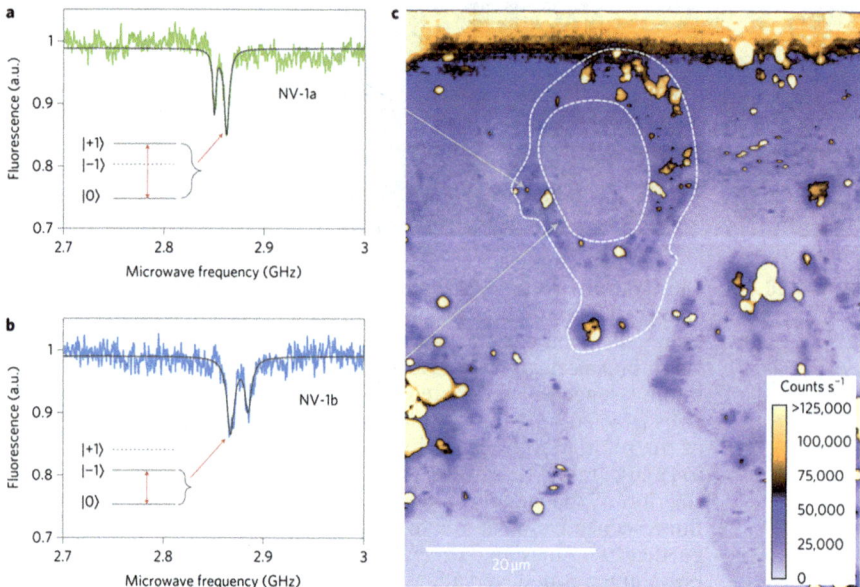

Figure 10.12   Confocal image of a HeLa-1 cell, with photoluminescence spectrally filtered around the NV emission (650–800 nm) (c). The nucleus and cell membrane are indicated with dashed lines for clarity. The optically detected magnetic resonance spectra of FNDa (a) and FNDb (b) show the different strain splitting between the two centres. Reprinted from ref. 76 with permission from Macmillan Publishers Ltd, Copyright 2011.

10.3.4   Ultra-Small FNDs

Imaging and sensing applications and also single-spin sensitive magnetometry[67] and Förster resonance energy transfer[77] require the nanodiamond layer to be thinned to exploit the strong dependence of photoluminescence signal on the interaction radius r ($1/r^6$ and $1/r^3$, respectively). The single-spin-sensitive imaging at the nanoscale led to an estimate that the NV⁻ centre had to be as close as 10 nm to an interrogated single spin for reliable detection.[67] In optical biological imaging applications, ultra-small diamonds are crucial to minimize disruption to the molecular trafficking under observation.[72] It is desirable to reduce the size of NDs to match the size of an average protein, *i.e.* 5 nm, which is achievable due to the progress in the production of ultra-small nanodiamonds.[78] It has also been reported that FNDs smaller than 10 nm retain their fluorescent properties characteristic of the NV centres. At the same time, the NV centre stability sets a size limit on the FND crystal. A 2.5 nm ultra-small nanodiamond may be incapable of hosting an NV centre in the core, as predicted theoretically.[79] The environmental susceptibility of the FND emission in small crystals, or NV centres in the nanometre proximity to a surface layer, has been reported on several occasions.[4] The question of the dependence of the stability of the NV centre on its distance to the diamond surface became important two decades ago owing to the progress towards the production of ultra-small NDs of detonation origin in view of the significant potential applications of low-cost detonation NDs.[57] Spectroscopic studies pointed to the origin of ND fluorescence as a graphitic layer that formed in the detonation process, manifested by the broadband photoluminescence,[80] which was blue shifted from the characteristic NV spectra (*cf.* Figure 10.8). The photoluminescence excitation spectra analysis of HTHP diamonds also supported the notion of NV centre instability in ultra-small NDs by showing a broadband component attributed to the amorphous carbon phase.[81] The obscuring background of the graphite layer to the overall fluorescence signal from FNDs was suppressed by employing a time-gated detection scheme making use of the difference in the emission lifetimes of graphite and NV⁻ (sub-nanosecond *versus* ~10–20 ns, respectively). As a result, the characteristic NV⁻ centre spectral signature was observed, thus providing unambiguous evidence for the existence of NV centres in detonation nanodiamonds.[82] Direct evidence for the background from the graphite layer was also reported by Bradac *et al.*[70] Thoroughly acid-cleaned detonation nanodiamonds were sparsely dispersed on a glass slide and interrogated individually, thus minimizing the graphite photoluminescent signal background, so that a signal originating from only one ND became observable. As a result, the characteristic NV⁻ spectral signatures from single centres were acquired.

An alternative top-down approach for the production of ultra-small NDs was reported by Tisler *et al.*,[59] and it can be generalized to the production of other photoluminescent nanomaterials. This approach was based on milling a millimetre-sized diamond crystal, following the NV centre implantation using the conventional high-energy electron irradiation and high-temperature annealing procedure. Oxidative etching in air/oxygen at high temperature is another top-down approach for ultra-small ND production. This method relies on the removal of carbon atoms from the ND surface at a controllable rate *via* oxidation reactions. Since an outer ND layer in the graphitized (sp^2) form is removed at a higher rate than that of the diamond sp^3 phase, this method was originally employed to remove sp^2 carbon from ND, as a surface-cleaning procedure.[58] This method was extended for *in situ* studies of the ND size evolution, as oxidative etching in air provides the least disruptive method of graphite cleaning and ND core size reduction, and NV⁻ emission in ND crystals smaller than 10 nm has been reported.[4]

10.3.5   Functionalization of FNDs

As-produced FNDs are poorly dispersible in water, which is commonly attributed to the sp^2 hybridized graphite on the nanoparticle surface, especially in the case of detonation nanodiamonds. One of the most established methods to make FNDs dispersible in water is to expose them to a repetitive chemical reflux under acidic (sulfuric and nitric acid) and basic (sodium hydroxide) conditions at elevated temperatures for long periods,[70] which removes the graphitic surface layer and reveals the sp^3 hybridized diamond

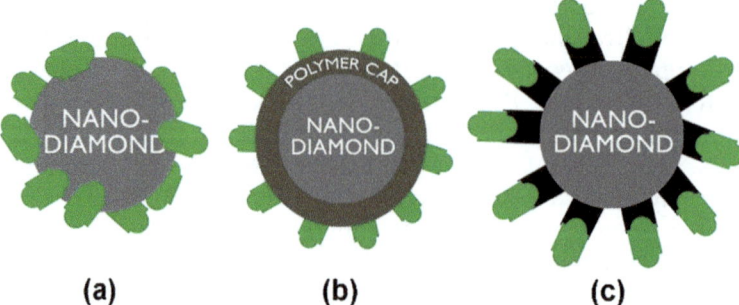

Figure 10.13   Broad classification of the architectures for nanodiamond biofunctionalization: (a) adsorption, (b) polymer/micelle encapsulation and
(c) covalent reaction. Reprinted from ref. 94 with permission from
IOP Publishing Limited.

surface. The procedure makes the FNDs colloidally stable due to electrostatic repulsion between individual FND particles with their large surface
charge (zeta potential ≈ −45 mV in 18 MΩ cm Milli-Q water).[45,83] However,
we observed that in the presence of trace amounts of salt (10 mM NaCl), the
FND colloid flocculated – a phenomenon that is attributed in the literature
to salt-induced charge screening.[45] This screening results in the loss of stabilizing electrostatic repulsion, which can be countered by introducing steric
stability *via* surface functionalization with macromolecules such as proteins,
polymers or dextran.[84] Such complexes may also perform an additional function of biomolecular targeting. Surface biofunctionalization using macromolecules such as nucleic acids,[85] drugs,[86] peptides[87] and proteins[45] have been
reported. The methods used for biofunctionalization include (i) adsorption –
primarily driven by van der Waals, hydrophilic or electrostatic forces,[88,89]
(ii) polymer or micelle encapsulation[61,90] and (iii) covalent biofunctionalization, as illustrated schematically in Figure 10.13. The more stable covalent
functionalization normally relies on the carbon-derived surface functional
groups such as carboxyl, ketone, aldehyde and anhydride groups, which
are activated to react with appropriate groups on a targeting macromolecule.[85,87,91,92] Biofunctionalization of the FNDs using high-affinity molecular
pairs as described earlier has also been reported by several groups.[45,93]

10.3.6   Limitations of FNDs for Optical Biomedical Imaging

FNDs can be employed as fluorescent molecular probes, especially for intracellular imaging, as its core is physically and chemically inert, whereas its
surface is amenable to biofunctionalization. Acid-cleaned FND surfaces are
immediately suitable for covalent biofunctionalization with a variety of reactive chemical functional groups. FND emission is photostable, exhibiting
non-fading emission upon continuous excitation, which makes them suitable for tracking discrete biomolecules labelled with FNDs. Properties of the

nanodiamond host such as non-cytotoxicity[73] and their availability in small sizes (down to 5 nm) extend the scope of their application to protein purification[95] and targeted delivery.

However, two major shortfalls limit the application of FNDs for cellular imaging. First, production of FND remains expensive and cumbersome, since it relies on high-energy electron or light-ion irradiation. These irradiation sources are costly and not readily available, and the production process is not easily scalable, despite some progress in this direction.[72] Annealing at relatively high temperatures in vacuum represents another complication in the FND production process that requires specialized equipment. Second, the FND emission is scaled down dramatically as the size of the crystalline host is decreased (as r^{-3}, where r = FND radius). This is exacerbated by the decreased density of emission states, as r is much smaller than the wavelength of light, which is very important owing to the large refractive index of the diamond host. Finally, the FND excitation takes place in the visible spectral region, ranging from 488 nm (argon ion laser) to 545 nm (helium–neon laser), where excitation at 532 nm (second harmonic, Nd:YAG laser) is often preferred. Radiation in this spectral range is highly absorbed in (live) biological tissue, in particular, by blood (*cf.* Figure 10.1(a)). Biological systems respond to this excitation by emitting fluorescence, termed autofluorescence, which is spectrally broad, extending through the visible to the near-infrared spectral range (tailing off at about 800 nm). This autofluorescence signal is spectrally overlapped with the broadband fluorescence of FND (*cf.* Figures 10.1(a) and 10.8), thus making spectral filtering of the FND signal inefficient. The cell autofluorescence ultimately limits the detection sensitivity of FNDs, precluding the use of ultra-small FNDs for imaging in live cells. More efficient mechanisms of suppression of the excitation light and autofluorescence are required in order to realize ultra-high-sensitivity optical biomedical imaging. These mechanisms and their realizations are discussed in the following sections.

10.4 Nanoruby

Despite the outlined advantages of NPs, such as photostability and surface-to-graft biocompatible moieties, many of the NP-based molecular probes reported in the literature so far and discussed in earlier sections have several drawbacks. For example, europium chelates require excitation at shorter wavelengths (up to 405 nm), which is highly absorbed by biological tissue and produces cell/tissue autofluorescence. The cytotoxicity of several types of NPs, including carbon nanotubes, metal oxides (*e.g.* ZnO) and semiconductor QDs[53] is still under debate. The imaging contrast of FNDs is limited by spectral overlap with biological tissue autofluorescence, and hence remains confined to niche applications.[71] Consequently, there is continuing interest in alternative nanomaterials that show prospects as fluorescent molecular probes. Here, we review the recently reported photoluminescent ruby nanoparticles developed by the Australian group, suitable for ultrasensitive imaging.[5]

Ruby is the Cr-doped form of aluminium oxide in its most stable crystallographic phase, corundum (α-Al$_2$O$_3$). It is photoluminescent due to the emission from Cr^{3+} substituted at Al sites and the emission spectrum in bulk material at 300 K contains two sharp zero-phonon lines (ZPLs) at 694.3 and 692.9 nm (labelled R1 and R2, respectively). The photoluminescence lifetime at room temperature, for Cr doping <1%, is of the order of milliseconds and the quantum yield of the photoluminescence has been determined as 90 ± 5%. With these characteristics, combined with the low cost and ease of obtaining synthetic ruby crystals, the material is attractive for photoluminescence labelling, provided that it can be produced as a colloidal NP suspension. In earlier studies where chemically synthesized nanocrystalline powders of Al$_2$O$_3$ doped with Cr were reported, they were poorly suited to photoluminescence labelling.

10.4.1 Production and Characterization of Nanorubies and Their Colloids

Laser ablation of bulk ruby material was demonstrated to produce colloidal nanorubies. It was carried out using a high-energy titanium/sapphire laser, with 100 fs pulses. The ablation was performed in air and the ablated material was immediately captured in water. Centrifugal separation allowed particle size selection, so that stable colloidal suspensions with a mean size of 12 nm were obtained. The optical properties of the synthesized material were investigated using a hybrid system comprised of an atomic force microscope (AFM) and a laser-scanning confocal microscope with a 532 nm continuous-wave laser used for excitation. This system was also employed to provide an independent verification of the particle size. The optical properties of 12 nm nanorubies, including excitation/emission spectra, quantum yield and photoluminescence lifetime, appeared to be comparable to those of bulk ruby, which demonstrated the nanoruby material's immunity to environmental perturbations, and therefore its attraction as a photoluminescent molecular probe.

The zeta potential of the as-synthesized nanoruby colloid, repeatedly recorded over a 6 month period, was between +30 and +40 mV. Such a high and remarkably stable zeta potential is an indication of good colloidal stability, attributed to the strong electrostatic repulsion of nanoruby particles. The ruby target was scrupulously cleaned prior to ablation, hence this high value of zeta potential was not due to contamination. For further confirmation of the stability, dynamic light scattering (DLS) measurements were conducted over a 1 week period and the hydrodynamic radii were found to remain unchanged within the measurement uncertainties. The stability of the water-based colloid was related to the amphoteric character of Al$_2$O$_3$ and is explained by the Al$_2$O$_3$ surface charge produced by ion complexation reactions, which depended on the solvent pH. Positively (AlOH$_2^+$) and negatively (AlO$^-$) charged moieties dominate at pH below and above the isoelectric point (p*I*), respectively.[96] This mechanism provided particles with good hydrophilic

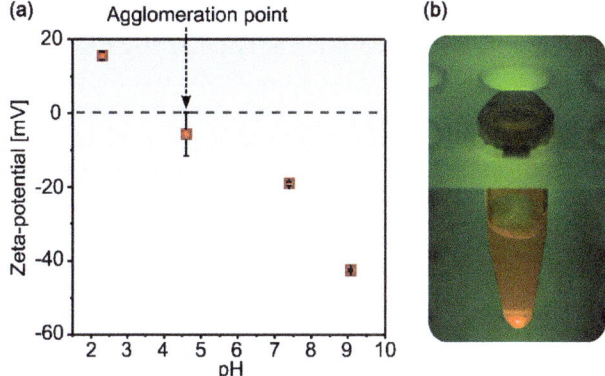

Figure 10.14 (a) Zeta potential of nanoruby particles as a function of pH in various buffers and (b) example nanoruby colloid (in an Eppendorf tube), displaying ruby-related photoluminescence under excitation using a 532 nm laser beam. Reprinted from ref. 5 with permission from John Wiley and Sons, Copyright 2013.

properties due to the electrostatic (electric double-layer repulsion) stabilization forces, which was in agreement with the classical Derjaguin–Landau–Verwey–Overbeek theory of colloidal interactions.

More importantly, the zeta potentials measured in a wide range of pH conditions revealed the remarkable stability of nanorubies in many anionic buffers, including 2-(*N*-morpholino)ethanesulfonic acid (100 mM, pH 2.3 and 4.6), phosphate (12 mM, pH 7.4) and borate (50 mM, pH 9.1). The results (Figure 10.14) show the dependence of the zeta potential on pH; it was negative between pH 4.6 (*ca.* p*I*) and 9 and reached −20 mV at a neutral pH of 7. The observed negative shift of the zeta potential (compared with that in water) was due to the chemisorption of buffer anions (*e.g.* PO_4^{3-} in phosphate-buffered saline (PBS)) to the positively charged $AlOH^{2+}$ moieties, such that the zeta potential becomes highly negative at pH >7, yielding excellent colloidal stability of the nanorubies in buffers.[97]

A lower estimate of the quantum yield of nanorubies was measured to be at least as high as 25 ± 2%, using a method based on the analysis of the average photoluminescence lifetime in particles displaying the R1 and R2 zero-phonon lines, compared with that found in the bulk of ruby. This is a remarkable result showing that the quantum yield was minimally affected by the interface and environmental quenchers, such as hydroxyl groups, present in aqueous media.

10.4.2 Nanorubies as Photoluminescent Molecular Probes

Nanorubies provide high optical imaging contrast due to their very long photoluminescence lifetime of ~3 ms, which is almost 10^6-fold of that of endogenous fluorophores in biological materials that generate a background autofluorescence signal. In order to suppress this autofluorescence background and also the background due to the backscattered excitation light, a time gating approach is employed. Time gating is a method in which pulsed excitation provides discrimination between different sources of fluorescence with distinct decay times. This is achieved by adjusting the delay time (gate) between the excitation pulse and the detection time window. Owing to this delay, the short-lived components of fluorescence can be suppressed. This is particularly useful in the case of multiple sources of fluorescence that have overlapping emission spectra, which cannot be separated by spectral filtering.

The feasibility of the time-gated imaging of photoluminescent nanorubies and the suppression of the undesirable short-lifetime autofluorescence background, typical of biological systems, was demonstrated using a conventional organic fluorescent dye Rhodamine 6G (fluorescence lifetime ~4 ns and peak emission at 551 nm), which showed complete suppression of signals from the dye. The as-produced nanorubies had sufficient brightness to be observed within cellular environments, such as Chinese hamster ovary (CHO-K1) cells, which allowed non-specific internalization of the NPs *via* endocytosis, as shown in Figure 10.15. Three samples were imaged: a control sample with untreated cells, cells incubated with nanoruby particles and cells incubated with nanoruby, which were also membrane labelled with QDs (Qdot 605, Invitrogen). Even in the absence of time gating (see Figure 10.15(a)), the nanorubies were sufficiently bright to be clearly visible on the autofluorescence background of the cell and individual microscale spots of

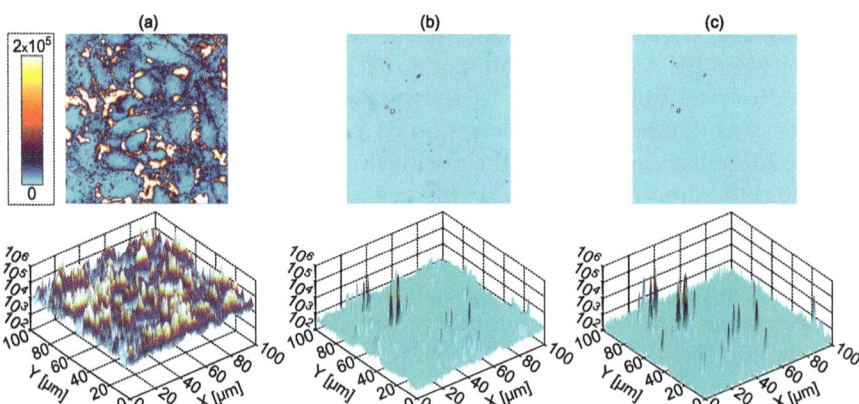

Figure 10.15   Photoluminescence laser scanning confocal microscopy images of CHO-K1 cells treated with both Qdot 605 quantum dots and as-synthesized nanoruby particles without time gating or spectral filtering (a), with a 650 nm long-pass filter (b) and with time gating enabled (c). In the case of the spectral filtering only, the staining from the QDs and autofluorescence are still evident, whereas this is completely suppressed in the case of time gating. Reprinted from ref. 5 with permission from John Wiley and Sons, Copyright 2013.

bright fluorescence, similar to those observed in bright-field transmission electron microscopy images. With the time gating (see Figure 15(b)), the cell autofluorescence was completely suppressed. The photoluminescence properties of the particles within cells were also characterized. The photophysical properties of the nanorubies also appeared to be unaffected by the cellular environment. Our results suggest that in the time-gated images of nanoruby in cells, *single* particles were observed.

Aluminium oxide, the host material of nanoruby, was shown to be the least toxic among other metal oxide nanoparticles, ZnO, TiO_2 and SiO_2, by assaying metabolic activity (MTT) in human foetal lung fibroblasts.[98] The toxicity was demonstrated to be dose dependent, with MTT activity reaching down to 80% of control during 48 h of exposure to 1.5 mg mL^{-1} alumina nanoparticles with a mean size of 190 nm. However, the nanoparticles were shown to exist in micrometre-sized aggregates during the cell incubation. In contrast to this report, our own observations suggested negligible toxicity, if any. The differences are likely to be due to differences in the characteristics of the particles studied, especially the size and concentration. Moreover, the toxicity was measured by means of a trypan blue-based assay, which measured the normal cellular function of expelling this extraneous chromophore.

Nanorubies can be conjugated with antibodies by straightforward physisorption. A model rabbit–anti-rabbit IgG immunoassay confirmed that such conjugated NPs can be targeted to biomolecules of interest. Collectively, these results demonstrate that nanoruby has promising potential as a molecular probe for high-contrast bioimaging and biosensing. At the same time, the excitation bands of nanoruby are situated in the visible spectral range and fall within the blood absorption spectral range. This limits the application scope of nanoruby to ultra-high-sensitivity imaging and sensing in cells, and also to biological tissue and systems devoid of blood. Nanomaterials such as upconversion nanoparticles circumvent this problem.

10.5   Upconversion Nanoparticles (UCNPs)

Photophysical properties unique to UCNP allow almost complete suppression of the biological tissue background and lead to ultimate single nanoparticle photodetection sensitivity.

10.5.1   Photophysics of Upconversion Nanomaterials

Upconversion photoluminescence is a process in which photons of longer wavelength are absorbed, resulting in the emission of light at shorter wavelength (higher energy). Like many other "anti-Stokes" processes (*e.g.* two-photon), the mechanism of upconversion is based on combining energies of multiple absorbed photons, by engaging intermediate energy levels. A special property of the UCNP(s) is that these intermediate energy levels are real, not virtual, as in the case of multiphoton photoluminescence that is widely used in multiphoton microscopy. This relaxes the requirement of simultaneous absorption of two and more photons, and significantly reduces the excitation intensities required to observe upconversion photoluminescence from 1×10^5 to 1×10^3 W cm^{-2}. Therefore, imaging and detection of upconversion particles can be achieved by using relatively low-intensity, inexpensive, continuous-wave 980 nm lasers,[3] as opposed to complex and expensive femtosecond lasers required to trigger multiphoton processes. This makes UCNPs well suited for biological applications, especially in the case of *in vivo* scenarios, where the maximum laser power is limited by their effect on the cells.[99] Example nanomaterials where upconversion photoluminescence was demonstrated include NaYF$_4$:Yb:Er,[99–101] NaYF$_4$:Y:Yb:Er,[102] and Y$_2$O$_3$:Yb:Er.[103] The most efficient in terms of the photon energy conversion and most widely studied UCNP is an inorganic (NaYF$_4$) nanocrystal co-doped with near-infrared light sensitizer ytterbium (Yb^{3+}) ions and activator erbium (Er^{3+}) or thulium (Tm^{3+}) lanthanide ions.

As an example, UCNP co-doped with Yb^{3+} and Tm^{3+} is excited at 980 nm and radiates at 800 nm (following absorption of two sequential photons) and 474 nm (three sequential photons) spectral bands (Figure 10.16(b)). The conversion efficiency, *i.e.* the emission/excitation (watts per watts) power ratio (η_{uc}), increases with the excitation intensity, saturating at excitation intensities of 10–1000 W cm^{-2}.[101,104] Until recently, the conversion efficiency of upconversion materials at the nanometre scale was very low and prevented practical applications. However, a breakthrough in UCNP synthesis[105,106] resulted in much brighter nanoparticles, where "brightness" is measured in terms of the product of η_{uc} and absorption cross-section. A combination of increased activator doping with higher excitation irradiance can enable single UCNP tracking, with sensitivities up to 70 times more than with QDs.[107]

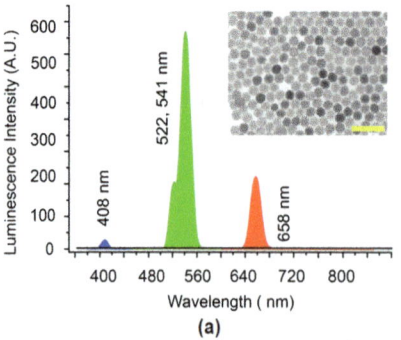
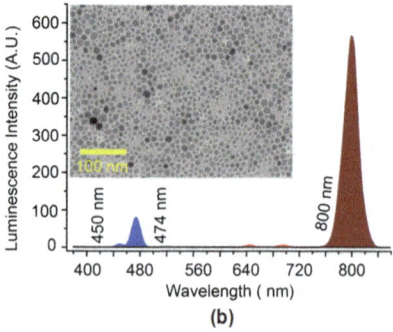

Figure 10.16   Photoluminescent spectra of (a) NaYF$_4$:Yb:Er; and (b) NaYF$_4$:Yb:Tm nanomaterials, with their corresponding transmission electron micrographs. Reproduced from ref. 6.

10.5.2 Production and Characterization of UCNPs

A two-step protocol established by Wang *et al.*[102] and Mai *et al.*[108] is the most widely used method for synthesizing UCNPs. This procedure yields UCNPs with sizes ranging from sub-10 nm NaLuF$_4$-based nanocrystals, co-doped with Gd^{3+}, Yb^{3+} and Er^{3+} (or Tm^{3+}),[109] to sub-micron-sized upconversion phosphors. This synthesis can be modified to produce a variety of UCNPs with different sizes and spectral and temporal upconversion properties.[110] η_{uc} represents a measurable quantity that reports on the quality of the produced nanoparticles. Our measurements carried out using an integrating sphere showed a dramatic increase in η_{uc} with size and transformation from the cubic (alpha) to hexagonal (beta) crystal phase, which confirmed the existing results extensively published in the literature. The 10 nm α-phase UCNPs had an η_{uc} of 0.03%, whereas the 60 nm β-phase UCNPs were characterized by an η_{uc} of 2%, and the saturation intensity was measured as ~70 W cm^{-2} for both samples.

Upconversion nanoparticles based on NaYF$_4$ host material require further processing to meet stringent demands for biological applications, one of these being the stability of an aqueous colloidal suspension to facilitate UCNP biofunctionalization. First, as-synthesized UCNPs have oleic groups on the nanoparticle surfaces that make this nanomaterial hydrophobic. The removal of the oleic groups is of little utility, because the resultant bare NaYF$_4$ surface is deprived of any chemical groups for reliable biochemical functionalization. Early attempts to replace the weakly bonded oleic groups with more suitable moieties, such as citric groups,[3] were ineffective because of their instability at pH $\neq$ 7 conditions. The other strategy is micelle encapsulation using amphiphilic polymers to produce water-dispersible UCNPs, similar to that used with FNDs.[24] Our own experiments using this protocol resulted in UCNPs that were stable in water and buffer solutions for 1 month.[46] However, it is also worth noting that the η_{uc} may be adversely affected by surface functionalization. This is because the high-energy vibronic modes of some functional groups, such as OH, provide non-radiative relaxation pathways for the upper excited states of Er^{3+} or Tm^{3+} in NaYF$_4$:Yb:Er(Tm).[110]

10.5.3 UCNP-Assisted Optical Biomedical Imaging

There are several advantages of UCNP in the context of biomedical optical imaging. First, the NaYF$_4$:Yb:Tm photoluminescence band (800 nm; see Figure 10.16(b)) falls into the biological tissue transparency window. Their emission in the biological tissue window experiences reduced absorption and scattering in comparison with the ultraviolet (UV), visible (VIS) and infrared (IR) spectral ranges. The biological tissue absorption is dominated by oxygenated and deoxygenated haemoglobin constituents of blood and proteins (*e.g.* melanin) in the UV/VIS range and by water in the IR range. The scattering extinction cross-section decreases slowly from the UV towards the IR region. The combined effect of the scattering/absorption of biological tissue is plotted in Figure 10.1(a). As can be seen, the absorption of the excitation light at 980 nm by water and haemoglobin in skin (Figure 10.1(a)) is compensated by the reduced tissue scattering at this wavelength. Second, UCNP excitation at 980 nm elicits very little autofluorescence from biological tissue. We have demonstrated that UCNPs topically applied both on freshly excised human skin and delivered into the dermis layer by a microneedle exhibit high imaging contrast, with the tissue-induced laser light scattering and autofluorescence background almost completely suppressed. The acquired epi-luminescence images of skin autofluorescence and UCNP distribution excited at 365 and 980 nm, respectively, are shown in Figure 10.17.

In order to counter the excitation light passing into the detection path, the detection band can be spectrally offset by as much as 330 nm from the excitation band, by choosing the $NaYF_4$:Yb:Er upconversion emission multiplet centred at 550 nm.[103] The other major advantage of UCNP lies in the 10^6-fold difference in the emission lifetimes of UCNP and endogenous fluorophores that make up the autofluorescence background ($\tau_{uc} \approx 1$ ms *versus* $\tau_{af} \approx 3$ ns) that allows complete suppression of the autofluorescence and excitation background by an optical time-gated approach alone, thus avoiding the use of expensive interference filters. This scheme is realized by setting a time

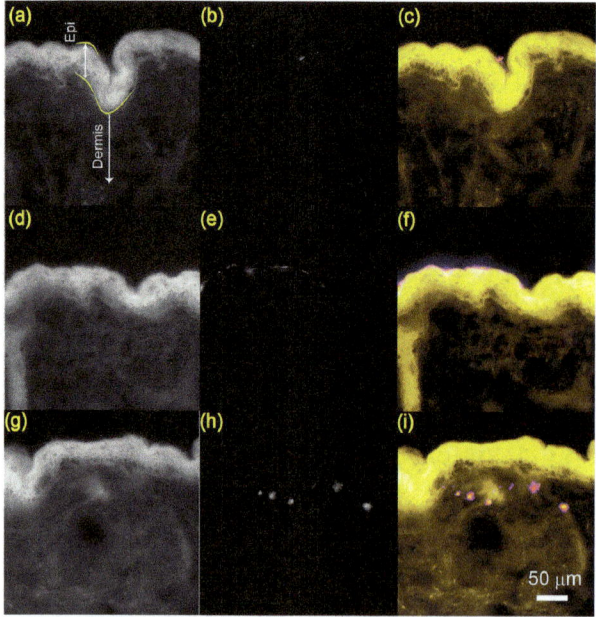

Figure 10.17 $NaYF_4$:Yb:Tm nanoparticle distribution in human skin in the case of (top two rows) topical application and (bottom row) microneedle treatment. (a, d and g) UV (365 nm)-excited autofluorescence images of skin; (b, e and h) images of UCNP excited by 980 nm laser radiation; (c, f and i) pseudo-colour overlaid images of (a), (d) and (g) showing UCNPs (purple) in skin folds and dermis (yellow), respectively.

delay (~10 μs) between the excitation laser pulse and photodetection of the sample response. The long τ_{uc} of UCNP also leads to enhanced efficiency of Förster resonance energy transfer (FRET) processes, used to investigate protein interactions.[111] This also makes UCNPs useful as sensitizers of singlet oxygen-generating proteins for use in theranostics applications.[112]

The promise of upconversion nanoparticles has recently been demonstrated by imaging of UCNP bioconjugates in cell cultures[3,46,109] and whole animal models[113,114] under background-free conditions. Figure 10.18 shows the results of the live-cell imaging of UCNPs by using laser-scanning photoluminescence confocal microscopy.[3] The laser-scanning modality allows high-sensitivity detection, since high intensity at the sub-femtolitre focal excitation volume, such as 10^5 W cm^{-2}, is easily achievable. Evading the signal cross-talks caused by the multiple scattering of photoluminescence photons in the biological samples is important for single nanoparticle imaging, which would otherwise be obscured by intense signals from unwanted remnant UCNP clusters. At the same time, the confocal imaging modality suffers from the long image acquisition times arising from the necessity to dwell for several τ_{uc}s on each acquisition pixel. As an example, consider $\tau_{uc} \cong 1$ ms, which requires setting an acquisition time of at least 5 ms per pixel, and takes ~30 min to capture a 512×512 pixel image – this is impractical in most applications.

Intense research activity has been focused on the optical tomography of UCNP-tagged tissue sites, where upconversion nanoparticles are deployed as molecular probes. In particular, whole-animal imaging has been reported on a number of occasions, where the UCNP-assisted target delivery and optical tomographic imaging of cancer tumour sites demonstrated imaging depths of up to 2 cm in transmission mode.[115]

This raises the question of the ultimate sensitivity achievable with UCNP imaging in biological tissue. In order to answer this question, we performed single UCNP imaging through a layer of hemolyzed blood using the full-field epi-luminescence optical microscopy configuration. We found that, provided the scattering of the biological sample is not overwhelming as in the

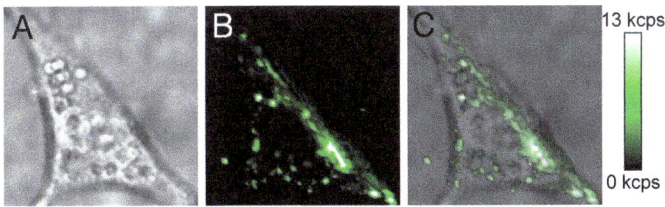

Figure 10.18   Live-cell imaging of UCNPs in NIH 3T3 murine fibroblasts using laser-scanning photoluminescence confocal microscopy. (A) Bright field image of a cell with endocytosed UCNPs; (B) upconverted photoluminescence following 980 nm excitation; (C) overlay. Scale bar, 10 μm. Reprinted from ref. 3 with permission from the National Academy of Sciences.

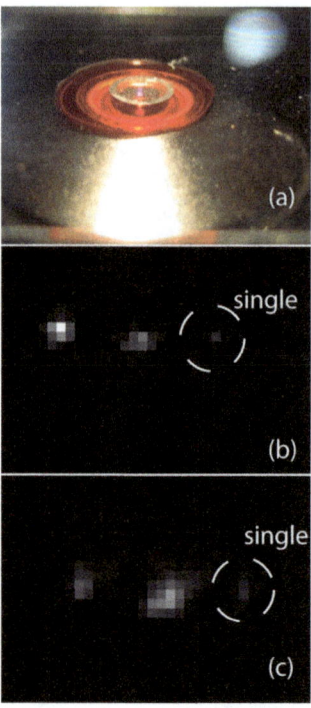

Figure 10.19   Optical epi-luminescence micrograph of a single UCNP particle deposited on a glass slide using a "blood immersion" objective lens shown in (a). (b and c) Images of the constellation of UCNPs acquired using a water and a blood immersion objective lens, respectively. A single UCNP particle, verified separately by correlative electron microscopy, is encircled. The epi-luminescence image acquisition time was 1 s.

case of hemolyzed blood, the sample absorption presented little problem for imaging, and no degradation of the imaging of a single (UCNP) nanoparticle was observed (Figure 10.19).

These studies also revealed shortcomings of the UCNP technology. These include limited penetration depth of the UCNP-assisted optical imaging in whole animals because of the requirement for the high excitation intensity that is usually realized by focusing, and is not possible in turbid biological tissue.[103] This point is illustrated in Figure 10.20, where a comparison between optical tomography imaging deploying an organic fluorescence dye and UCNP in an animal model is presented.[103] Although UCNP-assisted optical tomography provided much more accurate localization of the photoluminescence probe (due to the background-free imaging and non-linear dependence of the photoluminescence signal *versus* the excitation intensity), the sensitivity of the UCNP detection is inferior to that of the conventional infrared fluorescence dye. It is important to note that, unlike in the case of

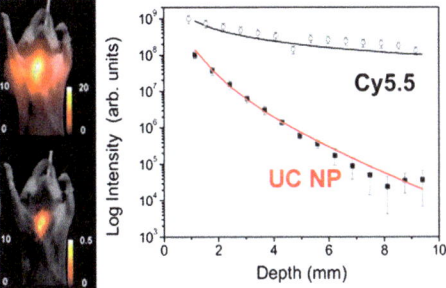

Figure 10.20 Left panel (top): Image of a mouse with a thin capillary tube inserted into the oesophagus. The tube is filled with Cy5.5 dye (excitation at 680 nm and emission at 712 nm). Left panel (bottom): The same mouse with UCNP instead of the dye (excitation at 980 nm and emission at 670 nm). The absence of autofluorescence in the bottom image is obvious. However, a much stronger excitation power is required in the case of the UCNP bioprobe. This is illustrated in the right panel, where the decay of the signal is shown as a function of the tissue imaging depth. The signal decays much faster for the UCNP, because it is proportional to the intensity to the power of 2.3. Reprinted from ref. 103 with permission from The Optical Society.

infrared fluorescence dye Cy5, the UCNP signal intensity is quickly degraded *versus* the imaging depth. This sets an inherent limit on the applicability of the UCNPs for deep tissue imaging, especially considering that the maximum intensity allowed at 980 nm for use in human studies and practices is 726 mW cm^{-2}.[104] Further, the spectral filtering efficiency of the UCNP photoluminescence is compromised by the broad angular distribution of unwanted background photons emerging from biological tissue (10^6-fold *versus* 10^3-fold[116]), although this can be mitigated by employing the time-gated detection method. These findings suggest that only some applications are able to capitalize fully on the key UCNP merits, while minimizing their limitations. One such application is the ultrasensitive imaging of UCNPs in thin tissue slices, such as skin, aimed at the demonstration of the background-free imaging of nanoparticles, paving the way towards ultrasensitive *in vivo* imaging of nanoparticle penetration in skin.

10.5.4 Specific Aspects of Bioconjugation of Upconversion Nanoparticles

As-synthesized UCNPs are surface coordinated with oleic groups that render this nanomaterial hydrophobic, *i.e.* not miscible with water and buffer solutions. In addition, NaYF$_4$ does not seem to provide appropriate surface anchoring points for strong docking of biomolecules or auxiliary moieties. Third, the UCNP conversion efficiency appeared to be highly susceptible to the surface functionalization. The early attempts to displace the weakly bonded oleic groups with more affine moieties, such as citric[3] and mercaptopropionic[117] groups, demonstrated only marginal improvements in the colloidal stability of UCNPs at neutral pH (pH 7), precluding biomedical applications, where the pH varies widely. This implies that the functional group attachment to the UCNP surface was governed by electrostatic or/and van der Waals forces, *i.e.* attachment by adsorption. The adsorption-based biofunctionalization protocol was also employed by Niedbala and co-workers to attach NeutrAvidin to the UCNP surface by a facile pot-mix reaction.[118] As can be expected, this approach was subject to desorption of the adsorbed molecules in biological solutions, soon resulting in particle agglomeration. A common solution is to cover the surface of nanoparticles with poly(ethylene glycol) (PEG), which prevents formation of aggregates by introducing steric hindrance.[119] Also, PEGylation represents the core of the so-called stealth technology preventing PEGylated nanoparticles injected intravenously into the bloodstream from rapid immune-mediated removal. However, PEG groups have disadvantages associated with poorly controlled polymer chains and, most importantly, PEG chemistry is expensive.

A radical solution to avoid the dependence on bioconjugation reactions based on UCNP surface anchoring involves coating the nanoparticles with an additional layer that is amenable to subsequent bioconjugation. Coating the UCNP nanoparticles with a silica shell[120,121] is an attractive approach owing to the stability of the shell over a range of pH and the maturity of the silica surface coating technology.[120] Another approach makes use of amphiphilic polymers that harbour moieties that are hydrophobic and hydrophilic in nature. This approach has been successfully demonstrated for the surface activation of QDs,[24] and partly accounted for the commercial success of QD-based procedures. An amphiphilic polymer was wrapped around the particles, exploiting the non-specific hydrophobic interactions between the alkyl chains of poly(maleic anhydride-*alt*-1-tetradecene) and the nanocrystal surfactant molecules (oleic acid).[46] This allowed the transfer of hydrophobically capped nanocrystals from organic to aqueous solution. Addition of bis(6-aminohexyl)amine resulted in cross-linking of the polymer chains around each nanoparticle, further stabilizing the alkyl chains.[24] Our initial experiments using this protocol resulted in UCNPs stable in water and buffer solutions for 1 month. Nanocrystals coated with COOH-amphiphilic polymer[24] are readily processed using a universal bioconjugation protocol, *i.e.* the EDC/sNHS reaction, which we demonstrated by attaching streptavidin to UCNP (unpublished work). A facile pot-mix reaction will lead to strong attachment of a targeting biomolecule of choice, if this molecule is biotinylated. A range of commercially available biotinylated ligands and antibodies, including antibodies to epitope tags, permits the versatile design of various bioconjugates that are linked together *via* the strong bond between streptavidin and biotin. An alternative approach that we have demonstrated is based on a high-affinity protein pair, barnase:barstar, that takes advantage of recombinant protein technology.[45]

10.6 Conclusion

The rapid development of photoluminescence nanotechnology in recent years has stimulated considerable interest in their application in the life sciences. In particular, several unique features of photoluminescent nanoparticles appeared very attractive for their implementation as diagnostic and therapeutic complexes. These include a physically and chemically stable nanoparticle core, and useful photophysical properties that are often distinct from those of the surrounding biological environment than can also be tuned by manipulating their production process. Such an important photophysical property of NPs as photoluminescence is usually stable upon continuous excitation and is tuneable (by varying the size, as in QDs, or active ion doping ratio, as in upconversion nanoparticles), and its range of lifetimes is very broad, from sub-nanoseconds (as in zinc oxide semiconductor QDs) to milliseconds (as in nanoruby). These features of photoluminescent NPs allow ultra-high-sensitivity imaging and sensing of these nanoparticles due to the possibility of discriminating their emission from the background comprised of backscattered excitation and autofluorescence from the biological environment. Another important advantage of NPs is their developed surface, which can be decorated with biocompatible functional moieties, form versatile docking stations. NPs can also serve as nanovehicles to host biologically significant modules, such as therapeutic, targeting and stealth modules for targeted delivery, and diagnostic modules that guide and monitor the effects of the NP-assisted therapy for pathological lesions. These properties provide the foundations for significant emerging areas in applied biomedical science, including (personalized) nanomedicine and theranostics.

We have reviewed several important types of photoluminescent nanoparticles, including well-established quantum dots (QDs), fluorescent nanodiamonds (FNDs), nanorubies and upconversion nanoparticles (UCNPs). The choice of these NPs was dictated by their emission properties with different photophysical origins, and these were discussed in the context of biomedical imaging applications, which offer a variety of important possibilities for nanomaterial deployment in bionanophotonics applications. The choice of nanomaterial, however, needs to be carefully scrutinized based on its physical, chemical and optical properties, where core chemical inertness, amenable surface chemistry and minimal cytotoxicity are essential. Optical properties remain the key factors determining the utility of these nanomaterials, where the NP brightness determined by its action cross-section is a necessary, but not sufficient, prerequisite. It appears that the imaging contrast of photoluminescent NPs against the crowded morphological and autofluorescence background of the cell needs to be taken into account. To this end, QDs and FNDs offer limited possibilities for achieving ultra-high contrast and ultra-high sensitivity, whereas nanorubies hold good promise by virtue of their narrow emission spectrum and long emission lifetime, albeit the excitation band situated in the visible spectral range may limit their application scope to imaging of cells and thin tissue slices. Despite the relatively low action cross-section of UCNPs, they offer unprecedented means for discriminating the scattered excitation and autofluorescence background, paving the way towards ultimate detection sensitivity at the single biomolecule level. At the same time, the non-linear dependence of the UCNP photoluminescence emission on the excitation intensity limits the imaging penetration depth in highly scattering biological tissue, and should be taken into account when deploying UCNPs for diagnostic imaging.

In order to apply photoluminescent NPs for the nanoparticle-assisted imaging of cellular processes, such as receptor-mediated endocytosis and associated molecular events, one needs to consider three key aspects: design of bright and photostable photoluminescent nanomaterials that are conspicuous among the background consisting of backscattered excitation light and cell autofluorescence; mild surface modification to permit facile interfacing with biomolecules; and modular engineering of the biomolecular complexes with targeting vectors (antibodies, mini-antibodies or peptides) firmly attached to the NP for targeted delivery to specific cellular or tissue sites.

Interfacing nanoparticles with biomolecules using bioconjugation procedures poses challenges owing to the poor dispersibility of (often hydrophobic) nanoparticles in aqueous colloids. Fluorescent nanodiamond represents a fortunate exception whose acid-cleaned surface exhibits biocompatible carbonyl groups that permit direct realization of a modular assembly of a fluorescent nanodiamond bioconjugate in a facile pot-mix reaction. Hydrophobic QDs and UCNPs require several additional procedures to render these particles mixable with buffer solutions and allow grafting of biocompatible groups, *e.g.* COOH groups, on their surface. This is achieved by complete surface reshaping with a shell comprising of, for example, an amphiphilic polymer prefabricated with COOH groups that render the UCNP surface amenable to bioconjugation. As-produced nanorubies display remarkable colloidal stability in water and buffers, making them amiable for bioconjugation, although covalent attachment of biocompatible surface groups remains a problem to be solved.

The biological properties of the nanoparticle biocomplexes are primarily determined by functional biomolecules attached to the NP surface, and the choice of biomolecules to be used to interact with a participating biological system is important. In this chapter, we considered several targeting vectors, including the mini-antibody HER2/neu and peptide somatostatin, to demonstrate the potential of photoluminescent nanoparticle labelling.

The development of photoluminescent nanoparticle bioconjugates capable of controllable interactions with targeted cells and tissues opens up new opportunities in several areas of the life sciences, including new imaging modalities allowing long time-lapse investigation of molecular trafficking in cells, cellular imaging at the unprecedented single-molecule level, *in vivo* diagnosis based on ultra-high-sensitivity optical imaging of whole organisms and targeted drug delivery. This chapter has outlined such exciting prospects for future developments.

Acknowledgements

We wish to acknowledge Russian Foundation for Basic Research (grant No. 13-04-40227), Russian Scientific Foundation (grant No. 14-24-00106), Ministry of Science and Education of Russian Federation (grant No. 2013-220-04-6237), for the support towards the cell technology, biochemistry and nanotechnology aspects of this work, respectively. VKAS and EMG acknowledge the funding of Macquarie University Research Fellowship and Australian Research Council Centre of Excellence for Nanoscale BioPhotonics (CE14010003), respectively.

References

1. H. Y. Li, Y. H. Li, J. Jiao and H. M. Hu, *Nat. Nanotechnol.*, 2011, **6**, 645–650.
2. A. M. Schrand, H. Huang, C. Carlson, J. J. Schlager, E. Ōsawa, S. M. Hussain and L. Dai, *J. Phys. Chem. B*, 2007, **111**, 2–7.
3. S. W. Wu, G. Han, D. J. Milliron, S. Aloni, V. Altoe, D. V. Talapin, B. E. Cohen and P. J. Schuck, *Proc. Natl. Acad. Sci. U. S. A.*, 2009, **106**, 10917–10921.
4. C. Bradac, T. Gaebel, C. I. Pakes, J. M. Say, A. V. Zvyagin and J. R. Rabeau, *Small*, 2013, **9**, 132–139.
5. A. M. Edmonds, M. A. Sobhan, V. K. A. Sreenivasan, E. A. Grebenik, J. R. Rabeau, E. M. Goldys and A. V. Zvyagin, *Part. Part. Syst. Charact.*, 2013, **30**, 506–513.
6. Z. Song, Y. G. Anissimov, J. Zhao, A. V. Nechaev, A. Nadort, D. Jin, T. W. Prow, M. S. Roberts and A. V. Zvyagin, *J. Biomed. Opt.*, 2013, **18**, 61215.
7. V. K. A. Sreenivasan, E. J. Kim, A. K. Goodchild, M. Connor and A. V. Zvyagin, *Nanomedicine*, 2012, 1551–1560.
8. A. L. Rogach, *Semiconductor Nanocrystal Quantum Dots: Synthesis, Assembly, Spectroscopy and Applications*, Springer, Berlin, 2008.
9. A. D. Yoffe, *Adv. Phys.*, 2001, **50**, 1–208.
10. B. O. Dabbousi, J. Rodriguez-Viejo, F. V. Mikulec, J. R. Heine, H. Mattoussi, R. Ober, K. F. Jensen and M. G. Bawendi, *J. Phys. Chem. B*, 1997, **101**, 9463–9475.
11. L. Cademartiri, E. Montanari, G. Calestani, A. Migliori, A. Guagliardi and G. A. Ozin, *J. Am. Chem. Soc.*, 2006, **128**, 10337–10346.
12. O. Schmelz, A. Mews, T. Basché, A. Herrmann and K. Müllen, *Langmuir*, 2001, **17**, 2861–2865.
13. A. Striolo, J. Ward, J. M. Prausnitz, W. J. Parak, D. Zanchet, D. Gerion, D. Milliron and A. P. Alivisatos, *J. Phys. Chem. B*, 2002, **106**, 5500–5505.
14. X. Michalet, F. F. Pinaud, L. A. Bentolila, J. M. Tsay, S. Doose, J. J. Li, G. Sundaresan, A. M. Wu, S. S. Gambhir and S. Weiss, *Science*, 2005, **307**, 538–544.
15. D. Gerion, F. Pinaud, S. C. Williams, W. J. Parak, D. Zanchet, S. Weiss and A. P. Alivisatos, *J. Phys. Chem. B*, 2001, **105**, 8861–8871.
16. C. B. Murray, D. J. Norris and M. G. Bawendi, *J. Am. Chem. Soc.*, 1993, **115**, 8706–8715.

17. M. Howarth, W. Liu, S. Puthenveetil, Y. Zheng, L. F. Marshall, M. M. Schmidt, K. D. Wittrup, M. G. Bawendi and A. Y. Ting, *Nat. Methods*, 2008, **5**, 397–399.

18. H. S. Yang, P. H. Holloway and S. Santra, *J. Chem. Phys.*, 2004, **121**, 7421–7426.

19. D. Douroumis, O. Obonyo, E. Fisher and M. Edwards, *Crit. Rev. Biotechnol.*, 2010, **30**, 283–301.

20. M. A. Hines and P. Guyot-Sionnest, *J. Phys. Chem.*, 1996, **100**, 468–471.

21. *Fluorescence Applications in Biotechnology and Life Sciences*, ed. E. M. Goldys, John Wiley & Sons, Hoboken, NJ, 2009.

22. V. Biju, T. Itoh, A. Anas, A. Sujith and M. Ishikawa, *Anal. Bioanal. Chem.*, 2008, **391**, 2469–2495.

23. W. C. W. Chan and S. M. Nie, *Science*, 1998, **281**, 2016–2018.

24. T. Pellegrino, L. Manna, S. Kudera, T. Liedl, D. Koktysh, A. L. Rogach, S. Keller, J. Radler, G. Natile and W. J. Parak, *Nano Lett.*, 2004, **4**, 703–707.

25. T. Rajh, O. I. Mićić and A. J. Nozik, *J. Phys. Chem.*, 1993, **97**, 11999–12003.

26. Y. Zhu, Z. Li, M. Chen, H. M. Cooper, G. Q. Lu and Z. P. Xu, *J. Colloid Interface Sci.*, 2013, **390**, 3–10.

27. Z. P. Li, B. B. Liu, X. L. Li, S. D. Yu, L. Wang, Y. Y. Hou, Y. G. Zou, M. G. Yao, Q. J. Li, B. Zou, T. Cui, G. T. Zou, G. R. Wang and Y. H. Liu, *Nanotechnology*, 2007, **18**, 255602.

28. A. N. Generalova, S. V. Sizova, T. A. Zdobnova, M. M. Zarifullina, M. V. Artemyev, A. V. Baranov, V. A. Oleinikov, V. P. Zubov and S. M. Deyev, *Nanomedicine*, 2011, **6**, 195–209.

29. C. T. Dameron, R. N. Reese, R. K. Mehra, A. R. Kortan, P. J. Carroll, M. L. Steigerwald, L. E. Brus and D. R. Winge, *Nature*, 1989, **338**, 596–597.

30. S. R. Stürzenbaum, M. Höckner, A. Panneerselvam, J. Levitt, J. S. Bouillard, S. Taniguchi, L. A. Dailey, R. A. Khanbeigi, E. V. Rosca, M. Thanou, K. Suhling, A. V. Zayats and M. Green, *Nat. Nanotechnol.*, 2013, **8**, 57–60.

31. K. Manzoor, S. Johny, D. Thomas, S. Setua, D. Menon and S. Nair, *Nanotechnology*, 2009, **20**, 13.

32. K. Manzoor, S. R. Vadera, N. Kumar and T. R. N. Kutty, *Mater. Chem. Phys.*, 2003, **82**, 718–725.

33. C. Shu, B. Huang, X. Chen, Y. Wang, X. Li, L. Ding and W. Zhong, *Spectrochim. Acta, Part A*, 2013, **104**, 143–149.

34. X. Y. Wu, H. J. Liu, J. Q. Liu, K. N. Haley, J. A. Treadway, J. P. Larson, N. F. Ge, F. Peale and M. P. Bruchez, *Nat. Biotechnol.*, 2003, **21**, 41–46.

35. T. A. Zdobnova, S. G. Dorofeev, P. N. Tananaev, R. B. Vasiliev, T. G. Balandin, E. F. Edelweiss, O. A. Stremovskiy, I. V. Balalaeva, I. V. Turchin, E. N. Lebedenko, V. P. Zlomanov and S. M. Deyev, *J. Biomed. Opt.*, 2009, **14**, 021004.

36. D. S. Lidke, P. Nagy, R. Heintzmann, D. J. Arndt-Jovin, J. N. Post, H. E. Grecco, E. A. Jares-Erijman and T. M. Jovin, *Nat. Biotechnol.*, 2004, **22**, 198–203.

37. R. Savla, O. Taratula, O. Garbuzenko and T. Minko, *J. Controlled Release*, 2011, **153**, 16–22.

38. R. Mahtab, H. H. Harden and C. J. Murphy, *J. Am. Chem. Soc.*, 2000, **122**, 14–17.

39. S. Pathak, S. K. Choi, N. Arnheim and M. E. Thompson, *J. Am. Chem. Soc.*, 2001, **123**, 4103–4104.

40. P. C. Weber, D. H. Ohlendorf, J. J. Wendoloski and F. R. Salemme, *Science*, 1989, **243**, 85.

41. S. M. Deyev, R. Waibel, E. N. Lebedenko, A. P. Schubiger and A. Pluckthun, *Nat. Biotechnol.*, 2003, **21**, 1486–1492.

42. V. K. Sreenivasan, T. A. Kelf, E. A. Grebenik, O. A. Stremovskiy, J. M. Say, J. R. Rabeau, A. V. Zvyagin and S. M. Deyev, *Proteomics*, 2013, 1437–1443.

43. J. H. Warner, E. Thomsen, A. R. Watt, N. R. Heckenberg and H. Rubinsztein-Dunlop, *Nanotechnology*, 2005, **16**, 175–179.

44. U. F. Aghayeva, M. P. Nikitin, S. V. Lukash and S. M. Deyev, *ACS Nano*, 2013, **7**, 950–961.

45. V. K. A. Sreenivasan, E. A. Ivukina, W. Deng, T. A. Kelf, T. A. Zdobnova, S. V. Lukash, B. V. Veryugin, O. A. Stremovskiy, A. V. Zvyagin and S. M. Deyev, *J. Mater. Chem.*, 2011, **21**, 65–68.

46. E. A. Grebenik, A. Nadort, A. N. Generalova, A. V. Nechaev, V. K. A. Sreeni-vasan, E. V. Khaydukov, V. A. Semchishen, A. P. Popov, V. I. Sokolov, A. S. Akhmanov, V. P. Zubov, D. V. Klinov, V. Y. Panchenko, S. M. Deyev and A. V. Zvyagin, *J. Biomed. Opt.*, 2013, **18**, 076004.

47. X. Y. Cheng, R. Gondosiswanto, S. Ciampi, P. J. Reece and J. J. Gooding, *Chem. Commun.*, 2012, **48**, 11874–11876.

48. L. Ruizendaal, S. P. Pujari, V. Gevaerts, J. M. J. Paulusse and H. Zuilhof, *Chem.–Asian J.*, 2011, **6**, 2776–2786.

49. N. J. Agard, J. A. Prescher and C. R. Bertozzi, *J. Am. Chem. Soc.*, 2005, **127**, 11196.

50. W. J. Jin, J. M. Costa-Fernandez, R. Pereiro and A. Sanz-Medel, *Anal. Chim. Acta*, 2004, **522**, 1–8.

51. N. J. Durr, T. Larson, D. K. Smith, B. A. Korgel, K. Sokolov and A. Ben-Yakar, *Nano Lett.*, 2007, **7**, 941–945.

52. T. A. Kelf, V. K. A. Sreenivasan, J. Sun, E. J. Kim, E. M. Goldys and A. V. Zvyagin, *Nanotechnology*, 2010, **21**, 285105.

53. A. M. Derfus, W. C. W. Chan and S. N. Bhatia, *Nano Lett.*, 2003, **4**, 11–18.

54. N. Pradhan, D. Goorskey, J. Thessing and X. Peng, *J. Am. Chem. Soc.*, 2005, **127**, 17586–17587.

55. L. Ye, K. T. Yong, L. W. Liu, I. Roy, R. Hu, J. Zhu, H. X. Cai, W. C. Law, J. W. Liu, K. Wang, J. Liu, Y. Q. Liu, Y. Z. Hu, X. H. Zhang, M. T. Swihart and P. N. Prasad, *Nat. Nanotechnol.*, 2012, **7**, 453–458.

56. C. Emmanuel, T. Nadhi, W. Y. William, L. C. Vicki and D. Rebekah, *Small*, 2006, **2**, 1412–1417.

57. V. Y. Dolmatov, M. V. Veretennikova, V. A. Marchukov and V. G. Sushchev, *Phys. Solid State*, 2004, **46**, 611–615.

58. S. Osswald, G. Yushin, V. Mochalin, S. O. Kucheyev and Y. Gogotsi, *J. Am. Chem. Soc.*, 2006, **128**, 11635–11642.

59. J. Tisler, G. Balasubramanian, B. Naydenov, R. Kolesov, B. Grotz, R. Reuter, J. P. Boudou, P. A. Curmi, M. Sennour, A. Thorel, M. Borsch, K. Aulenbacher, R. Erdmann, P. R. Hemmer, F. Jelezko and J. Wrachtrup, *ACS Nano*, 2009, **3**, 1959–1965.

60. A. Krueger, M. Ozawa, G. Jarre, Y. Liang, J. Stegk and L. Lu, *Phys. Status Solidi A*, 2007, **204**, 2881–2887.

61. B. R. Smith, M. Niebert, T. Plakhotnik and A. V. Zvyagin, *J. Lumin.*, 2007, **127**, 260–263.

62. I. I. Vlasov, A. S. Barnard, V. G. Ralchenko, O. I. Lebedev, M. V. Kanzyuba, A. V. Saveliev, V. I. Konov and E. Goovaerts, *Adv. Mater.*, 2009, **21**, 808.

63. A. Gruber, A. Drabenstedt, C. Tietz, L. Fleury, J. Wrachtrup and C. von Borczyskowski, *Science*, 1997, **276**, 2012–2014.

64. L. Rondin, G. Dantelle, A. Slablab, F. Grosshans, F. Treussart, P. Bergonzo, S. Perruchas, T. Gacoin, M. Chaigneau, H. C. Chang, V. Jacques and J. F. Roch, *Phys. Rev. B*, 2010, **82**, 115449.

65. T. L. Wee, Y. K. Tzeng, C. C. Han, H. C. Chang, W. Fann, J. H. Hsu, K. M. Chen and Y. C. Yull, *J. Phys. Chem. A*, 2007, **111**, 9379–9386.

66. A. V. Zvyagin and N. B. Manson, in *Ultrananocrystalline Diamond*, ed. O. Shenderova and D. Gruen, William Andrew Publishing, Oxford, 2nd edn, 2012, pp. 327–354.

67. J. R. Maze, P. L. Stanwix, J. S. Hodges, S. Hong, J. M. Taylor, P. Cappellaro, L. Jiang, M. V. G. Dutt, E. Togan, A. S. Zibrov, A. Yacoby, R. L. Walsworth and M. D. Lukin, *Nature*, 2008, **455**, 644–647.

68. F. Cichos, C. von Borczyskowski and M. Orrit, *Curr. Opin. Colloid Interface Sci.*, 2007, **12**, 272–284.

69. P. A. Frantsuzov and R. A. Marcus, *Phys. Rev. B*, 2005, **72**, 155321.

70. C. Bradac, T. Gaebel, N. Naidoo, M. J. Sellars, J. Twamley, L. J. Brown, A. S. Barnard, T. Plakhotnik, A. V. Zvyagin and J. R. Rabeau, *Nat. Nanotechnol.*, 2010, **5**, 345–349.

71. C.-C. Fu, H.-Y. Lee, K. Chen, T.-S. Lim, H.-Y. Wu, P.-K. Lin, P.-K. Wei, P.-H. Tsao, H.-C. Chang and W. Fann, *Proc. Natl. Acad. Sci. U. S. A.*, 2007, **104**, 727–732.

72. Y.-R. Chang, H.-Y. Lee, K. Chen, C.-C. Chang, D.-S. Tsai, C.-C. Fu, T.-S. Lim, Y.-K. Tzeng, C.-Y. Fang, C.-C. Han, H.-C. Chang and W. Fann, *Nat. Nanotechnol.*, 2008, **3**, 284–288.

73. S. J. Yu, M. W. Kang, H. C. Chang, K. M. Chen and Y. C. Yu, *J. Am. Chem. Soc.*, 2005, **127**, 17604–17605.

74. A. M. Schrand, S. A. C. Hens and O. A. Shenderova, *Crit. Rev. Solid State Mater. Sci.*, 2009, **34**, 18–74.

75. *Nanodiamonds: Applications in Biology and Nanoscale Medicine*, ed. D. Ho, Springer, New York, 2010.

76. L. P. McGuinness, Y. Yan, A. Stacey, D. A. Simpson, L. T. Hall, D. Maclaurin, S. Prawer, P. Mulvaney, J. Wrachtrup, F. Caruso, R. E. Scholten and L. C. L. Hollenberg, *Nat. Nanotechnol.*, 2011, **6**, 358–363.

77. Y. Y. Chen, H. L. Shu, Y. Kuo, Y. K. Tzeng and H. C. Chang, *Diamond Relat. Mater.*, 2011, **20**, 803–807.

78. A. Kruger, F. Kataoka, M. Ozawa, T. Fujino, Y. Suzuki, A. E. Aleksenskii, A. Y. Vul' and E. Osawa, *Carbon*, 2005, **43**, 1722–1730.

79. A. S. Barnard and M. Sternberg, *J. Phys. Chem. B*, 2005, **109**, 17107–17112.

80. P. H. Chung, E. Perevedentseva and C. L. Cheng, *Surf. Sci.*, 2007, **601**, 3866–3870.

81. K. Iakoubovskii and G. J. Adriaenssens, *Phys. Rev. B*, 2000, **61**, 10174–10182.

82. B. R. Smith, D. W. Inglis, B. Sandnes, J. R. Rabeau, A. V. Zvyagin, D. Gruber, C. J. Noble, R. Vogel, E. Osawa and T. Plakhotnik, *Small*, 2009, **5**, 1649–1653.

83. N. Gibson, O. Shenderova, T. J. M. Luo, S. Moseenkov, V. Bondar, A. Puzyr, K. Purtov, Z. Fitzgerald and D. W. Brenner, *Colloidal stability of modified nanodiamond particles*, 2009, **18**, 620–626.

84. S. Kuchibhatla, A. S. Karakoti and S. Seal, *J. Mater.*, 2005, **57**, 52–56.

85. K. Ushizawa, Y. Sato, T. Mitsumori, T. Machinami, T. Ueda and T. Ando, *Chem. Phys. Lett.*, 2002, **351**, 105–108.

86. K. V. Purtov, A. I. Petunin, A. E. Burov, A. P. Puzyr and V. S. Bondar, *Nanoscale Res. Lett.*, 2010, **5**, 631–636.

87. A. Kruger, Y. J. Liang, G. Jarre and J. Stegk, *J. Mater. Chem.*, 2006, **16**, 2322–2328.

88. L. C. L. Huang and H. C. Chang, *Langmuir*, 2004, **20**, 5879–5884.

89. K. K. Liu, F. Chen, P. Y. Chen, T. J. F. Lee, C. L. Cheng, C. C. Chang, Y. P. Ho and J. I. Chao, *Nanotechnology*, 2008, **19**, 10.

90. Y. Y. Hui, B. L. Zhang, Y. C. Chang, C. C. Chang, H. C. Chang, J. H. Hsu, K. Chang and F. H. Chang, *Opt. Express*, 2010, **18**, 5896–5905.

91. K. Anke, *Chem.-Eur. J.*, 2008, **14**, 1382–1390.

92. Y. Liu, Z. N. Gu, J. L. Margrave and V. N. Khabashesku, *Chem. Mater.*, 2004, **16**, 3924–3930.

93. F. Neugart, A. Zappe, F. Jelezko, C. Tietz, J. P. Boudou, A. Krueger and J. Wrachtrup, *Nano Lett.*, 2007, **7**, 3588–3591.

94. V. K. A. Sreenivasan, A. V. Zvyagin and E. M. Goldys, *J. Phys.: Condens. Matter*, 2013, **25**, 194101.

95. V. S. Bondar, I. O. Pozdnyakova and A. P. Puzyr, *Phys. Solid State*, 2004, **46**, 758–760.

96. K. W. Goyne, A. R. Zimmerman, B. L. Newalkar, S. Komarneni, S. L. Brantley and J. Chorover, *J. Porous Mater.*, 2002, **9**, 243–256.

97. N. Bogdan, F. Vetrone, G. A. Ozin and J. A. Capobianco, *Nano Lett.*, 2011, **11**, 835–840.

98. X. Zhang, L. Yin, M. Tang and Y. Pu, *Biomed. Environ. Sci.*, 2011, **24**, 661–669.

99. M. Yu, F. Li, Z. Chen, H. Hu, C. Zhan, H. Yang and C. Huang, *Anal. Chem.*, 2009, **81**, 930–935.

100. J. Zhao, Z. Lu, Y. Yin, C. McRae, J. A. Piper, J. M. Dawes, D. Jin and E. M. Goldys, *Nanoscale*, 2012, 825–1236.

101. R. H. Page, K. I. Schaffers, P. A. Waide, J. B. Tassano, S. A. Payne, W. F. Krupke and W. K. Bischel, *J. Opt. Soc. Am. B*, 1998, **15**, 996–1008.

102. C. Wang, H. Tao, L. Cheng and Z. Liu, *Biomaterials*, 2011, **32**, 6145–6154.
103. C. Vinegoni, D. Razansky, S. A. Hilderbrand, F. W. Shao, V. Ntziachristos and R. Weissleder, *Opt. Lett.*, 2009, **34**, 2566–2568.
104. A. Nadort, V. K. A. Sreenivasan, Z. Song, E. A. Grebenik, A. V. Nechaev, V. A. Semchishen, V. Y. Panchenko and A. V. Zvyagin, *PLoS One*, 2013, **8**, e63292.
105. F. Wang, R. R. Deng, J. Wang, Q. X. Wang, Y. Han, H. M. Zhu, X. Y. Chen and X. G. Liu, *Nat. Mater.*, 2011, **10**, 968–973.
106. H. X. Mai, Y. W. Zhang, R. Si, Z. G. Yan, L. D. Sun, L. P. You and C. H. Yan, *J. Am. Chem. Soc.*, 2006, **128**, 6426–6436.
107. J. Zhao, D. Jin, E. P. Schartner, Y. Lu, Y. Liu, A. V. Zvyagin, L. Zhang, J. M. Dawes, P. Xi, J. A. Piper, E. M. Goldys and T. M. Monro, *Nat. Nanotechnol.*, 2013, **8**, 729–734.
108. H. X. Mai, Y. W. Zhang, L. D. Sun and C. H. Yan, *J. Phys. Chem. C*, 2007, **111**, 13721–13729.
109. Q. Liu, Y. Sun, T. S. Yang, W. Feng, C. G. Li and F. Y. Li, *J. Am. Chem. Soc.*, 2011, **133**, 17122–17125.
110. F. Wang, Y. Han, C. S. Lim, Y. H. Lu, J. Wang, J. Xu, H. Y. Chen, C. Zhang, M. H. Hong and X. G. Liu, *Nature*, 2010, **463**, 1061–1065.
111. P. R. Selvin, *Annu. Rev. Biophys. Biomol. Struct.*, 2002, **31**, 275–302.
112. K. E. Mironova, G. M. Proshkina, A. V. Ryabova, O. A. Stremovskiy, S. A. Lukyanov, R. V. Petrov and S. M. Deyev, *Theranostics*, 2013, **3**, 831–840.
113. Q. Q. Zhan, J. Qian, H. J. Liang, G. Somesfalean, D. Wang, S. L. He, Z. G. Zhang and S. Andersson-Engels, *ACS Nano*, 2011, **5**, 3744–3757.
114. T. Y. Cao, Y. Yang, Y. A. Gao, J. Zhou, Z. Q. Li and F. Y. Li, *Biomaterials*, 2011, **32**, 2959–2968.
115. A. D. Ostrowski, E. M. Chan, D. J. Gargas, E. M. Katz, G. Han, P. J. Schuck, D. J. Milliron and B. E. Cohen, *ACS Nano*, 2012, **6**, 2686–2692.
116. F. Leblond, S. C. Davis, P. A. Valdes and B. W. Pogue, *J. Photochem. Photobiol., B*, 2009, **98**, 77–94.
117. D. Li, B. A. Dong, X. Bai, Y. Wang and H. W. Song, *J. Phys. Chem. C*, 2010, **114**, 8219–8226.
118. H. J. M. A. A. Zijlmans, J. Bonnet, J. Burton, K. Kardos, T. Vail, R. S. Niedbala and H. J. Tanke, *Anal. Biochem.*, 1999, **267**, 30–36.
119. J. C. Boyer, M. P. Manseau, J. I. Murray and F. C. J. M. van Veggel, *Langmuir*, 2010, **26**, 1157–1164.
120. S. A. Osseni, S. Lechevallier, M. Verelst, C. Dujardin, J. Dexpert-Ghys, D. Neumeyer, M. Leclercq, H. Baaziz, D. Cussac, V. Santran and R. Mauricot, *J. Mater. Chem.*, 2011, **21**, 18365–18372.
121. R. S. Niedbala, H. Feindt, K. Kardos, T. Vail, J. Burton, B. Bielska, S. Li, D. Milunic, P. Bourdelle and R. Vallejo, *Anal. Biochem.*, 2001, **293**, 22–30.

Protein-Integrated Electronic Nanodevices and Systems

FANBEN MENG[a], HONG WANG[*a], AND XIAODONG CHEN[*a]

[a]School of Materials Science and Engineering, Nanyang Technological University, 50 Nanyang Avenue, Singapore 639798
*E-mail: chenxd@ntu.edu.sg, wanghong@ime.ac.cn

11.1 Biooptoelectronic Devices

Combining the advantages of high capacity and parallelism of photonics and the versatility of electronics,[1] optoelectronic devices exhibit great potential to impact the development of many applications, such as sensing, computing, and signal communication and processing. Biomacromolecules, especially proteins with specific functions in photosynthesis processes *in vivo*,[2] offer researchers a rich natural source to explore optoelectronic integrated computing circuits.

11.1.1 Proteins as Functional Elements for Optoelectronic Devices

Bacteriorhodopsin (BR), a chromoprotein, serves as a light-driven proton pump.[3] It has been found in the purple membrane of *Halobacterium salinarum*, where BR is organized into trimers. Each protein is composed of seven transmembrane helices with a retinal chromophore that is covalently

RSC Smart Materials No. 16
Bio-Synthetic Hybrid Materials and Bionanoparticles: A Biological Chemical Approach Towards Material Science
Edited by Alexander Böker and Patrick van Rijn
© The Royal Society of Chemistry 2015
Published by the Royal Society of Chemistry, www.rsc.org

bound in the central region. By absorption of visible green light (~570 nm), a BR molecule undergoes a multistep reaction cycle including seven intermediates with different isomerized states of retinal.[4] It exhibits high efficiency of light conversion, long-term stability of structure, and desirable photoelectric and photochromic properties. More interestingly, it is capable of forming thin films or gels and maintaining its biological activity on solid supports. Therefore, BR has emerged as an outstanding material for biooptics and bioelectronics and has attracted much interest in recent years.

To demonstrate the electronic properties of BR, junctions with a format of metal–protein(s)–metal were successfully prepared by several nanofabrication techniques, such as vesicle fusion with soft Au deposition of the top electrode,[5] direct incorporation within a metal nanogap[6] and embedding in a polymer matrix coated on an electrode.[7] Typical *I–V* characteristics of metal–BR–metal junctions was observed as shown in Figure 11.1(a), and Figure 11.1(b) illustrates a representative photovoltage response of BR-based molecular junctions. Compared with other light sources with different wavelengths, only the green light can trigger a clear excitation response with high conductivity, as expected. Experimental and theoretical studies demonstrated that the presence of retinal or a similar p-electron system in the protein is a key factor to support efficient electron transport, and a photoresponse may be associated with the isomerization of the retinal group under suitable light irradiation. Furthermore, complex Boolean computing of OR, AND, NOR and NAND logic gates has been achieved based on the photovoltaic characteristics of BR molecules.

In addition to BR, other chromoproteins and related biosystems, such as photosynthetic reaction centres and photosystem I/II have also been incorporated into integrated electronic chips. All photo-triggered behaviours have been extensively studied in BR integrated electronic devices to demonstrate the applicability of photosensitive biosystems as active optoelectronic materials.

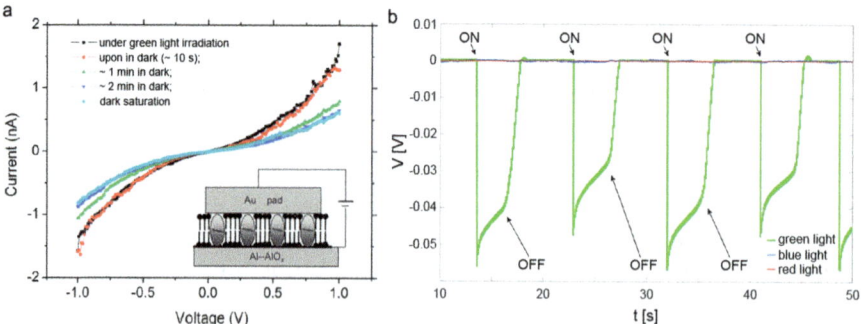

Figure 11.1   (a) *I–V* curves for a BR monolayer junction after equilibration in the dark. Inset: schematic of the BR junction (the proteins are incorporated in the fused lipid bilayers as a separate component of the array). (b) Different behaviour of BR under light irradiation by three different LEDs. Parts (a) and (b) reprinted with permission from ref. 5 and 6, respectively.

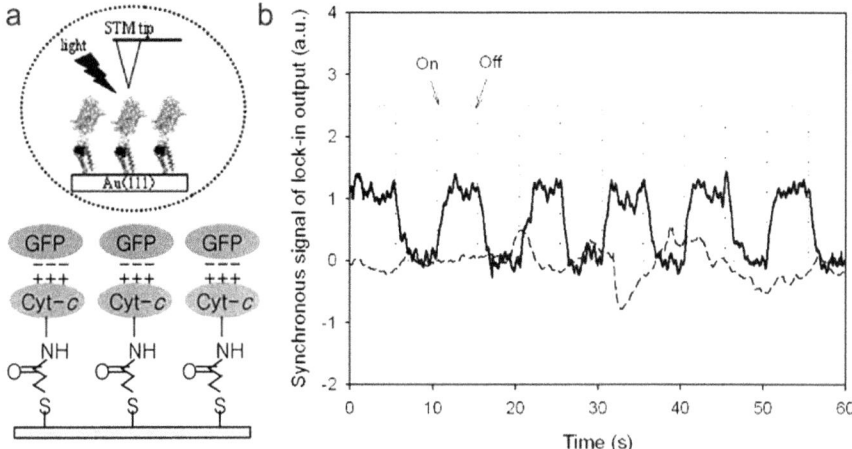

Figure 11.2 (a) Schematic of heteroprotein film consisting of GFP and cytochrome *c* with thiol groups on an Au surface. (b) The photocurrent of the heteroprotein layer excited by a blue laser (488 nm, solid line) and a green laser (532 nm, dashed line). Parts (a) and (b) reprinted with permission from ref. 8 and 9, respectively.

11.1.2 Biophotodiodes Consisting of Multi-Protein Heterolayers

As discussed above, based on active optical switching, BR-based devices successfully demonstrated the optoelectronic application of proteins. Further, the advantages of two kinds of proteins were combined and a heteroprotein layer was fabricated to achieve more useful optoelectronic functions.[8] With a system consisting of photosensitive green fluorescent protein (GFP) and an electro-active cytochrome family as a sensitizer and an electron acceptor, respectively [Figure 11.2(a)], the heteroprotein layer can adopt the functions of a biophotodiode. The photocurrent can only be generated by electron transfer from excited GFP to the cytochrome family, and therefore a unidirectional electron flow as characteristic of a diode is achieved. The photoswitching and rectified photocurrent properties of such a multiprotein heterolayer upon repeated on–off switching of blue laser illumination at a certain bias can be observed, as shown in Figure 11.2(b), indicating future applications of such biomolecular devices in optoelectronics.[9]

11.2 Protein Structures Combined with Molecular Electronic Components

Electron transfer through proteins is a fundamental element of many biochemical reactions from respiration to photosynthesis, which is prominent in diverse metabolic cycles.[10] Motivated by the unique properties of proteins that carry out complicated physiological or biochemical functions, such as support, storage, transport, signalling and catalysis, many attempts have been made to integrate proteins into solid-state junctions, in order to study their electronic properties and introduce proteins as a new type of conductive material.

11.2.1 Proteins as Electronic Conductors

A major challenge is to integrate proteins between two electrical contacts in solid state and retain natural structures and functions.[11] Recently, such solid-state protein monolayer junctions have been successfully fabricated by both scanning probe techniques at the single (or few) molecule(s) level and a large-area protein monolayer junction [Figure 11.3(a)].

Many representative types of proteins, such as azurin (AZ) (a redox protein with blue copper centre), bacteriorhodopsin (BR) (a membrane protein as the light-driven proton pump) and bovine serum albumin (BSA), have been achieved on these junctions with maintaining their native conformation,

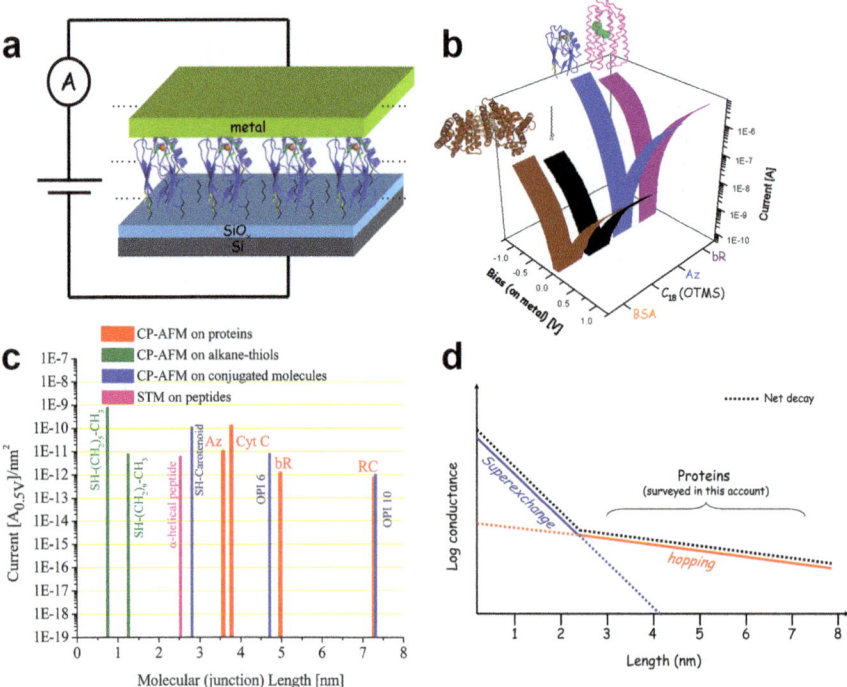

Figure 11.3 (a) Scheme of solid-state protein monolayer junction. (b) *I–V* characteristics of protein monolayer junctions (BR, AZ, BSA) and an organic monolayer junction (OTMS). (c) Current densities, under 0.5 V applied bias, measured by conductive atomic force microscopy (c-AFM) through saturated molecules (green), conjugated molecules (blue), peptides (magenta) and proteins (red). (d) Illustrated description of the size dependence of molecular resistance. Reprinted with permission from ref. 11.

leading to high-yield electrical transport [Figure 11.3(b) and (c)].[12] As a result, proteins with the physiological function of charge transfer provide more efficient electron transport than a protein such as BSA without such functions. This monolayer of BSA, a protein that is not known to be electroactive, still exhibited higher efficiency of electron transport than a monolayer of saturated organic carbon chains. Remarkably high current densities were measured, indicating that proteins are not insulators, and especially redox proteins should be viewed as active electronic materials.

The remarkably efficient electron transfer of proteins in their physiological conditions is well known,[13] and leave the question of how to interpret the electron transport in solid-state junctions. By studying the length-dependent electron transfer, the direct tunnelling efficiency may obviously decrease across protein molecules with such large size [Figure 11.3(d)], and therefore the observed electron transport in proteins may be dominated by hopping. With further temperature-dependent investigations of junctions formed by cytochrome *c*, BR or AZ, a more generalized mechanism was proposed. Tunnelling and hopping coexists with multiple transport paths through proteins and each dominates in different temperature regimes. Meanwhile, the pivotal function of their active centres in the solid-state electron transport process has been further clarified.[14–17]

Although the mechanisms behind the highly efficient electron transfer of protein molecules are still not fully understood, the remarkable measured current densities indicate that proteins should not be viewed as insulators. Based on their functions and observed electronic properties, proteins are promising candidates as electronic natural biomaterials for fabricating integrated functional devices and systems.

11.2.2 Protein-Based Memory Resistive (Memristive) Devices

Ferritin is the main intracellular iron storage protein, containing a spherical shell with a diameter of ~12 nm and an inorganic hydrous ferric oxide ($Fe^{III}O-OH$) mineral core inside.[18] Owing to the oxidase activity, an apoferritin shell has often been used as a template for nanoparticle synthesis of metal oxides. Additionally, a highly stable structure of the protein can be observed, which can withstand a pH range of 2.0–12.0 and high temperatures up to 80 °C. Therefore, it is a desirable bionanoparticle for fabricating electronic devices. Ferritin also exhibits the properties of a protein conductor, as demonstrated by measurements with c-AFM.[19] Meanwhile, the mechanism of iron uptake and release *in vivo* shows biological "memory" storage processes within protein nanocages, providing a potential target to develop bionanoparticle-based memory devices.

The memristor (short for memory resistor), with electrical resistive switching effects, has been named as the fourth fundamental two-terminal circuit element following the resistor, the capacitor and the inductor.[20] It has been proposed as a candidate for the next generation of non-volatile memory devices, combining the advantages of Flash and DRAM.[21] Thus far, various man-made materials, such as using metal oxides, silicon or carbon, and polymer composite materials have been used to fabricate electronic devices with memristive properties. Recently, the memristor has been applied in biosystems. Natural ferritin has been incorporated into gold nanoelectrode pairs, resulting in reproducible memristive behaviour.

As shown in Figure 11.4(a), ferritin molecules were covalently assembled as a monolayer within a chemically fabricated Au nanogap.[22] Two terminal I–V characteristics of the ferritin-based nanodevices exhibited reversible and repeatable programmable resistance, switching between the high-conductivity ON state and low-conductivity OFF state (Figure 11.4). Such hysteretic curves are a feature of memristors, similar to the bipolar switching of devices made of titanium dioxide thin films, the first physical model of a memristor reported by Williams and co-workers.[20] Such bipolar memristive behaviour has been demonstrated to be related to the inorganic iron oxide core with a protein shell. This hypothesis was confirmed by the results from c-AFM measurements of individual Fe cores.[23] Redox processes of solid-state electrochemistry in metal oxides were proposed as one of the possible mechanisms for the memristive behaviour.

Furthermore, these ferritin-based nanodevices with reversible conductance could be used for non-volatile memory based on the results of

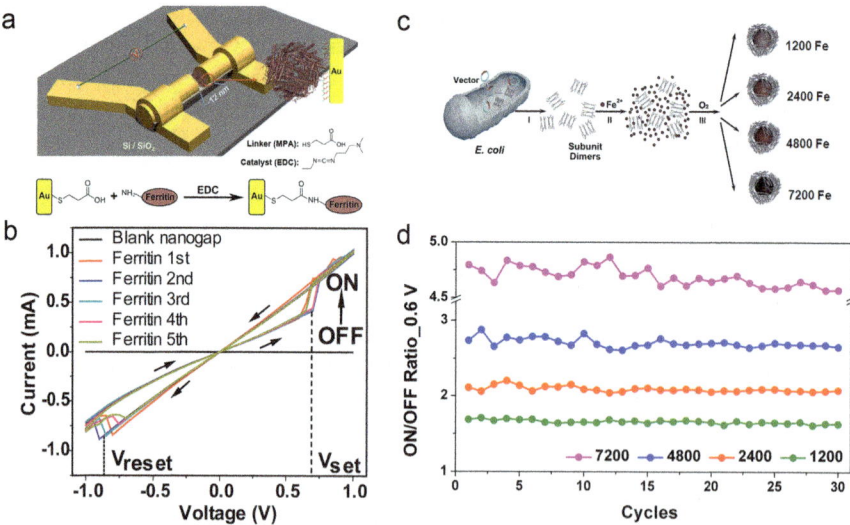

Figure 11.4   (a) Schematic covalent assembly of ferritin within nanogap device. (b) Representative I–V characteristics of ~12 nm gaps before (black curve) and after immobilization of ferritin (coloured curves). (c) Scheme to obtain AfFtn with different Fe loadings (1200, 2400, 4800 and 7200): I, protein purification; II, ratio-controlled loading; III, self-assembled cage formation. (d) Endurance performance of the AfFtn-based memristive nanodevices with different Fe loadings. Parts (a) and (b) and parts (c) and (d) reprinted with permission from ref. 22 and 24, respectively.

write–read–erase cycle tests. More importantly, the hysteretic resistive switching can be modulated by loading variable amounts of iron, which is achieved by nanodevices functionalized with archaeon *Archaeoglobus fulgidus* ferritin (AfFtn) produced in *Escherichia coli* by bioengineering [Figure 11.4(c)].[24] Similar programmable memristive behaviour was observed to be enduring and stable for each type of ferritin-based device [Figure 11.4(d)]. Meanwhile, the difference in resistance between ON and OFF states became increasingly apparent with higher Fe loadings inside the protein cage, indicating tuneability of electronic devices with rapidly developing bioengineering.

Following these studies, both kinds of silk protein, fibroin[25] and sericin,[26] have been investigated for non-volatile memory application with the electronic properties of bipolar memristive switching. This work provides solid proof that bionanoparticles can be utilized in the fabrication of functional electronic circuits.

11.3 Protein–Inorganic Structures for Nanowire Applications

A nanowire is a wire structure with a diameter of the order of nanometres and typically with a length/width ratio of >20. Various types of nanowires, including conductor, semiconductor and insulator materials, have been prepared. Nanowires can be used to create next-generation electronic and optoelectronic devices, in the sensing of proteins and chemicals, and even for exploring cell interiors. Several physical and chemical strategies have been proposed for the preparation of nanowires.[27] On the other hand, natural biomaterials, especially proteins, provide great potential for use as scaffolds in the building of nanowires owing to their well-defined and exactly tailored nanostructures.[28,29]

11.3.1 Proteins as Templates for Fabricating Inorganic Nanowires

Compared with conventional physical and chemical strategies such as lithographic methods, microcontact printing and chemical synthesis, bio-template-directed synthesis is attractive for preparing nanowires owing to its inherent advantages.[28,30,31] First, such an approach can reduce energy consumption and avoid the use of some toxic solvents, leading to environmentally friendly processing. In addition, the properties, such as the size, shape, chemistry and crystal structure of inorganic nanowires synthesized by this method can be exquisitely controlled. Furthermore, proteins have great potential to generate inorganic nanowires with highly specific or multiple functions. So far, various inorganic nanowires, including conductors, semiconductors and insulators, have been synthesized with the use of biomaterials as a template, such as a virus as a template for the synthesis of magnetic, metal and semiconductor nanowires and bacteria as a template for the synthesis of superlattices CdS nanowires.[28-32] In this section, typical processes involving the use of proteins as templates in the synthesis of nanowires are introduced.

Tobacco mosaic virus (TMV) is a commonly used template for the synthesis of metal nanowires with diameters down to only a few atoms and micrometre lengths.[28] TMV is a stable tube-shaped complex with an outer diameter of 18 nm, a central channel with a diameter of 4 nm, and lengths of 300, 600, 900 nm, and so on. Here, typical processes for metal nanowires deposition using TMV as template are introduced with nickel as an example. First, the virus surface was activated with Pd(II) or Pt(II). After dialysis against deionized water, the virus was incubated with 0.9 mM $PdCl_4^{2-}$ or $PtCl_4^{2-}$ in 1 M Cl$^-$ at pH 5, and Pd or Pt clusters with diameters larger than 4 nm could be adsorbed on the outer virus surface. Then, the virus particles were dispersed in an electroless deposition bath that contained an Ni(II) reductant under the conditions of pH 5–8 at 25 °C for 3 min. Finally, after removal of the samples from the bath and drying, Ni nanowires with a diameter of the virus central channel of 3 nm and a length of more than 500 nm could be formed, as shown in Figure 11.5.[28] In addition to metal nanowires, compound nanowires can also be prepared by protein-assisted synthesis. Lysozyme has been used as a morphology-controlling agent for the synthesis of bismuth sulfide and oxide nanowires.[30] Single-crystal bismuth sulfide nanowires with

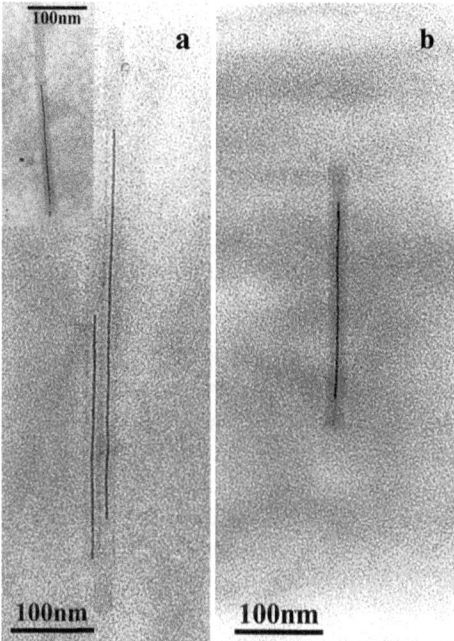

Figure 11.5 Transmission electron microscope (TEM) image of TMV after Pd(II) activation, followed by deposition of (a) Ni and (b) Co. Reprinted with permission from ref. 28.

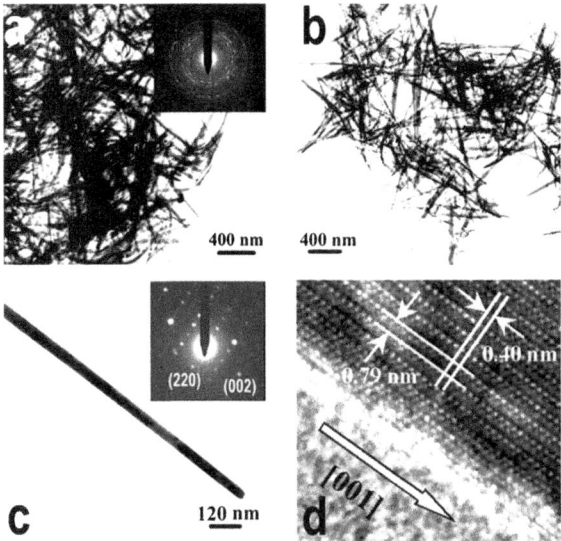

Figure 11.6 (a and b) TEM images, (c) single-crystal selected area electron diffraction (SAED) patterns and (d) high-resolution TEM image of the bismuth sulfide nanowire obtained by protein-assisted synthesis. Reprinted with permission from ref. 30.

diameters in the range 10–50 nm and lengths up to micrometres could be obtained from $Bi(NO_3)_3 \cdot 5H_2O$, thiourea and lysozyme as reactants at 160 °C under hydrothermal conditions, as shown in Figure 11.6.[30] Additionally, the proposed protein-assisted growth of nanowires could be extended to a variety of compounds. Further extension of protein template synthesis would be of great importance in future developments in nanotechnology, bioinspired materials and biointegrated systems.

11.3.2 Protein–Inorganic Nanowires

Nanowires are promising building blocks for future electronic and optoelectronic devices.[33] However, the assembly of orderly nanowires always shows high error frequencies and low assembly rates.[34] Recently, the biological assembly of inorganic nanowires for nanoelectronics applications has attracted great interest.[34–39] Some interesting approaches such as metal nanowires connected by protein connectors and protein-driven assembly of semiconductor nanowires in a desired network have been developed.[34–36] It is worth mentioning that Kotov's group assembled semiconductor cadmium telluride (CdTe) in a crossbar and side-to-end mode for a prototype of logical circuit applications.[34] For the nanowire network assembly, thioglycolic acid (TGA)- or L-cysteine (LCY)-stabilized CdTe nanowires were first prepared. Various protein molecules, such as bovine serum albumin (BSA), anti-BSA antibody [immunoglobulin G (IgG)], biotin–streptavidin (SA) and biotin

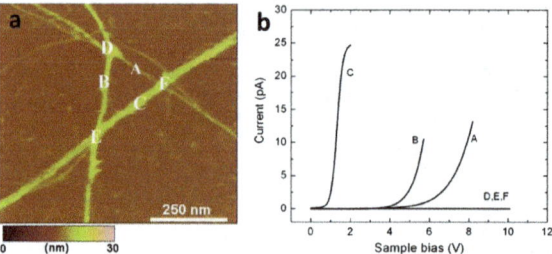

Figure 11.7   (a) AFM image of NW–TGA–BSA/NW–MSA–IgG deposited on a silicon substrate; (b) *I–V* characteristics of the CdTe nanowires on a silicon substrate. Reprinted with permission from ref. 34.

hydrazide (BH), could be utilized as the connectors to bridge the nanowires. Then, conjugated complexes could be prepared by the sulfo-NHS (*N*-hydroxy-sulfosuccinimide) and EDC [1-ethyl-3(3-dimethylaminopropyl)carbodiimide hydrochloride] reaction. A typical image of a biologically assembled CdTe–TGA–BSA/CdTe–MSA–IgG (MSA = mercaptosuccinic acid) crossbar is shown in Figure 11.7(a).[34] The potential of the protein-assembled nanowires for applications in electronics was also explored by c-AFM.

As shown in Figure 11.7(b), when the c-AFM probe was positioned on top of nanowires and bias was applied, diode-like behaviour could be obtained for the middle points on the sides (points A, B, C), whereas there was no current going through the cross points (points D, E, F). The non-conductive nature of the nanowire junctions indicated that the protein molecules assembled between the inorganic nanowires. The protein molecules existing between the semiconductor nanowires provide an opportunity for building hybrid nanodevices. For future bioinspired nanocircuit applications, large-area bottom-up assemblies of nanowire arrays are needed for further study.

11.4   Protein-Integrated Electrochemically Active Systems and Devices

11.4.1   Electrochemically Gated Protein Devices

As demonstrated with respect to the applicability of proteins as electronic elements, redox proteins capable of accepting and transporting electrons by switching redox states of their active centres provide us with a useful protein source to develop functional electronic systems with tuneable conductive properties.[40] As became well known, powerful electrochemical techniques are most commonly used to control the redox switching of individual molecules. Therefore, a method for the electrochemical modulation of molecular devices based on redox proteins was proposed by adding a third electrode in electrolytes, applying a gate potential through the double layer developed at the electrodes/electrolyte interfaces, and

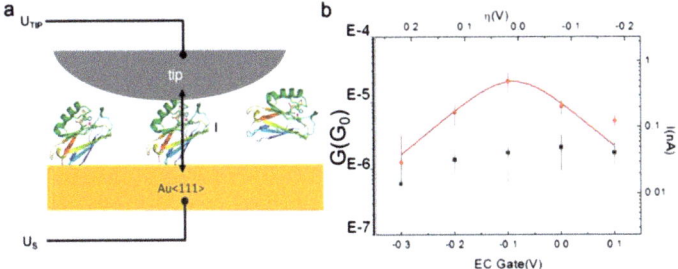

Figure 11.8 (a) Scheme of single-azurin junction formation by the ECSTM–Break Junction (BJ) method. (b) Conductance as a function of electrochemical gate potential for azurin (red circles) and non-redox Zn azurin junctions (black squares). Reprinted with permission from ref. 42.

measuring the gate-dependent transport. An electrochemically gated three-terminal arrangement can also circumvent the limitation of placing a third electrode in close proximity to the molecular electrical contact when measuring the gate effect of molecular devices.[41]

As an example, azurin junctions were formed by a self-assembled protein monolayer on an Au substrate with a probe of an electrochemical scanning tunnelling microscope (ECSTM) [Figure 11.8(a)]. A large on/off ratio of up to 20 was demonstrated with the modulation of redox gating, suggesting the field-effect transistor-like behaviour of the protein junctions (Figure 11.8(b)). Electrochemically gated protein-integrated devices offer an excellent platform to understand charge transport with complex protein molecules. Although more experimental and theoretical efforts are necessary to elucidate fully the mechanisms involved, this novel research field is still open to explore more possibilities in bioelectronics regarding structure-dependent protein device operations.

11.4.2 Protein-Based Electrochemical Memory Devices

Redox proteins have also emerged as building blocks of biointegrated devices for robust memory devices with visible electrochemical output.[43] A single redox protein can act as an individual memory element due to the three distinct conducting states corresponding to the electrochemical processes of the redox protein [Figure 11.9(a)] as follows: (1) *writing*: application of an oxidation potential, causing electron transfer from the Au electrode to protein molecules and positive charges stored; (2) *erasing*: after application of a reduction potential, neutralizing the positively charged protein molecules; and (3) *reading*: application of an open-circuit potential (OCP), leading protein molecules to an equilibrium state, in which the direction of electron transfer is decided by the previous redox state of the protein, and indicating the memory effect of such electrochemical biodevices, as shown in Figure 11.9(b), which was measured by the chrono-amperometry method.

Furthermore, more kinds of redox protein were introduced as heterolayers to develop more complicated and advanced biointegrated devices with multistate memory [Figure 11.9(b)].[44] The electrochemical behaviours of recombinant protein heterolayers present two pairs of redox peaks, suggesting multistates for the write, read and erase functions and enhancement of the memory density for single devices. The proof-of-concept of the redox protein-based electrochemical memory device provides possibilities to realize and produce commercially biointegrated devices for information storage.

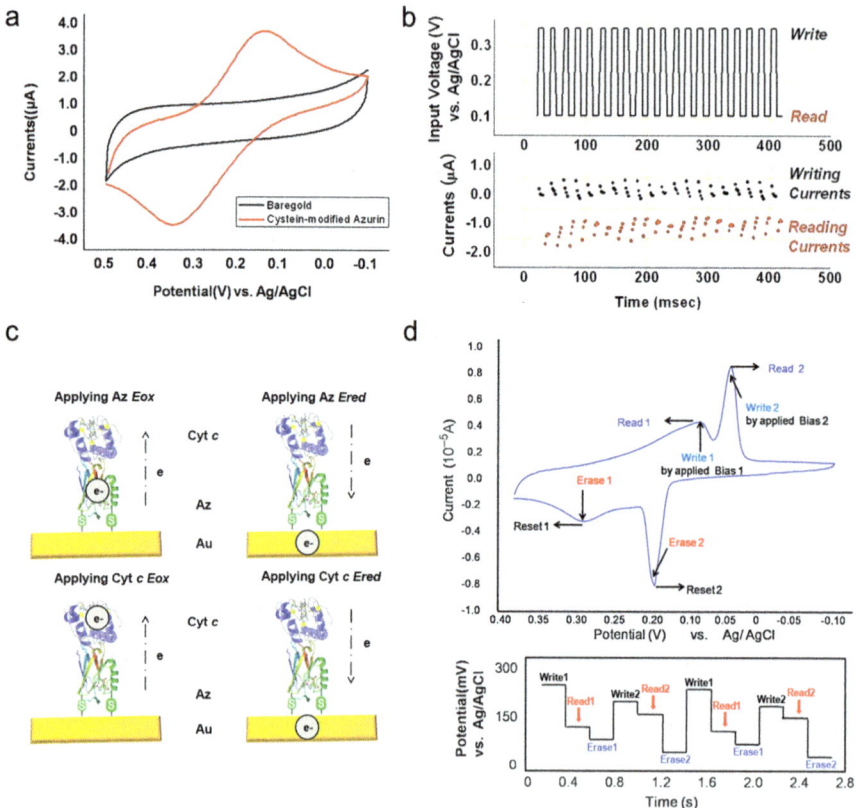

Figure 11.9 (a) Cyclic voltammogram of cysteine-modified azurin on an Au substrate; (b) the corresponding OCP amperometry experiment; (c) schematic electron-transfer mechanism of the protein heterolayer of cytochrome *c*–azurin on an Au surface; (d) cyclic voltammogram of such a protein heterolayer and the memory function characteristics of a multistate protein-based biomemory device. Parts (a) and (b) and parts (c) and (d) reprinted with permission from ref. 43 and 44, respectively.

11.5 Floating Gate Memory Devices

11.5.1 Introduction of Floating Gate Memory

The floating gate memory, first reported by Kahng and Sze in 1967,[45] is a field-effect transistor with a structure similar to that of the conventional metal–oxide–semiconductor field-effect transistor (MOSFET). The typical device structure, shown in Figure 11.10(a), mainly contains gate, source and drain electrodes, a semiconductor layer, a blocking insulator layer, a floating gate (charge-trapping layer), a tunnelling insulator layer, and a substrate. The typical *I–V* characteristics of a floating gate memory are shown in Figure 11.10(b); by applying a program or erase voltage to the gate electrode, the threshold voltage of the transistor can be modulated. The floating gate memory is widely used in non-volatile data storage systems and recently has also shown great potential for neuronal computational elements in neural network applications.[46] Nowadays, the use of proteins for electronic devices fabrication is a promising technology since proteins are attractive for constructing nanostructures for device applications and proteins are typically renewable and environmentally friendly.[47] Proteins have been widely used as a charge-trapping layer or as a template to fabricate a charge-trapping layer in floating gate memory devices.[48]

11.5.2 Protein as Charge Trapping Layer for Floating Gate Memory

A floating gate memory cell with discrete nanoparticles as charge-trapping layer has attracted much attention owing to its great potential for scaling down.[49] Metal nanoparticles and metal nanoparticle-incorporated block copolymer micelles are always used as a discrete charge-trapping layer for floating gate memory devices.[50] At the same time, protein molecules are potential candidates for the charge-trapping layer of floating gate memory because of their ability of construct uniform nanostructures.[51] Ferritin, which is an iron-storing protein with an outer diameter of 12 nm and a shell thickness of 2 nm, is a widely used protein as a charge-trapping layer in a

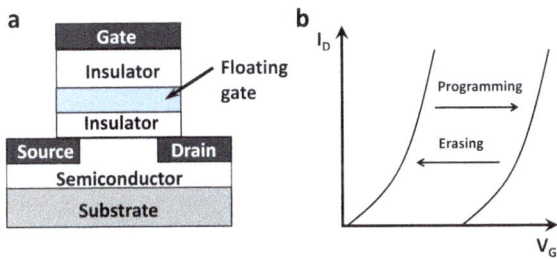

Figure 11.10   (a) Schematic of a floating gate memory device; (b) schematic of typical *I–V* curves of a floating gate memory device.

floating gate memory.[51,52] For example, Cho's group developed an organic field-effect transistor (OFET) memory with ferritin nanoparticle multilayers as the charge-trapping layer; a typical device structure is shown in Figure 11.11(a).[51] Ferritin multilayer films could be prepared by layer-by-layer (LbL) deposition of cationic poly(allylamine hydrochloride) (PAH) and anionic ferritin nanoparticles. From the TEM image [Figure 11.11(b)], a uniformly ferritin nanoparticle layer with a density of 1.8×10^{11} cm^{-2} can be observed. With a PAH–ferritin bilayer number of 10, a large memory window of 20 V [Figure 11.11(c)], a high program/erase current ratio of 10^4 and excellent endurance and data retention [Figure 11.11(d) and (e)] can be obtained for the organic floating gate memory. Bioinspired memory devices that employ protein nanoparticles as charge-trapping layer would be a basis for the exploitation of the protein-based floating gate memory.

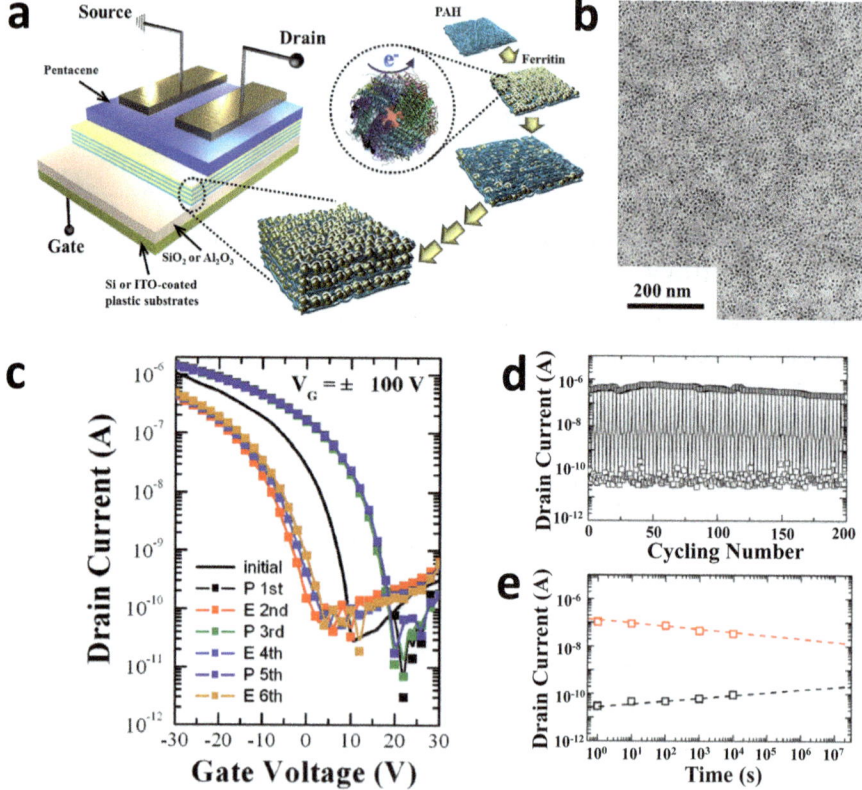

Figure 11.11   (a) Schematic of organic field-effect transistor-based memory device with layer-by-layer assembled (PAH–ferritin NP)$_n$ multilayered gate dielectrics; (b) TEM image of ferritin layer; (c) typical transfer curves for the memory device using (PAH–ferritin NP)$_{10}$ gate dielectrics; (d) endurance and (e) retention tests for memory device with (PAH–ferritin NP)$_{40}$ insulator layer. Reprinted with permission from ref. 51.

11.5.3 Protein as Template for Fabricating the Charge-Trapping Layer of Floating Gate Memory

The floating gate memory with nanocrystals or quantum dots embedded between the block and tunnelling insulator layer is promising for use as a flash memory owing to the further scaling down potential, lower operating voltage and excellent non-volatile information retention time.[53] The traditional methods for preparing nanodots in a floating gate memory include chemical vapour deposition, ion implantation, annealing of silicon-rich oxide, aerosol nanocrystal formation and self-assembled metal nanoparticles. However, the size distribution and the density of the nanodots fabricated by these methods are not very uniform. Several proteins can be used as templates for synthesizing nanoparticles and the uniformity of proteins provides a promising biological path to fabricate uniformly distributed nanodots by a biomineralization process.[54] Ferritin is commonly utilized as a template to fabricate uniform nanodots for the floating gate memory. Ferritin is a cage-shaped protein with an inner cavity and protein shell of about 7 and 12 nm, respectively, and can be used to construct various types of inorganic nanoparticles, such as Fe, Ni, Co and lead selenide.[55,56] Here, we introduce the typical procedures for use of ferritin as a template to fabricate Fe nanodots for high-performance floating gate memory application. As Ohara *et al.* proposed, in a typical experiment, 5 nm HfO_2 or HfSiO was used as a tunnelling insulator layer.[56] For ferritin deposition, HfO_2 or HfSiO was first cleaned by SPM cleaning followed by UV–O_3 treatment, then ferritin solution containing biomineralized iron bionanodots was spin coated on the HfO_2 or HfSiO. After ferritin absorption, the outer shell of ferritin was removed by UV–O_3 treatment. Uniformly distributed Fe nanodots were observed. Finally, the device was obtained by deposition of a 20 nm blocking insulator layer of SiO_2 and Al electrodes. By applying a programming and erasing pulse voltage, a memory window larger than 10 V and a retention time of more than 10^4 s could be achieved for the biomineralized iron nanodot-based floating gate memory.

11.6 Fuel Cells

A fuel cell is a device that can convert the chemical energy from a fuel into electricity through chemical reactions. A typical fuel cell consists of an anode, a cathode and an electrolyte. Direct current electricity can be produced when electrons are drawn from the anode to the cathode through an external load.[57] There are many types of fuel cells and they are usually classified according to the electrolyte utilized in the cell. Recently, protein molecules integrated into fuel cells have attracted much attention and have become an important interdisciplinary research field since proteins are renewable and the structure of proteins can be effectively controlled. Especially redox proteins that exchange electrons upon oxidation or reduction of specific substrates have been widely developed.[58] Connection of redox proteins to electrodes allows

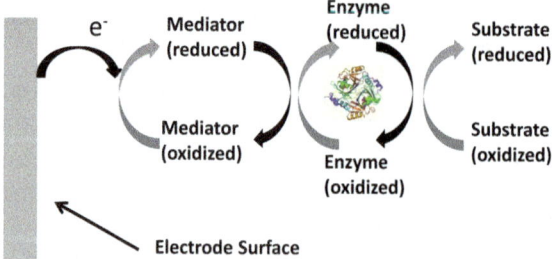

Figure 11.12   Schematic of a mediated electron transfer-based bioelectrode for bio-fuel cells.

bioelectrocatalytic electrodes to be used as active components in fuel cells. For example, oxidation of glucose by molecular oxygen has been utilized as a process for generating electrical energy. Electrical power can be generated through the bioelectrocatalytic oxidation of glucose with construction of a system consisting of a glucose oxidase (GOx) enzyme electrode and bioelectrocatalytic oxygen-reducing laccase as anode and cathode.[59] Various proteins, especially enzymes, have been used in fuel cells. Such enzyme-based fuel cells can not only be used for producing clean energy, but also yield a new generation of implanted devices for various bioapplications. In this section, we mainly introduce the enzymes utilized in fuel cells.

Enzymes are typically employed as biocatalysts in biofuel cells.[57,58] Most of the enzymes are proteins that are highly selective catalysts and responsible for the thousands of metabolic processes that sustain life. Enzymes are commonly used as electron shuttles in a fuel cell. Depending on the electrical connection between the electrode–enzyme pair, enzymatic biofuel cells can be divided into two main types: mediated electron transfer- and direct electron transfer-based fuel cells. In mediated electron transfer devices, as shown in Figure 11.12, mediator molecules participate in the catalytic reaction by reacting with enzyme and transferring the electrons to/from the electrode surface.[58] A metal-based mediator such as ferrocene and osmium complexes, and also azine mediators, are always employed in the mediated electron transfer enzymatic fuel cells. On the other hand, direct electron transfer enzymatic fuel cells have attracted much attention owing to the simplicity of fabrication. As shown in Figure 11.13, some enzymes are able to interact with an electrode due to the short electron tunnelling distance.[57,60] For example, haem-containing redox proteins including cytochrome *c* and bifunctional haem enzymes such as cellobiose dehydrogenase, D-fructose dehydrogenase and alcohol dehydrogenase have shown efficient direct electron transfer. In addition, modification of the electrode surface can promote direct electron transfer.[60–62] Deposition of conductive polymers such as polypyrrole, polyaniline, polyacetylene and carbon nanotubes on the electrode surface can significantly enhance the electrocatalytic properties of enzymes. Tremendous

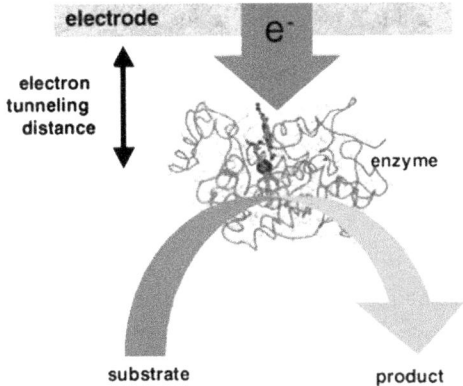

Figure 11.13 Schematic of direct electron transfer enzymatic biofuel cell. Reprinted with permission from ref. 57.

progress has been made in the area of enzymatic biofuel cells, *e.g.* the output power of enzymatic biofuel cells with glucose as fuel is nearly 1000 μW.[62] However, the most important constraint for application of enzymatic biofuel cells is the short lifetime. The lifetime of enzymatic fuel cells is always insufficient because enzymes in solution easily lose their three-dimensional structure or denature. Therefore, increasing the stability of enzymes at the electrode surface by improving electron transfer pathways and using novel immobilization methods represents some of the most important work on enzymatic biofuel cells needed in the future.

11.7   Summary and Critical Discussion

The utilization of biomolecules as functional elements represents an attractive strategy for producing functional electronic devices and systems. Proteins perform specific functions in all physiological and biochemical processes *in vivo*, offering us untapped resources to fulfil the requirements of biointegrated electronic devices. Although the diversity of natural proteins is vast, tailoring their structures and properties to fit the variable and complex requirements of electronic functionality is achievable by leveraging on the rapidly developing biotechnology. For instance, by genetic engineering, more than 10^6 variants of an individual protein may be created. Consequently, the employment of protein as functional elements in electronic devices and systems rather than traditional chemical molecules and semiconductors creates additional choices. However, thus far, the stability and durability of these protein-integrated electronic nanodevices and systems are still not comparable to those of traditional semiconductor electronics, leaving a lot of room for the further optimization and improvement of current fabrication techniques and operating conditions.

References

1. H. J. Caulfield and S. Dolev, *Nat. Photonics*, 2010, **4**, 261.
2. M. H. B. Stowell, T. M. McPhilips, D. C. Rees, S. M. Soltis, E. Abresch and G. Feher, *Science*, 1997, **276**, 812.
3. D. Oesterhelt and W. Stoeckenius, *Nat., New Biol.*, 1971, **233**, 149.
4. N. Hampp, *Chem. Rev.*, 2000, **100**, 1755.
5. Y. Jin, N. Friedman, M. Sheves and D. Cahen, *Adv. Funct. Mater.*, 2007, **17**, 1417.
6. A. Dimonte, S. Frache, V. Erokhin, G. Piccinini, D. Demarchi, F. Milano, G. De Micheli and S. Carrara, *Biomacromolecules*, 2012, **13**, 3503.
7. M. Prasad and S. Roy, *IEEE Trans. Nanobioscience*, 2012, **11**, 410.
8. J.-W. Choi and M. Fujihira, *Appl. Phys. Lett.*, 2004, **84**, 2187.
9. B. Lee, S. Takeda, K. Nakajima, J. Nohd, J.-W. Choi, M. Hara and T. Nagamunea, *Biosens. Bioelectron.*, 2004, **19**, 1169.
10. R. A. Marcus and N. Sutin, *Biochim. Biophys. Acta*, 1985, **811**, 265.
11. I. Ron, I. Pecht, M. Sheves and D. Cahen, *Acc. Chem. Res.*, 2010, **43**, 945.
12. I. Ron, L. Sepunaru, S. Itzhakov, T. Belenkova, N. Friedman, I. Pecht, M. Sheves and D. Cahen, *J. Am. Chem. Soc.*, 2010, **132**, 4131.
13. R. Winkler, A. J. Di Bilio, N. A. Farrow, J. H. Richards and H. B. Gray, *Pure Appl. Chem.*, 1999, **71**, 1753.
14. L. Sepunaru, I. Pecht, M. Sheves and D. Cahen, *J. Am. Chem. Soc.*, 2011, **133**, 2421.
15. L. Sepunaru, N. Friedman, I. Pecht, M. Sheves and D. Cahen, *J. Am. Chem. Soc.*, 2012, **134**, 4176.
16. N. Amdursky, I. Pecht, M. Sheves and D. Cahen, *J. Am. Chem. Soc.*, 2012, **134**, 18221.
17. N. Amdursky, I. Pecht, M. Sheves and D. Cahen, *J. Am. Chem. Soc.*, 2013, **135**, 6300.
18. E. C. Theil, *Annu. Rev. Biochem.*, 1987, **56**, 289.
19. D. G. Xu, G. D. Watt, J. N. Harb and R. C. Davis, *Nano Lett.*, 2005, **5**, 571.
20. D. B. Strukov, G. S. Snider, D. R. Stewart and R. S. Williams, *Nature*, 2008, **453**, 80.
21. R. Waser and M. Aono, *Nat. Mater.*, 2007, **6**, 833.
22. F. Meng, L. Jiang, K. Zheng, C. F. Goh, S. Lim, H. H. Hng, J. Ma, F. Boey and X. Chen, *Small*, 2011, **7**, 3016.
23. M. Uenuma, T. Ban, N. Okamoto, B. Zheng, Y. Kakihara, M. Horita, Y. Ishikawa, I. Yamashita and Y. Uraoka, *RSC Adv.*, 2013, **3**, 18044.
24. F. Meng, B. Sana, Y. Li, Y. Liu, S. Lim and X. Chen, *Small*, 2014, **10**, 277.
25. K. Hota, M. K. Bera, B. Kundu, S. C. Kundu and C. K. Maiti, *Adv. Funct. Mater.*, 2012, **22**, 4493.
26. H. Wang, F. Meng, Y. Cai, L. Zheng, Y. Li, Y. Liu, Y. Jiang, X. Wang and X. Chen, *Adv. Mater.*, 2013, **25**, 5498.
27. V. Schmidt, J. Wittemann, S. Senz and U. Gösele, *Adv. Mater.*, 2009, **21**, 2681.

28. M. Knez, A. Bittner, F. Boes, C. Wege, H. Jeske, E. Maiss and K. Kern, *Nano Lett.*, 2003, **3**, 1079.
29. M. Dickerson, K. Sandhage and R. Naik, *Chem. Rev.*, 2008, **108**, 4935.
30. F. Gao, Q. Lu and S. Komarneni, *Chem. Commun.*, 2005, 531.
31. J. Juarez, A. Cambon, A. Topete, P. Taboada and V. Mosquera, *Chem.–Eur. J.*, 2011, **17**, 7366.
32. J. Slocik, S. Kim, T. Whitehead, D. Clark and R. Naik, *Small*, 2009, **5**, 2038.
33. Y. Cui and C. Lieber, *Science*, 2001, **291**, 851.
34. Y. Wang, Z. Tang, S. Tan and N. Kotov, *Nano Lett.*, 2005, **5**, 243.
35. K. Ainslie, G. Sharma, M. Dyer, C. Grimes and M. Pishko, *Nano Lett.*, 2005, **5**, 1852.
36. N. Birenbaum, B. Lai, C. Chen, D. Reich and G. Meyer, *Langmuir*, 2003, **19**, 9580.
37. Y. Maeda and H. Matsui, *Soft Matter*, 2012, **8**, 7533.
38. Y. Leng, H. Wei, Z. Zhang, Y. Zhou, J. Deng, Z. Cui, D. Men, X. You, Z. Yu, M. Luo and X. Zhang, *Angew. Chem., Int. Ed.*, 2010, **122**, 7401.
39. D. Men, Z. Zhang, Y. Guo, D. Zhu, L. Bi, J. Deng, Z. Cui, H. Wei and X. Zhang, *Biosens. Bioelectron.*, 2010, **26**, 1137.
40. Q. J. Chi, O. Farver and J. Ulstrup, *Proc. Natl. Acad. Sci. U. S. A.*, 2005, **102**, 16203.
41. S. Guo, J. M. Artés and I. Díez-Pérez, *Electrochim. Acta*, 2013, **110**, 741.
42. J. M. Artés, I. Díez-Pérez and P. Gorostiza, *Nano Lett.*, 2012, **12**, 2679.
43. J.-W. Choi, B.-K. Oh, Y. J. Kim and J. Min, *Appl. Phys. Lett.*, 2007, **91**, 263902.
44. T. Lee, S.-U. Kim, J. Min and J.-W. Choi, *Adv. Mater.*, 2010, **22**, 510.
45. D. Kahng and S. Sze, *Bell Syst. Tech. J.*, 1967, **46**, 1288.
46. Q. Lai, L. Zhang, Z. Li, W. Stickle, R. Williams and Y. Chen, *Adv. Mater.*, 2010, **22**, 2448.
47. H. Tao, D. Kaplan and F. Omenetto, *Adv. Mater.*, 2012, **24**, 2824.
48. I. Yamashita, K. Iwahori and S. Kumagai, *Biochim. Biophys. Acta*, 2010, **1800**, 846.
49. K. Yamada, S. Yoshii, S. Kumagai, A. Miura, Y. Uraoka, T. Fuyuki and I. Yamashita, *J. Appl. Phys.*, 2007, **46**, 7549.
50. K. Ohara, I. Yamashita, T. Yaegashi, M. Moniwa, M. Yoshimaru and Y. Uraoka, *Appl. Phys. Express*, 2009, **2**, 095001.
51. B. Kim, Y. Ko, J. Cho and J. Cho, *Small*, 2013, **9**, 3784.
52. I. Yamashita, K. Iwahori and S. Kumagai, *Biochim. Biophys. Acta*, 2010, **1800**, 846.
53. K. Ohara, B. Zheng, M. Uenuma and Y. Ishikawa, *Appl. Phys. Express*, 2011, **4**, 085004.
54. A. Miura, Y. Uraoka, T. Fuyuki, S. Yoshii and I. Yamashita, *J. Appl. Phys.*, 2008, **103**, 074503.
55. Y. Tojo, A. Miura, Y. Uraoka, T. Fuyuki and I. Yamashita, *Jpn. J. Appl. Phys.*, 2009, **48**, 04C190.
56. K. Ohara, Y. Uraoka, T. Fuyuki, I. Yamashita, T. Yaegashi, M. Moniwa and M. Yoshimaru, *Jpn. J. Appl. Phys.*, 2009, **48**, 04C153.

57. S. Barton, J. Gallaway and P. Atanassov, *Chem. Rev.*, 2004, **104**, 4867.
58. M. Cooney, V. Svoboda, C. Lau, G. Martin and S. Minteer, *Energy Environ. Sci.*, 2008, **1**, 320.
59. I. Willner, *Science*, 2002, **298**, 2407.
60. M. Falk, Z. Blum and S. Shleev, *Electrochim. Acta*, 2012, **82**, 191.
61. R. Bullen, T. Arnot, J. Lakeman and F. Walsh, *Biosens. Bioelectron.*, 2006, **21**, 2015.
62. T. Miyake, S. Yoshino, T. Yamada, K. Hata and M. Nishizawa, *J. Am. Chem. Soc.*, 2011, **133**, 5129.

Subject Index